D0023521

Science in the Enlightenment

Other Titles in ABC-CLIO's

History of Science

Series

The Scientific Revolution, William E. Burns

FORTHCOMING

Contemporary Science, Eric Swedin

Science in the Ancient World, Russell Lawson

Science in the Early Twentieth Century, Jacob Hamblin

Science in the Enlightenment

An Encyclopedia

William E. Burns

Santa Barbara, California
Denver, Colorado
Oxford, England

Library of Congress Cataloging-in-Publication Data

Burns, William E.
Science in the Enlightenment : an encyclopedia / William E. Burns
p. cm. (ABC-CLIO's history of science series)
Includes bibliographical references and index.
ISBN 1-57607-886-8 (hardcover : alk. paper)–ISBN 1-57607-887-6 (eBook)
1. Science—History—Encyclopedias. I. Title. II. Series.
Q121.B87 2003
509′.003—dc22 2003011342

07 06 05 04 03 10 9 8 7 6 5 4 3 2 1

This book is available on the World Wide Web as an eBook. Visit abc-clio.com for details.

ABC-CLIO, Inc.
130 Cremona Drive, P.O. Box 1911
Santa Barbara, California 93116-1911

This book is printed on acid-free paper.
Manufactured in the United States of America

Dedicated to Tom, Jeni, Nathan, Carrie, and Ellen,
bearers of Enlightenment

Contents

Science in the Enlightenment: An Encyclopedia

A

B

C

Foreword

The universe of science became vast precisely in the eighteenth century. It was as if some invisible barrier had been broken, and like the novel, science found an audience, a new set of practitioners, textbooks, societies, amateurs, and specializations. Reading through the correspondence of one of the first journalists to popularize Newtonian science, Prosper Marchand, a Huguenot refugee in the Dutch Republic, one letter written from Surinam around 1720 announced, "I will abjure my Cartesianism if you will abjure your Newtonianism." When someone in Surinam can joke about philosophical positions, you know that the intellectual universe has expanded. William Burns's dictionary (to use the eighteenth-century term) attempts to walk us through the landscape of the new scientific world.

Notice that universities, while present in the index, do not swell the pages. The remarkable thing about the spread of scientific learning in the period concerns its catholicity. If anything, most universities were behind, not in the forefront of, the latest science. Certainly, it was possible to learn as much about Newton's science in a London coffeehouse where lecturers regularly gave courses as at the Sorbonne. The gaps between Britain and the Continent were noticeable by 1720, and the fault lay largely with the clergy. Whether Jesuits or not, the Catholic clergy had clung to Aristotle as long as they could, finally embraced Descartes in the 1690s, and then dug in their heels. Thus the academies and informal societies became very important in Catholic Europe as did the so-called philosophes. The term is simply French for philosopher, but in the eighteenth century, people like Voltaire and his lover, Madame du Châtelet, made the new science into a rallying cry for Enlightenment. Thus they earned the title of philosophe, as did Georges-Louis Leclerc de Buffon and Erasmus Darwin. They pushed the philosophical limits by taking up the materialism that could be extracted from the new science. With their interests we see a maturing of biology and medicine as well as philosophical daring.

Most of the devotees of science concentrated not on philosophical outrage but on experiments and gadgets. Scientific instruments became part of any lecture and even many households. To be cultivated meant having scientific books, or a microscope, or an orrery (a metal gadget that displayed the planets) in one's home. It also sometimes meant having the courage to inoculate a child against smallpox. The academies provided a comfort zone for the progressive advocates of science, and, not surprisingly, they figure prominently in this dictionary. In London and Haarlem the main scientific societies were private and received no state support. In Paris, Berlin, and St. Petersburg they were state institutions where pensions and royal patronage were expected and coveted. Careers could be made in both settings, as the life of Pierre-Louis Moreau de Maupertuis will reveal. Yet the character of the continental societies differed from the English and Dutch varieties. The first was more formal, while the private societies struggled and expanded their membership to just about anyone who could pay the dues and showed some interest.

But science also had its detractors throughout the period covered by the dictionary. All the major religions were made nervous by one or another aspect, with liberal Protestants in England and Germany being the least anxious. Halle gave them a home, and so too did Cambridge. But among conservative Anglicans, Hutchinsonianism made great strides, and it was openly hostile to Newtonianism. As can be seen, the various natural philosophies had become "isms" with fierce battles often waged on behalf of one or another. In that respect science might be said to have been more exciting, provocative, and controversial than it is in our day. Sample these pages and witness for yourself the array of novelty and controversy that the vast universe of eighteenth-century science offered to an eager public. You will not be disappointed.

Margaret C. Jacob
University of California at Los Angeles

Preface and Acknowledgments

Enlightenment and science are two concepts often yoked together in Western—and world—culture. The eighteenth century in Europe, the classic "Age of Enlightenment," also saw significant scientific advances, advances that continued through the French Revolution of 1789 and into the romantic period of the early nineteenth century. For the most part, eighteenth-century science has received less attention from historians than has either the scientific revolution of the seventeenth century or the "second scientific revolution" of the nineteenth century.

The Scientific Revolution of the Seventeenth Century

The Enlightenment was preceded by the scientific revolution, the transformation of natural knowledge in the sixteenth and seventeenth centuries that created modern science. Many aspects of Enlightenment science and the society and institutions in which it was carried on have their roots in the preceding century. The most important eighteenth-century scientific societies were founded in the 1660s (*see* Royal Academy of Sciences; Royal Society). The Copernican view of the planets, including Earth, traveling around the Sun at the center rather than Earth at the center had defeated the traditional Ptolemaic view of Earth at the center with the Sun and planets rotating around it (*see* Astronomy). The subordination of scientific to religious authority, most memorably exemplified in the trial and condemnation of Galileo in 1633, had greatly diminished by the start of the eighteenth century, even if the Catholic Church continued to forbid the teaching of heliocentric astronomy as fact. Medieval and Renaissance Aristotelianism had been displaced through most of Europe by the natural philosophies that dominated the eighteenth century (*see* Cartesianism; Newtonianism). The philosophy of the ancient physician Galen and the doctrine of humors that had dominated medieval and Renaissance medicine had been eliminated from most medical schools and replaced by more mechanistic medicine (*see* Boerhaave, Hermann; Medicine). Science had also increasingly distanced itself from magical thinking. Alchemists had mostly (but not entirely) given up the quest for the universal solvent and the philosophers' stone that would turn base metals to gold and had become chemists (*see* Böttger, Johann Friedrich; Chemistry), while astronomy had definitively separated itself from astrology. New tools for scientists had been invented or dramatically improved (*see* Microscopes; Telescopes; Thermometers). This new science had been introduced in many universities, particularly the more recently founded ones (*see* Universities; University of Edinburgh; University of Halle; University of Leiden), and in the curricula of many other educational institutions (*see* Education).

The French Predominance

One thing that changed in the transition from the scientific revolution to Enlightenment science was the relative role of different European nations and their scientific institutions. Although the leading scientific nation of the late seventeenth century, the age of Newton, had been England, its leading role was quickly yielded to France in the early eighteenth century. French scientists possessed by far the most formidable apparatus for the support of science, as a benefit of residence in Europe's most powerful and wealthy state. Paris, the home of the Royal Academy of Sciences as well as many other scientific associations and institutions, was the capital of European science into the Napoleonic period of the early nineteenth century (*see* Napoleonic Science). France's leading role persisted through the political changes of the French Revolution of 1789, although some old institutions were eliminated and new ones created and some important French scientists were executed by the new regime (*see* Condorcet, Marie-Jean-Antoine-Nicolas Caritat, Marquis de; French Revolution; Lavoisier, Antoine-Laurent; Museum of Natural History; Royal Academy of Sciences; Royal Society of Medicine). French increasingly became the language of international scientific communication, although it did not entirely displace Latin in this period (*see* Periodicals). The French state also led in the conscription of scientific, medical, and technical knowledge for state purposes (*see* Berthollet, Claude-Louis; Coudray, Angelique Marguerite Le Boursier du; Education; Technology and Engineering; War). The power of the French government also protected its scientists from the sometimes heavy hand of the Catholic Church, which was still capable of blocking scientific initiative in the Iberian Peninsula and sometimes in Italy. In its ability to finance and mount large-scale scientific expeditions or scientific components to expeditions launched for other purposes, France's only rival was Great Britain (*see* Egyptian Expedition; Exploration, Discovery, and Colonization; Maupertuis, Pierre-Louis Moreau de; Metric System; Transits), and for most of the century the French state possessed a far more effective intellectual and institutional organization for analyzing data gathered from expeditions. One French weakness—the inferiority of its scientific-instrument industry to that of Britain—could be partially remedied simply by purchasing British equipment (*see* Cassini Family; Instrument Making). French science also benefited from the concentration of so many important minds in the capital of Paris, as opposed to the British or German model, where scientists were more diffused throughout the country (*see* Arcueil, Society of). This was not always an advantage, however, as French science lacked the diversity of approaches characteristic of German or British scientists.

Of course, none of France's advantages would have mattered so much if the country had not produced a plethora of outstanding and innovative scientists. Outstanding French scientists of the eighteenth century, all of whom were members of the Royal Academy of Sciences, include Antoine-Laurent Lavoisier, the founder of modern chemistry; Pierre-Simon de Laplace, who perfected the Newtonian view of the solar system; and Charles-Augustin de Coulomb. French science quickly recovered from the damage wrought by the Revolution with the creation of new institutions. In the early nineteenth century surviving scientists such as Laplace and the chemist Claude-Louis Berthollet continued their careers, and they were joined by younger scientists taking advantage of French military supremacy and the fluid social situation the Revolution and its attack on traditional hierarchies had produced. French scientific supremacy was maintained by the brilliant group of zoologists at the Museum of Natural History, Georges Cuvier, Étienne Geoffroy Saint-Hilaire, and Jean-Baptiste-Pierre-Antoine de Monet de Lamarck, as well as the chemist Joseph-Louis Gay-Lussac and the versatile André-Marie Ampère. France also benefited from its ability to attract outstanding foreign scientists to Paris, a tradition dating back to the founding of the Royal Academy of Sciences in the seventeenth century. Examples include the mathematician Joseph-Louis Lagrange and the versatile American-born Benjamin Thompson, Count Rumford.

The Diffusion of Science throughout Europe and the World

Complementing the French predominance was the spread of science beyond its original core in France, England, the Dutch Republic, Germany, and Italy to nations that had hitherto been marginal to scientific development. Perhaps the most dramatic example is Scotland, politically united with England to form Great Britain in 1707. Scotland played a very modest role in the scientific revolution, producing few scientists and lacking scientific institutions. Beginning in the mid-eighteenth century it produced a number of important scientists of European renown, perhaps the most important among them the chemist Joseph Black, and sustained continuing scientific institutions (*see* Royal Society of Edinburgh; University of Edinburgh). Scottish science's unique character was partly due to the strength of its universities and particularly their medical schools, which distinguished it from its English neighbor, in which the universities played little scientific role and university medical education was nearly entirely moribund. Sweden was another nation that emerged from scientific obscurity to scientific prominence in the eighteenth century and produced one of Europe's most powerful scientists in the botanist Carolus Linnaeus, as well as influential scientific societies (*see* Royal Swedish Academy of Science). Russia, although it did not produce any scientists of the importance of Black or Linnaeus, also saw the spread of scientific institutions (*see* Imperial Academy of Sciences of St. Petersburg; Lomonosov, Mikhail Vasilyevich). Both new and old centers of science benefited from the mobility of many eighteenth-century scientists, who were willing to move either toward or away from scientific capitals. Particularly in medicine and its numerous allied fields, Scotland became an exporter of scientific talent (*see* Brown, Robert; Hunter Family), enriching English science and enabling Scottish universities to enroll far more students than Scotland would ever be able to provide careers for. Another rising scientific power, Switzerland, was a poor country that produced much more scientific talent (particularly in mathematics) than it could afford to support, but became a prominent source of recruitment for the St. Petersburg and Berlin Academies as well as German universities (*see* Bernoulli Family; Euler, Leonhard; Haller, Albrecht von).

The rise of these peripheral centers of science contrasted with the decline of the Dutch Republic, a leading scientific nation in the seventeenth century that produced no scientists of the first rank after the physician and chemist Hermann Boerhaave. England remained second only to France as a scientific power, benefiting from the wealth and power of the British state, the skill of its manufacturers, and the expansion of the British Empire (*see* Banks, Sir Joseph; Cook, James; Industrialization; Instrument Making). However, the stubborn British refusal to adopt the mathematical methods pioneered on the European continent isolated England from the mainstream of advanced mathematical science until the very end of this period and ensured that, despite the greatness of English scientists like Henry Cavendish, London would not rival Paris as a center of scientific theory. Italy, although the primacy it had enjoyed during the age of Galileo was long gone, produced a number of important scientists during the Enlightenment, often of an experimental rather than theoretical bent (*see* Galvani, Luigi; Spallanzani, Lazzaro; Volta, Alessandro Giuseppe Antonio Anastasio). Italian science was also strongly represented among university faculties and somewhat more welcoming to women (*see* Bassi, Laure Maria Catarina). Iberia remained peripheral to European science in the eighteenth century as it had been for most of the scientific revolution, but by the late eighteenth century was determinedly trying to catch up (*see* Royal Botanical Expedition).

Although centralized France was a partial exception, the major countries of Europe also saw science spread outside the political capitals and university towns that had nurtured it in the seventeenth century. The most striking example is England, where the weakness of its universities was compensated by a remarkable series of scientific institutions in industrial and commercial towns in the north (*see* Derby Philosophical Society; Lunar Society of Birmingham; Manchester Literary and Philosophical Society). In the late eighteenth century these northern towns, where scientists, physicians, engineers, and industrialists traveled in the same social circles, made it possible for English scientists to have careers independent of the Royal Society in London or the English universities. The scientific style of these provincial centers was more pragmatic and technologically oriented as well as prominently including Protestant Dissenters rather than the members of the established Church of England who dominated London science and the English universities. Perhaps the most notable English Dissenting provincial scientist was the chemist Joseph Priestley, although eventually his religious and political radicalism forced him to emigrate to the newly formed United States. John Dalton the atomist was another Dissenting scientist whose career was almost entirely provincial.

Germany, then organized in the form of the radically decentralized Holy Roman Empire, also saw the domestic spread of science. The capital of Prussia, Berlin, emerged from scientific obscurity to scientific prominence with the rise of the Berlin Academy, which recruited many French and Swiss scientists as well as Germans. Germany's decentralization also contributed to its strong university system, which was much more scientifically dynamic than its French and English counterparts (*see* Blumenbach, Johann Friedrich; Crell, Lorenz Florens Friedrich von; Gauss, Carl Friedrich; Haller, Albrecht von; Kant, Immanuel; Klaproth, Martin Heinrich; Lichtenberg, Georg Christoph; Mayer, Johann Tobias; Reil, Johann Christian; University of Halle; Wolff, Christian). Though located in provincial towns, German and Scottish universities and academies rose in importance for science, while princely, royal, and papal courts declined. Royal patrons of science usually founded or supported scientific institutions rather than patronizing individual scientists (*see* Nationalism).

The culture of Western science in the eighteenth and early nineteenth centuries extended over a far larger geographic area outside Europe than ever before (*see* Egyptian Expedition; Exploration, Discovery, and Colonization; Humboldt, Alexander von). Knowledge gathered from all over the world, not just the European corner of it, assumed a central role in more disciplines (*see* Astronomy; Botany; Meteorology; Oceanography; Transits; Volcanoes; Zoology). The ability to gather, analyze, and coordinate information from a variety of global sources assumed a new importance, particularly in astronomy and botany, the most important Enlightenment life science other than medicine (*see* Maupertuis, Pierre-Louis Moreau de). Carolus Linnaeus was a scientist of genius, and Joseph Banks was not, but both derived enormous influence from their far-flung network of followers and disciples gathering plants from all over the globe (*see* Botanical Gardens; Kew Gardens; Sloane, Sir Hans). The physical evidence of non-European phenomena also had to be displayed to European scientists as much as possible (*see* Museum of Natural History; Museums). But not all science produced outside Europe was controlled from European centers. Communities of practitioners and students of Western science arose in colonial and non-Western contexts, in many cases taking an increasingly self-confident stance in relation to European science (*see* American Philosophical Society; Bartram Family; Colonial Science; Japan, Western Science in). The newly independent United States, while lagging far behind the major European states in the capacity to identify, nurture, and support scientific talent, was able to launch scientific initiatives of its own (*see* Lewis and Clark Expedition; Michaux Family) while also providing an additional source of employment for European scientists. Just a few scientists of colonial origin in this period achieved European celebrity, often with a more practical and less mathematically sophisticated scientific bent than the Parisian mainstream (*see* Franklin, Benjamin; Thompson, Benjamin [Count Rumford]).

Science, Enlightenment, and Romanticism

The social and cultural importance of science continued to rise in the eighteenth century. Thinkers of the Enlightenment (*see* Alembert, Jean Le Rond d'; Condillac, Étienne Bonnot de; Condorcet, Marie-Jean-Antoine-Nicolas Caritat, Marquis de; Diderot, Denis; The Enlightenment; Kant, Immanuel; Voltaire) found in science an alternative to the traditional, religious culture of Europe, although only a few radicals sought to use scientific materialism to overthrow the religious view of the world entirely (*see* La Mettrie, Julien Offroy de; Materialism). Religion mostly made peace with science, although some religious leaders either rejected it entirely or were attracted to scientific "heresies" (*see* Hutchinsonianism). Some devout scientists attempted to create new syntheses of the most advanced science with religion (*see* Boscovich, Ruggiero Giuseppe; Priestley, Joseph). Scientific developments seemed related more closely than ever to obvious technological advances, contributing to the Enlightenment ideal of progress (*see* Ballooning; Lightning Rods; Longitude Problem; Smallpox Inoculation). However, science was also invoked frequently to support Europe's traditional religion and distribution of power (*see* Nationalism; Natural Theology; Race; Religion; Sexual Difference).

Science is carried out not solely by the great scientists, but also by an ever growing army of lesser figures. An important part of the history of Western science has been the rise of the profession of science and its associated cultural prestige. The eighteenth century was very important for this transformation. Scientific academies on the model of the French Royal Academy of Sciences, which paid their academicians, provided a way of practicing science as a profession, as distinct from carrying it out as a supplement to a basically medical, clerical, or academic career (*see* Academies and Scientific Societies). They also offered publication opportunities and frequently sponsored prize competitions, which could be lucrative. The impact of this development was limited, however, as the new academies had few paid academicians, often paid too little to support a scientist and his household, and frequently gave prizes to established scientists rather than younger scientists starting out.

Another important development for the creation of the scientific profession was the rise of state sponsorship of specific scientific projects and scientific institutions with narrowly defined utilitarian goals (*see* Botanical Gardens; Böttger, Johann Friedrich; Observatories). The idea that science was economically useful went back to the scientific revolution and had been expressed most influentially by the hero of many Enlightened thinkers, the seventeenth-century English statesman and philosopher Francis Bacon. However, the eighteenth century saw a vastly increased coordination between science and state power. A similar development took place in medicine, where it became possible for more physicians and surgeons to devote substantial portions of their careers to research rather than practice or teaching (*see* Hospitals; Royal Society of Medicine). Not only governments saw science as pragmatically useful. Scientists, particularly in Great Britain, also contributed to the expansion of private capitalist enterprise in the era of the Industrial Revolution, and engineers and scientists often worked closely together (*see* Black, Joseph; Industrialization; Technology and Engineering).

Scientific professionalization narrowed the range of voices participating in science, as a pronounced division between "amateur" and "professional" scientists began to emerge. The ideal of the "amateur" scientist persisted longer in England, lacking both the German and Scottish university culture and the French scientific bureaucracy (*see* Cavendish, Henry). However, even the Royal Society, a bastion of scientific amateurs, was feeling the pressure of professionalization by the early nineteenth century (*see* Banks, Sir Joseph; Davy, Sir Humphry). Some entire populations found themselves barred from the new scientific professions, the most obvious example being women (*see* Germain, Sophie). The growing scientific presence in medicine was also related to the marginalization of women's traditional roles in health care, particularly in the English-speaking world (*see* Midwives). Scientific professionalization along with the growing volume of knowledge that a budding scientist had to absorb also led to a growing gap between working scientists and scientific popularizers (*see* Popularization).

Although scientists were growing somewhat more removed from the general population, science and scientific innovation, somewhat marginal activities in the medieval and Renaissance world and even in the scientific revolution, began to attract more broad cultural attention in the Enlightenment (*see* Literature; Periodicals). Europe became covered with a dense network of scientific institutions. Although the practice of European science remained heavily dominated by male Christians, its audience at least broadened considerably in the Enlightenment (*see* Jewish Culture; Women). The idea of the heroic scientist, a powerful archetype in Western culture, emerged in the eighteenth century, particularly in the cult of Isaac Newton. Scientists, particularly those with a flair for showmanship, could both inspire and capitalize on popular celebrity (*see* Buffon, Georges-Louis Leclerc de; Davy, Sir Humphry; Franklin, Benjamin). Even scientists of less creativity could make careers for themselves as popularizers and scientific showmen, particularly early in the eighteenth century when the distinction between the two roles was less great (*see* Desaguliers, John Theophilus; Hauksbee, Francis; Poliniere, Pierre; Whiston, William). Enlightenment and the belief in progress provided powerful motivations for the popularization of science, another area where the eighteenth century built on the accomplishments of the scientific revolution (*see* Châtelet, Gabrielle-Émilie du; Encyclopedias; Popularization).

The attitude toward science held by romantic thinkers, including both scientists and nonscientists, was more complicated than that of their Enlightenment predecessors and contemporaries (*see Naturphilosophie;* Romanticism). Some forms of romanticism, a movement emerging in the last decades of the eighteenth century and the first of the nineteenth, distrusted the scientific enterprise entirely (*see* Literature). Others endorsed a nonmechanical science, rejecting to varying degrees the tradition of mathematized and abstract science sometimes associated with the Newtonian tradition (*see* Goethe, Johann Wolfgang von). Romantics often viewed mainstream science as going astray through an obsession with mathematical abstraction as opposed to observation. Less mathematized sciences, such as electricity and optics, attracted many scientific romantics (*see* Davy, Sir Humphry; Ørsted, Hans Christian; Ritter, Johann Wilhelm). In the life sciences, romantic scientists tended to emphasize the degree to which living things were formed on a single plan and different organs were homologous to each other (*see* Geoffroy Saint-Hilaire, Étienne). Some romantic medical men were attracted to grand theories that offered hope for understanding all diseases (*see* Beddoes, Thomas; Brunonianism). Of course, many scientists strongly opposed romantic trends in science, finding them overly abstract and philosophical rather than firmly grounded in empirical studies (*see* Berzelius, Jöns Jakob; Cuvier, Georges).

Physics and Astronomy

The subsequent development of many sciences that had been "revolutionized" in the sixteenth and seventeenth centuries was more sedate in the eighteenth century, as physics and astronomy were marked more by conceptual development than sudden, revolutionary shifts comparable to Copernicus's development of the Sun-centered universe or Isaac Newton's formulation of the law of universal gravitation. Enlightenment mathematical physics and astronomy cannot be simply reduced, as some older histories have it, to the development of "Newtonianism," but there can be little doubt that Copernican astronomy and Newtonian dynamics were not seriously challenged, but rather further refined and expanded (*see* Astronomy; Mechanics; Optics; Physics). French, Swiss, and, by the nineteenth century, German mathematicians played the major role in developing the increasingly sophisticated mathematics used to tackle physical problems (*see* Alembert, Jean Le Rond d'; Bernoulli Family; Euler, Leonhard; Gauss, Carl Friedrich; Lagrange, Joseph-Louis; Laplace, Pierre-Simon de; Mathematics; Probability). This work culminated in Lagrange's *Analytical Mechanics* (1787) and Laplace's five-volume treatment of the solar system, *Treatise on Celestial Mechanics* (1799–1825).

More dramatic, if not revolutionary, developments in astronomy were associated with observation (*see* Bradley, James; Comets; Herschel Family; Meteors and Meteorites; Telescopes; Transits). By far the most dramatic, if not necessarily the most important, was William Herschel's discovery of the new planet Uranus in 1781, striking evidence of the progress of science, as no planet had been "discovered" by an individual in previous history. The predicted return of Halley's comet in 1758 also attracted great publicity and attention, as did the transits of Venus in 1761 and 1769. The first asteroids were discovered in the early nineteenth century, and after several false starts and wrong turns scientists established the nature of meteors, or "shooting stars," and their connection with meteorites, or lumps of metal that fell from the sky.

Most working astronomers neither expounded the workings of the solar system nor made spectacular discoveries. Astronomy was largely a pragmatic discipline in the eighteenth century, concerned primarily with solving practical problems of cartography and navigation (*see* Cassini Family; Longitude Problem; Maskelyne, Nevil; Mayer, Johann Tobias; Metric System; Observatories). As such, it had some remarkable triumphs, such as the development of the method of lunar distances for determining the longitude, the most accurate maps produced to date, and the creation of the metric system.

One major exception to the generally undramatic development of eighteenth-century physics was the rise of electrical science, beginning with the experimental displays of the early-eighteenth-century demonstrators (*see* Desaguliers, John Theophilus; Electricity; Hauksbee, Francis; Polinière, Pierre). Although all of the demonstrators were acquainted with some mathematics, the kind of "experimental physics" they were doing was differentiated in the world of science from "mathematical physics." Stunning effects, particularly in the early stages of a science's development, were more important than sophisticated equations. The really dramatic rise of electrical studies began around the middle of the century, with the development of the Leiden jar, literally the most stunning electrical effect of all. The jar was capable of delivering a much larger electrical spark than had ever been known before. Its rise to prominence was quickly followed by the development of Benjamin Franklin's popular electrical theory (*see* Franklin, Benjamin; Lightning Rods; Nollet, Jean-Antoine).

Important developments in late-eighteenth-century electrical science (and other areas of "experimental physics," like magnetism and optics) involved more rigorous applications of quantitative rather than purely qualitative analysis to electrical phenomena, as the methods of mathematical physics were applied to the phenomena produced by experimental physicists. This mathematization was mostly the work of French savants (*see*

Coulomb, Charles-Augustin de). A predominantly experimental Italian tradition also bore fruit in the late eighteenth century, thanks to the inventive genius of Alessandro Volta and the discovery of "animal electricity" by Luigi Galvani (*see* Galvani, Luigi; Volta, Alessandro Giuseppe Antonio Anastasio). Volta's battery meant that electricity no longer had to be studied in the form of brief electrical sparks, but could now be studied as a flowing "current." This led to the development of electromagnetism and electrodynamics in the hands of a diverse group of scholars (*see* Ampère, André-Marie; Ørsted, Hans Christian), as well as the application of electricity to chemistry (*see* Davy, Sir Humphry).

The Chemical Revolution

One science whose "revolution" is often regarded as taking place in the eighteenth rather than the seventeenth century is chemistry. Early-eighteenth-century chemistry had still not entirely freed itself from the vestiges of alchemy and the making of gold (*see* Böttger, Johann Friedrich; Stahl, Georg Ernst) and was handicapped by an overly complex and unsystematic tradition of chemical names (*see* Chemical Nomenclature). Mid-eighteenth-century chemists advanced the frontiers of chemical knowledge with the refinement of analysis and the discovery of new chemical substances but made few important conceptual innovations (*see* Macquer, Pierre Joseph; Marggraf, Andreas Sigismund; Scheele, Carl Wilhelm).

The "chemical revolution" began with the discovery, mostly carried out by British experimenters, that air was not a unified "element" but composed of different substances (*see* Black, Joseph; Cavendish, Henry; Hales, Stephen; Priestley, Joseph). Most historians of science oppose the reduction of the "chemical revolution" to a single event, but traditionally it is associated with Lavoisier's overthrow of the phlogiston theory of combustion in the late eighteenth century. Phlogistonism had seventeenth-century roots, but was associated in the eighteenth century with the German medical professor Georg Ernst Stahl. Phlogistonists held that combustion was associated with the release of phlogiston from a body while it was burning. Lavoisier, building on the work of previous French and British chemists, demonstrated experimentally that combustion was caused by the combination of a body with a recently discovered gas he referred to by the invented term *oxygen*. Lavoisier's chemical innovations were accepted relatively quickly in France (*see* Berthollet, Claude-Louis), but it took some time for the "French chemistry," which was associated with a new and much more rational system of chemical names, to be accepted internationally (*see* Crell, Lorenz Florens Friedrich von; Klaproth, Martin Heinrich).

Lavoisier's chemical revolution was followed in the early nineteenth century by events nearly as dramatic. Chemical and electrical science were brought together, most dramatically in the use of current electricity from the newly invented battery to decompose water into "oxygen" and "hydrogen," another chemical name introduced by Lavoisier (*see* Davy, Sir Humphry). Chemical substances were defined in terms of atoms forming compounds in numerical ratios (*see* Atomism; Avogadro, Amedeo; Berzelius, Jöns Jakob; Dalton, John), and significant discoveries were made concerning the behavior of gases (*see* Gay-Lussac, Joseph-Louis).

One subject examined by chemists, mathematical physicists, and experimental physicists was heat, around which clustered many mysteries. Joseph Black established the concepts of latent and specific heat, and the development of the calorimeter provided a means for the measurement of heat, but its essential nature was still an open question. Was it a substance, which eventually came to be named caloric, or was it the result of the movement or interior vibration of the particles of a substance (*see* Calorimeters; Lavoisier, Antoine-Laurent; Thompson, Benjamin [Count Rumford])? Eighteenth- and early-nineteenth-century scientists did not succeed in settling this question, which remained an open one until the development of nineteenth-century thermodynamics.

Sciences of the Earth

Another science that underwent revolutionary change in the eighteenth and early nineteenth centuries was geology (*see* Earthquakes; Geology; Hutton, James; Volcanoes; Werner, Abraham Gottlob). The intellectual roots of Enlightenment geology can be traced to the seventeenth century, but eighteenth-century geologists both hugely increased the science's stock of empirical information and elaborated exciting new theories, which in turn inspired more empirical work. Disagreements raged over the relative importance of water, Earth's internal heat, and volcanoes in the shaping of Earth. Theories of Earth produced in the eighteenth century made more use of time and often ascribed greater age to it than had been traditionally allowed on the basis of the biblical account (*see* Buffon, Georges-Louis Leclerc de; Cuvier, Georges). The use of fossils to establish the sequence of earthly events was an innovation of this period. Like most Enlightenment science, geology was believed to have practical goals—the foremost continental European geological theorist, Abraham Gottlob Werner, taught at a mining academy, not a university. Not only was the content of geology changing, but it was also rising in prominence among sciences generally. One of the first and most successful societies devoted to a particular science was the Geographical Society of London founded in 1807. The later conflict between "Genesis and geology," which influenced so many nineteenth-century intellectual developments, had roots in the Enlightenment.

Oceanography, a new science in the eighteenth century, and meteorology did not see as many revolutionary developments as did geology, but both sciences accumulated data and produced new theories. The eighteenth century is particularly important in the history of meteorology for the development or refinement of instruments (*see* Hygrometers; Rain Gauges; Thermometers). Both oceanographers and meteorologists expanded the geographical range of their sciences, with observations from all over the world by European seafarers, explorers, and colonists.

Botany and the Struggle for Classification

In no science was the flood of new information reaching European scientists more marked than in botany (*see* Banks, Sir Joseph; Bartram Family; Colonial Science; Exploration, Discovery, and Colonization; Kew Gardens; Lewis and Clark Expedition; Michaux Family; Royal Botanical Expedition). The thousands of new plants and plant descriptions reaching European centers made a workable system of plant classification, a goal toward which scientists had been striving since the sixteenth century, an even more urgent necessity. The most celebrated botanist of the eighteenth century was renowned above all as the inventor of a workable and easily learned system of classification. The Swedish scientist Carolus Linnaeus invented and disseminated through a remarkable group of disciples what came to be known as the "sexual system of Linnaeus," based on the number and arrangement of a plant's sexual organs (*see* Botany; Linnaeus, Carolus). Linnaeus's system, relatively easy to learn and apply, was at first a great success, particularly influential in the English-speaking world (*see* Darwin, Erasmus; Linnean Society). It was rivaled by the tradition of "natural classification" that dominated French botany (*see* Botanical Gardens). Natural classifiers attacked Linnaeus's reliance on sexual characteristics as reductionistic and "artificial." In the early nineteenth century, natural classification displaced the Linnaean system among professional botanists, even in the Linnaean stronghold of England, although it lingered among popularizers (*see* Brown, Robert).

Classification and discovery were not the only aspects of plant science enhanced during the Enlightenment. Plant sexuality, still considered controversial at the dawn of the eighteenth century, became universally accepted among scientists, as well as better understood. The inner workings of plants, how they absorbed nutrition from the air and ground, were mapped out for the first time (*see* Hales, Stephen; Plant Physiology). Nor were the uses of plants in human society forgotten, and the cultivation of agricultural crops was increasingly viewed in scientific terms and the subject of scientific work (*see* Agriculture; Botanical Gardens). The "new agriculture" of the Englishman Jethro Tull and other agricultural innovations were justified in scientific terms and often aroused more interest from scientists than working farmers or agricultural landowners. Another area where science particularly affected agriculture was in the project of acclimating crops among different parts of Europe's far-flung empires.

The Beginnings of Biology

Zoological science lagged far behind botany for most of the history of Western science, and has also received far less attention from historians. The science of animals began to make significant strides in the Enlightenment. Although Linnaeus classified animals as well as plants, the most important zoologists for the entire period were French, including Linnaeus's contemporary and rival the Comte de Buffon. The French tradition of zoological excellence was carried on by the remarkable group at the Museum of Natural History in the revolutionary and Napoleonic periods (*see* Cuvier, Georges; Geoffroy Saint-Hilaire, Étienne; Lamarck, Jean-Baptiste-Pierre-Antoine de Monet de; Museum of Natural History). In a development with parallels in botany and chemistry, animal anatomy, or "comparative anatomy," changed from a basically medical science practiced by physicians and aimed at aiding in the understanding of human anatomy to a more autonomous intellectual enterprise. Georges Cuvier, the first major comparative anatomist not to have received a medical education, was also responsible for broadening the science to take into consideration extinct species as recorded in fossils as well as existing species (*see* Fossils). The notion of extinction itself was controversial for much of this period, particularly as so much of the world remained unexplored that seemingly extinct species could still be undetected somewhere. The question of extinction was linked to that of the fixity of species, a doctrine upheld by Cuvier against his colleagues at the Museum of Natural History, Jean-Baptiste-Pierre-Antoine de Monet de Lamarck and Étienne Geoffroy Saint-Hilaire, believers in the possibility that species could change. Although it is a common mistake to evaluate Enlightenment and romantic zoologists solely on their anticipations of the theory of evolution, lively debates over the age of Earth and the extinction or transformation of species did result in the introduction of some evolutionary ideas, particularly in the work of Lamarck. Although pragmatic concerns did not play as large a role in animal science as in plant science, they were not forgotten. It was this period that saw the institutionalization of veterinary science with the founding of schools and chairs in the subject.

Life and the nature, or even the existence, of the distinction between the living and the nonliving presented many problems for Enlightenment scientists, physicians, and philosophers (*see* Diderot, Denis; Materialism; Physiology; Polyps; Spallanzani, Lazzaro; Vitalism). The question of whether animals reproduced by forming a new individual from scattered parts (epigenesis) or whether each began as a fully formed miniature in its mother's egg or father's sperm (preformation) provoked passionate controversy (*see* Bonnet, Charles; Embryology; Haller, Albrecht von). The idea of the spontaneous generation of living things from nonliving material, which had been debated during the scientific revolution, prompted more disagreement during the Enlightenment. The English Catholic priest and scientist John Turberville Needham experimentally proved spontaneous generation, while more careful and rigorous work of the Italian priest and scientist Lazzaro Spallanzani experimentally disproved it. The creation of the distinction between living and nonliving things was also partly due to the romantic movement; romantic scientists were the first to group the life sciences in an exclusive category called "biology" (*see Naturphilosophie;* Romanticism).

The life sciences in this period remained closely tied to medicine, and most innovations in physiology and anatomy were the work of physicians and other medical practitioners. Some of the most important scientists of the Enlightenment, in chemistry and physics as well as the life sciences, were physicians who taught in medical institutions (*see* Black, Joseph; Blumenbach, Johann Friedrich; Boerhaave, Hermann; Haller, Albrecht von). Below the level of the scientific elite, a number of physicians and surgeons participated in a broad range of sciences (*see* Beddoes, Thomas; Darwin, Erasmus; Reil, Johann Christian). The alliance between science and medicine had a long history in Europe, but it was fundamentally reshaped in the era of the Enlightenment. Some conceptual areas that had previously been viewed mostly in religious or other nonmedical and nonscientific terms were defined as medical matters in this period (*see* Madness; Masturbation; Psychology; Toft Case). Some attempts were made to remodel medical institutions on more scientific lines, with intellectual advance privileged over the immediate needs of patients (*see* Hospitals; Royal Society of Medicine; Surgeons and Surgery). Even new and dubious medical schools of thought, fundamentally challenging the views of Europe's medical elite, couched their theories in scientific terms (*see* Brunonianism; Mesmerism and Animal Magnetism). Although in some areas medicine drew closer to science, in others the two were moving further apart. Many scientific areas that had traditionally been the realm of medical personnel acquired a more independent status (*see* Anatomy; Botany; Chemistry). Many conceptual advances in the understanding of the human body and the diseases that could afflict it were the work of Enlightenment physicians (*see* Bichat, Marie-François-Xavier). However, they did not always lead to significant advances in patient care, aspects of which may have even declined in the period (*see* Midwives).

Science in the Enlightenment takes as its starting point the reorganization of the French Royal Academy of Sciences in 1699, which established the structure of the eighteenth century's greatest scientific institution and its numerous imitators. It concludes with the careers of the astounding group of scientists born between 1765 and 1780 and shaped by the dramatic changes of the French Revolution and the romantic movement (*see* Ampère, André-Marie; Avogadro, Amedeo; Berzelius, Jöns Jakob; Bichat, Marie-François-Xavier; Brown, Robert; Cuvier, Georges; Dalton, John; Davy, Sir Humphry; Gauss, Carl Friedrich; Gay-Lussac, Joseph-Louis; Geoffroy Saint-Hilaire, Étienne; Germain, Sophie; Humboldt, Alexander von; Ørsted, Hans Christian; Ritter, Johann Wilhelm).

I would like to thank Margaret Jacob, the Founders Library of Howard University, the Folger Shakespeare Library, and the Library of Congress.

Topic Finder

Instruments and Devices

Ballooning
Calorimeters
Hygrometers
Instrument Making
Leiden Jars
Lightning Rods
Microscopes
Rain Gauges
Telescopes
Thermometers

Language and Communication

Chemical Nomenclature
Education
Encyclopedias
Illustration
Literature
Metric System
Periodicals
Popularization

People

Colonial America and the United States

Bartram Family
Franklin, Benjamin
Michaux Family

France

Alembert, Jean Le Rond d'
Ampère, André-Marie
Berthollet, Claude-Louis
Bichat, Marie-François-Xavier
Buffon, Georges-Louis Leclerc de
Cassini Family
Châtelet, Gabrielle-Émilie du
Condillac, Étienne Bonnot de
Condorcet, Marie-Jean-Antoine-Nicolas Caritat, Marquis de
Coudray, Angelique Marguerite Le Boursier du
Coulomb, Charles-Augustin de
Cuvier, Georges
Diderot, Denis
Gay-Lussac, Joseph-Louis
Geoffroy Saint-Hilaire, Étienne
Germain, Sophie
La Mettrie, Julien Offroy de
Lagrange, Joseph-Louis
Lamarck, Jean-Baptiste-Pierre-Antoine de Monet de
Laplace, Pierre-Simon de
Lavoisier, Antoine-Laurent
Macquer, Pierre Joseph
Maupertuis, Pierre-Louis Moreau de
Nollet, Jean-Antoine
Poliniere, Pierre
Voltaire

Mathematics
Mechanics
Medicine
Meteorology
Oceanography
Optics
Physics
Physiology
Plant Physiology
Probability
Psychology
Surgeons and Surgery
Technology and Engineering
Zoology

Scientific Institutions, General

Academies and Scientific Societies
Botanical Gardens
Hospitals
Museums
Observatories
Universities

Scientific Institutions, Specific

American Philosophical Society
Arcueil, Society of
Berlin Academy
Bologna Academy of Sciences
Derby Philosophical Society
Imperial Academy of Sciences of St. Petersburg
Kew Gardens
Linnean Society
Lunar Society of Birmingham
Manchester Literary and Philosophical Society
Museum of Natural History (Paris)
Royal Academy of Sciences
Royal Society
Royal Society of Edinburgh
Royal Society of Medicine (France)
Royal Swedish Academy of Science
University of Edinburgh
University of Halle
University of Leiden

Theories and Ideologies

Atomism
Brunonianism
Cartesianism
The Enlightenment
Freemasonry
Hutchinsonianism
Materialism
Mesmerism and Animal Magnetism
Nationalism
Natural Theology
Naturphilosophie
Newtonianism
Phlogiston
Religion
Romanticism
Vitalism

Topics of Investigation

Agriculture
Comets
Earthquakes
Electricity
Fossils
Heat
Longitude Problem
Madness
Masturbation
Meteors and Meteorites
Polyps
Race
Sexual Difference
Toft Case
Transits
Volcanoes

Academies and Scientific Societies

Academies and scientific societies dominated institutional science in the eighteenth century more than ever before or since. Virtually every male scientist of importance was a member or correspondent of at least one scientific society. (Women, with very few exceptions, were barred from membership.) Society memberships were an important way of establishing one's standing as a member of the scientific community, and were frequently listed after the author's name on the title pages of books.

The formal chartering of a scientific society often occurred after years of organization and informal meetings of local savants, rather than being simply created by royal fiat. Societies and academies differed from these informal gatherings in that they had written rules, a fixed meeting schedule, a defined membership, and their own quarters. (A curious exception was the scientific society of the Holy Roman Empire, the Leopoldina, a relic from the seventeenth century that met wherever its president happened to live.) They usually had libraries and collections and were often charged with governmental functions such as overseeing observatories or advising the government on scientific and technological questions. Major societies, and many minor ones, also sponsored periodical publications and prize competitions. Scientific societies were only part of Europe's vast and growing array of learned institutions, and often a scientific society existed as one part of a large academy devoted to other topics as well, such as literature and industry. Groups specifically and exclusively devoted to science were usually found where there was a large concentration of resources, such as the national capitals of Europe's most powerful states. Other specialized societies, such as societies for medicine, technology, and economic development, also had scientific interests.

The two main patterns for scientific societies came from the oldest: Britain's Royal Society and France's Royal Academy of Sciences. Institutions following the Royal Society model dominated the British Isles, Britain's American colonies, and the Dutch Republic, whereas academies following the French model dominated the European continent. The Royal Society did not have a fixed number of members, was open to full membership by nonscientists, and lacked an internal hierarchy among its fellows, all of whom were theoretically equal. The Royal Academy was limited in membership, had salaried members, and divided them into a hierarchy of classes. The Royal Society did not pay its fellows, and supported its day-to-day activities through dues and other internal sources, whereas the Royal Academy was supported by the French state.

The first few decades of the eighteenth century saw new scientific societies spread beyond the original centers in Paris and London. The Berlin Academy was founded in 1700 and was followed by the Bologna Academy of Sciences in 1714, the Imperial Academy of Sciences of St. Petersburg in 1724, Sweden's Uppsala Academy in 1728, and the Royal Swedish Academy of Science in Stockholm in 1739. The London, Paris, Berlin, St. Petersburg, and Stockholm societies were the "big five" of the eighteenth century, internationally recognized as the most important and prestigious. The French provinces also produced several new academies in this period. The Montpellier Academy, closely associated with the medical faculty of the University of Montpellier, was founded in 1706 and was the only French provincial academy formally acknowledged as an equal by the Paris academy. Another important provincial academy was the Bordeaux Academy, founded in 1712 and known for its pioneering annual prize competitions in physics and natural history.

Communication and cooperation between scientific societies also grew in the first half of the eighteenth century, a process in which the Royal Society and the Imperial Academy of St. Petersburg were the leaders. Communication was mostly in the form of correspondence and the exchange of publications. Serious cooperation in joint scientific endeavors began in the 1750s and 1760s, with the transit observations and expeditions.

The second half of the eighteenth century saw a wave of society creation, as the leaders of most large European and colonial cities believed a society was essential to their cities' prestige. Several societies were formed in Germany: The Göttingen Academy, closely connected with its university, in 1752, followed by the Erfurt Academy in 1754 and the Bavarian Academy of Sciences in Munich in 1759. The Mannheim Academy, founded in 1763, is particularly noteworthy for its offshoot, the Meteorological Society of Mannheim founded in 1780, which had an ambitious program for worldwide weather observation employing standardized instruments and reporting forms. This was the last great cooperative effort of the scientific societies in the eighteenth century. Other institutions founded in this period were the Turin Society in 1759 (upgraded to an academy with paid members in 1783), the Padua Academy in 1759, the American Philosophical Society in 1768, the Brussels Academy in 1772, the Lisbon Academy in 1779, and the Boston Academy of Arts and Sciences (despite its name, a society-type organization) in 1780. Madrid and Vienna were the only European capitals to lack scientific societies, but the Academy of Natural Sciences and Arts of Barcelona was founded in 1770. The British Isles saw the founding of the Royal Society of Edinburgh, chartered in 1783, and the Royal Irish Academy in 1785. The North of England also had several less formal and state-connected bodies, notably the Manchester Literary and Philosophical Society.

Although many societies were showing a loss of vitality by the late eighteenth century, what ended the age of scientific societies was the French Revolution, which boldly abolished the Royal Academy of Sciences along with every other academy in Paris and the provinces in 1793. The Royal Academy was restored two years later in the form of the First Section of the Institute of France, but the provincial academies were not revived until after the fall of Napoléon in 1815, and never regained their old vitality. The Revolution and the Napoleonic Wars with their accompanying devastation also forced several continental scientific societies to close their doors, either temporarily, as Turin did between 1792 and 1801, or permanently, as Bologna did in 1804. A more long-term menace was the rise of smaller specialized societies. They had a long history, but were on the increase by the close of the eighteenth century. The various Linnaean societies for applying Carolus Linnaeus's botanical classification system had been either absorbed into the dominant society culture, as in England, or destroyed, as in France, but specialized societies continued to be created in the early nineteenth century, including France's Society of Arcueil, an elite but short-lived group specializing in physics and chemistry, and Britain's Geological Society, founded in 1807 by British geologists who believed the Royal Society was insufficiently attentive to their science. Despite Royal Society opposition, the new society proved a great success.

See also American Philosophical Society; Arcueil, Society of; Berlin Academy; Bologna Academy of Sciences; Derby Philosophical Society; Imperial Academy of Sciences of St. Petersburg; Linnean Society; Lunar Society of Birmingham; Manchester Literary and Philosophical Society; Periodicals; Royal Academy of Sciences; Royal Society; Royal Society of Edinburgh; Royal Society of Medicine; Royal Swedish Academy of Science; Transits.

References

McClellan, James E., III. *Science Reorganized: Scientific Societies in the Eighteenth Century.* New York: Columbia University Press, 1985.

Pyenson, Lewis, and Susan Sheets-Pyenson. *Servants of Nature: A History of Scientific Institution, Enterprises, and Sensibilities.* New York: W. W. Norton, 1999.

Agriculture

The so-called agricultural revolution of the eighteenth century sprang from scientific and nonscientific sources, but the period beginning from the mid-eighteenth century saw increasing scientific concern with agricultural productivity. Scientists in Europe and outside it exchanged seeds and information in the attempt to develop or acclimate new crops. The hope of agricultural improvement had a long history in the development of Western science, but despite some experimental work there had been little direct application of science to the problems of farmers. The most influential publication on agriculture in the eighteenth century, *The New Horse Houghing Husbandry* (1731) by Jethro Tull (1674–1741), was written by a farmer rather than a scientist. Influences on Tull's new theories included the science of his day, classical farming literature dating back to the ancient Romans, and his observations of viticulture in the south of France. Tull's new farming relied on thorough hoeing to make the resources of the soil more available to the plants and the use of a seed drill he invented for sowing, rather than scattering. Tull's more questionable ideas included opposition to manuring, which he regarded as valueless, and to crop rotation.

One area with a high level of interest in both science and agricultural development was Scotland, home of an early agricultural society, the Society of Improvers in the Knowledge of Agriculture, active between 1723 and 1745. One of the earliest works of agricultural chemistry was *Principles of Agriculture and Vegetation* (1757) by the Scottish professor Francis Home (1719–1813). It was translated into French and German, but its influence was limited by its old-fashioned chemistry, like Tull's, still based on the Aristotelian four elements.

The second half of the century saw the creation of many new institutions concerned with agriculture. Some were societies specifically devoted to agriculture; others included its improvement as part of an overall mission of economic development. There were also many agricultural periodicals founded in this period (agriculture was second only to medicine in the number of journals devoted to it), including the *Journal of Agriculture; Commerce and Finance,* which ran from 1763 to 1783; and *Annals of Agriculture,* which ran from 1784 to 1815, edited by the British agricultural writer Arthur Young (1741–1820).

Tull's work was brought to France in 1750 by Henri-Louis Duhamel du Monceau (1700–1782), himself from a landowning family. This was not simply a matter of translation, as Tull was a very obscure writer and his science had to be brought in line with current French thinking. The indefatigable Duhamel du Monceau also set up an experimental farm to test Tull's and other theories, and maintained a correspondence with other experimental farmers. The results of his work were published in 1762 in the two-volume *Elements of Agriculture,* frequently translated and reprinted.

The previous year had seen the foundation of the Paris-based Royal Society of Agriculture, the centerpiece of a vast state-led effort to promote agricultural enlightenment through the creation of a network of provincial associations. The later Committee on Agriculture of the Finance Ministry, founded in 1785, was a more bureaucratic body. Its secretary was Antoine-Laurent Lavoisier, himself an active experimental farmer. The French government also attempted to improve animal husbandry by encouraging the foundation of the world's first school of veterinary medicine at Lyon in 1762, and a second at Alfort, on the outskirts of Paris, in 1766. By the late eighteenth century, the idea of agricultural improvement through institution building had spread over Europe. In 1797, a Hungarian nobleman, György Festetics, founded the Georgikon, a school for agricultural technology, on his estate in Keszthely, hoping to improve its productivity.

The leading French agronomist in the decades before the French Revolution was the chemist Antoine-Augustin Parmentier (1737–1813), admitted as a member of the Royal Society of Agriculture in 1773 and best remembered as the great promoter of the potato in France. Unlike Duhamel du Monceau, Parmentier was laboratory oriented. Building on the work of Jacopo Bartolomeo Beccari (1682–1766), a Bolognese who had first broken down flour into gluten and starch, Parmentier launched an exhaustive series of chemical analyses of common food products, including bread and milk. His work on the potato was first put forth in *Chemical Analysis of Potatoes* (1773). Parmentier vigorously promoted the potato as a supplemental food that could be grown in soil inhospitable to grain rather than a dietary staple, having to overcome many prejudices against it. He gave famous dinners for members of the French elite (most of whom associated potatoes with poor peasants) in which every course was potato based. In the Napoleonic period, Parmentier was one of the many scientists working on the extraction of sugar from grapes and beets to substitute for British-controlled cane sugar.

British agricultural societies originally formed spontaneously rather than as part of a governmental effort. The London-based Society for the Encouragement of Arts, Commerce, and Manufactures, founded in 1754, offered prizes for agricultural innovations as well as industrial. The Society of Arts, as it was known, also supported agricultural innovators in the American colonies, after the American Revolution, with the founding of agricultural societies in Philadelphia and South Carolina in 1785. One of the most important provincial societies with an exclusively agricultural focus was the Bath and West and Southern Counties Society, founded in 1777. The society had connections with local scientific circles. It purchased a ten-acre plot of land for an experimental farm in 1779, although the effort came to nothing. In 1805, it set up a chemical laboratory for soil analysis.

The British government began to follow the French example of direct involvement in agricultural improvement in the 1790s, founding the London Veterinary College in 1792 and a board of agriculture in 1793. Beginning in 1803 and ending in 1812, the board sponsored a series of annual lectures by Sir Humphry Davy on agricultural chemistry, focusing on soil analysis and plant nutrition. The lectures were published in 1813 as *Elements of Agricultural Chemistry,* which was frequently reprinted and translated into French, Italian, and German.

The degree to which all this intellectual activity affected actual farming outside experimental farms is obscure. Many farmers distrusted "book farming" and thought the changes recommended by agricultural improvers to be too risky.

See also Botany; Plant Physiology; Technology and Engineering.

References

Ambrosoli, Mauro. *The Wild and the Sown: Botany and Agriculture in Western Europe, 1350–1850.* Translated by Mary McCann Salvatori. Cambridge: Cambridge University Press, 1997.

Gillispie, Charles Coulston. *Science and Polity in France at the End of the Old Regime.* Princeton: Princeton University Press, 1980.

Russell, Edward John. *A History of Agricultural Science in Great Britain, 1620–1954.* London: Allen & Unwin, 1966.

Alchemy

See Böttger, Johann Friedrich; Chemistry; Stahl, Georg Ernst.

Alembert, Jean Le Rond d′ (1717–1783)

Jean Le Rond d'Alembert was the only important mathematical scientist among the philosophes of the French Enlightenment. He received a fine mathematical education at the College of Four Nations in Paris, as well as training in the Cartesian physics he later rejected. After graduation he tried law, then medicine, but eventually decided on mathematics. He read his first paper to the Royal Academy of Sciences in 1739. Of minor interest in itself, it began d'Alembert's determined campaign to enter the academy. He was admitted in 1741. D'Alembert continued to present mathematical papers and published the work that made his reputation, *Treatise on Dynamics,* in 1743. In rational mechanics, the *Treatise on Dynamics* included what became known as "d'Alembert's principle" (although d'Alembert did not clearly formulate it), a way of reducing problems in dynamics to static terms, and a refinement of the Newtonian concept of force. Although d'Alembert affected to despise Cartesianism, he avowed his admiration for Descartes himself, and his early Cartesian education strongly influenced his physical theories. For d'Alembert, mechanics was a rational, deductive science, a branch of mathematics working from first principles, not an experimental one working from observed phenomena. D'Alembert's disdain for experiment and experimenters was legendary, and there is no evidence that he ever performed one.

The *Treatise on Dynamics* involved d'Alembert in a fierce feud, the first of many in his career, with another member of the academy, Alexis-Claude Clairaut (1713–1765), who was setting out a theory of dynamics at the same time. The two often worked on similar problems. Along with Leonhard Euler, both announced the inadequacy of Newton's inverse square law of gravitation to account for the motions of the Moon in 1747 and 1748. D'Alembert believed that the problem might be a magnetic attraction between Earth and the Moon, but Clairaut demonstrated in 1751 that the problem was a mathematical error shared by the three investigators rather than a flaw in Newton. D'Alembert acquired another enemy in 1746, when his paper on the causes of wind won a prize from the Berlin Academy that Daniel Bernoulli thought he himself deserved. D'Alembert and Bernoulli also quarreled over the mathematics of a vibrating string. In the course of solving this problem, d'Alembert solved the wave equation, a major feat. This paper was also submitted to the Berlin Academy and helped to cement an alliance between d'Alembert and the king of Prussia, Frederick the Great (r. 1740–1786), who invited d'Alembert to be president of the academy in 1752. D'Alembert, who found it hard to exist outside Paris, graciously declined. He continued to have great influence over the Berlin Academy, though, even as his relations with Euler, its mathematical star, grew worse. Euler often broadened and gave more clear and rigorous treatment to the mathematical ideas he shared with d'Alembert, which led to mutual accusations of plagiarism. The two were reconciled when d'Alembert visited Prussia in 1763 and recommended that Euler be appointed president of the Berlin Academy.

Outside the world of pure science, d'Alembert was active in the editing of the *Encyclopédie,* for which he wrote the famous

"Preliminary Discourse," a classic statement of the Enlightenment program first published separately in 1751, as well as most of the mathematical articles. The "Preliminary Discourse" made a great sensation, and the adulation he received began to turn d'Alembert's attention from mathematics to literature. He was already a lion of the Parisian salons due to his charm and delightful conversation (which contrasted with the heaviness and authoritarianism of his mathematical and scientific writings). As a philosophe, d'Alembert hoped that the highly mathematical version of natural science he espoused would furnish a basis for the reorganization of all knowledge. He shared the violent anticlericalism of his close friend Voltaire, along with a common dismay at the rise of materialism in the community of philosophes—a dismay that contributed to d'Alembert's decision to quit the editorship of the *Encyclopédie* in 1759, leaving it to the increasingly materialistic Denis Diderot.

By 1764, d'Alembert had abandoned original mathematical work after a bout of severe illness. He spent most of his labors advancing the cause of the philosophes in the intellectual arenas of Paris. He had been admitted to the French Academy, the dominant state-sponsored institution in the field of French literature, in 1754, and became its perpetual secretary in 1772. His power in Parisian intellectual life and his dominant influence over the Berlin Academy meant that he still influenced science as a patron and friend of younger mathematicians, the most notable being Pierre-Simon de Laplace, Joseph-Louis Lagrange, and the Marquis de Condorcet, Marie-Jean-Antoine-Nicolas de Caritat.

See also Diderot, Denis; Encyclopedias; The Enlightenment; Euler, Leonhard; Mathematics; Probability.

Reference

Hankins, Thomas. *Jean d'Alembert: Science and the Enlightenment.* Oxford: Oxford University Press, 1970. Reprint, New York: Gordon and Breach, 1990.

American Philosophical Society

The first enduring American scientific society was founded late in 1768 by a merger between two Philadelphia groups, the predominantly Quaker American Society for the Promotion of Useful Knowledge (founded 1766) and the predominantly Anglican American Philosophical Society (founded 1767). The official name of the new society was a compromise, "American Philosophical Society for the Promotion of Useful Knowledge." It was a typical scientific society of the type modeled on Britain's Royal Society, with a large membership and a small body of officers. It was supposed to be supported by admission fees and dues, which often went unpaid.

Two events in 1769 put the American Philosophical Society on the scientific map. One was the election of Benjamin Franklin as its president in abstentia. Franklin held that position until his death in 1790. The other was the society's participation in the collection of astronomical observations of the transit of Venus. The society applied for and received a grant from the Pennsylvania legislature for this purpose, and coordinated and collected the observations of astronomers throughout the British colonies in North America. The data gathered occupied a major part of the first volume of the society's *Transactions,* published in 1771. Copies were sent to a number of European scientific societies and distributed in Europe by Franklin. They made a favorable impression on many European astronomers. Subsequent volumes of the *Transactions* appeared intermittently. The society also engaged in activities relating to economic development.

The American Philosophical Society suffered from the general decline of American science after the American Revolution, although Franklin's return to Philadelphia from France resulted in the erection of a new building, Philosophical Hall. The surveyor and astronomer David Rittenhouse (1732–1796) succeeded Franklin as president. The society sponsored and was involved in raising the funds for André Michaux's abortive expedition through the American West in 1793. America's first natural history museum was established in the society's rooms in 1794 and would later be the principal repository of the zoological specimens from the Lewis and Clark Expedition. On his death, Rittenhouse was succeeded by Thomas Jefferson (1743–1826), and the society was politically identified with Jeffersonian republicanism. Distinguished American members included Joseph Priestley and the first four presidents of the United States; foreign members included the Marquis de Lafayette (1757–1834) and one woman, the Russian princess Yekaterina Dashkova (1744–1810). The society exists to the present day.

See also Academies and Scientific Societies; Bartram Family; Colonial Science; Franklin, Benjamin; Lewis and Clark Expedition; Michaux Family; Transits.

References

Greene, John C. *American Science in the Age of Jefferson.* Ames: Iowa State University Press, 1984.

McClellan, James E., III. *Science Reorganized: Scientific Societies in the Eighteenth Century.* New York: Columbia University Press, 1985.

Stearns, Raymond Phineas. *Science in the British Colonies of America.* Urbana: University of Illinois Press, 1970.

Ampère, André-Marie (1775–1836)

The French scientist André-Marie Ampère is best remembered as the founder of electrodynamics, but he had a wide range of scientific interests. As a youth, Ampère was largely self-taught in the sciences. His happy childhood was ended by a series of disasters—his father was executed by the Jacobins in 1793, and his first wife and beloved elder sister died young. His bad personal luck would continue throughout his life—his second marriage would be a disaster and his relations with his son and daughter were fraught with conflict.

In 1802 he became a physics teacher for the school of the department of Ain. In 1804 he moved to Paris and was appointed a tutor in the Polytechnic School, where he was promoted to professor in 1815. In 1808 he was appointed inspector for the Imperial University, a position that meant he spent every summer traveling through France, inspecting local schools. Ampère's original scientific work in this period was mostly in mathematics, probability, and the calculus of variations.

Although respected in the scientific community of Napoleonic Paris, Ampère was something of an outsider. He was a religious Catholic and interested in issues of metaphysics that most French scientists ignored. His admission to the First Section of the Institute in 1814 was based on pure mathematical work on the classification of partial differentials. This work was not congenial to him and was out of the main lines of French mathematical research. Despite his preference for science over pure mathematics, he was not a member of the elite physics and chemistry group, the Society of Arcueil. He did not accept the dominant school of "Laplacian" physics, focused on short-range interactions between particles, preferring fluid theories. Although Ampère corresponded with Sir Humphry Davy on chemical matters, he did not have the laboratory skills that were needed to be accepted as an equal by chemical leaders. As a chemist, he independently arrived at Amedeo Avogadro's hypothesis of identical volumes of a gas containing the same number of particles, and in 1816 set forth a classification scheme for the elements based on their chemical properties. Ampère also became a close friend of another outsider in French science, Augustin-Jean Fresnel (1788–1827), and was one of the few French scientists to support his wave theory of light.

Ampère's great scientific opportunity came with Hans Christian Ørsted's discovery of the relation of magnetic and electrical force. The Laplacians treated electricity and magnetism as completely different realms, and did not immediately realize the significance of Ørsted's discovery. Ampère plunged into a course of experiments, discovering the attractive and repulsive forces between electrical currents. Ampère's work established that electricity and magnetism were the same force, rather than being two different "fluids." He set forth his theory in mathematical form in *Memoir on the Mathematical Theory of Electrodynamic Phenomena,* published in slightly different forms in 1826 and 1827. This work is considered to have founded electrodynamics, an achievement recognized in 1881 by the Paris Congress of Electricians, which designated the unit of electric current the ampere.

After publication, Ampère lost interest in electrodynamics. His health began to deteriorate in 1829, and most of his intellectual energy went into an elaborate treatise on the classification of the sciences, *Essay on the Philosophy of Sciences* (two volumes, 1834–1843).

See also Electricity; Napoleonic Science; Ørsted, Hans Christian; Physics.

Reference

Hofmann, James R. *André-Marie Ampère.* Cambridge: Cambridge University Press, 1996.

Anatomy

Anatomy progressed slowly rather than by dramatic leaps in the eighteenth century. The larger features of the human body had been delineated by the late seventeenth century, and further progress depended on ever more meticulous dissection and the use of the microscope. The "iatromechanical" model of the body as a machine, dominant in early-eighteenth-century medicine, encouraged anatomical investigation. (By contrast, the leading vitalist Georg Ernst Stahl treated anatomy as unimportant; what determined health was the condition of the soul, which animated the body.) Physicians' need to distinguish the learned medicine they studied from the "empirical" medicine practiced by lesser practitioners and quacks also motivated them to acquire anatomical knowledge in scholarly surroundings such as universities. Surgeons, seeking to raise the status of their discipline, also sought anatomical knowledge.

Private schools spread as venues for the teaching of anatomy, particularly in England, where university medical education was moribund. Among the first to give anatomy lessons in London, which became the center of English medical training due to its anatomy schools, was William Cheselden (1688–1752). Cheselden, a surgeon and fellow of the Royal Society, was also author of *The Anatomy of the Humane Body* (1713), frequently reprinted to the end of the century. The most significant of these schools was the one run by William Hunter and carried on by his nephew Matthew Baillie in the late eighteenth and early nineteenth centuries.

Anatomy at Europe's leading medical school for most of the period, the University of Edinburgh, was in the hands of the longest-lived professorial dynasty in history, the three generations of Alexander Monros. The first Alexander Monro (1697–1767), a founder of Edinburgh's medical greatness, was appointed to the chair of surgery and anatomy in 1726. An Alexander Monro held the chair continuously until 1846, when the first Alexander's grandson retired. The first Alexander was best known as the author of the frequently reprinted *The Anatomy of the Human Bones and Nerves* (1741).

The pioneer of morbid anatomy, anatomy of the diseased rather than the healthy body, was the Padua professor Giovanni Battista Morgagni (1682–1771). His *Sites and Causes of Diseases* (1761) drew on hundreds of dissections. Morgagni described the anatomical phenomena associated with circulatory conditions such as angina pectoris and myocardial degeneration. He helped make anatomy central to medicine by showing the link between diseases and problems in specific organs. *Sites and Causes of Diseases* had an immediate impact and was translated into English and German. Morgagni's work was followed up in Baillie's *Morbid Anatomy of Some of the Most Important Parts of the Human Body* (1793). This is one of the many anatomy books that benefited from the remarkable skill of copperplate engravers, who produced the most detailed, accurate, and elegant anatomical illustrations yet.

Another Italian who became known for his

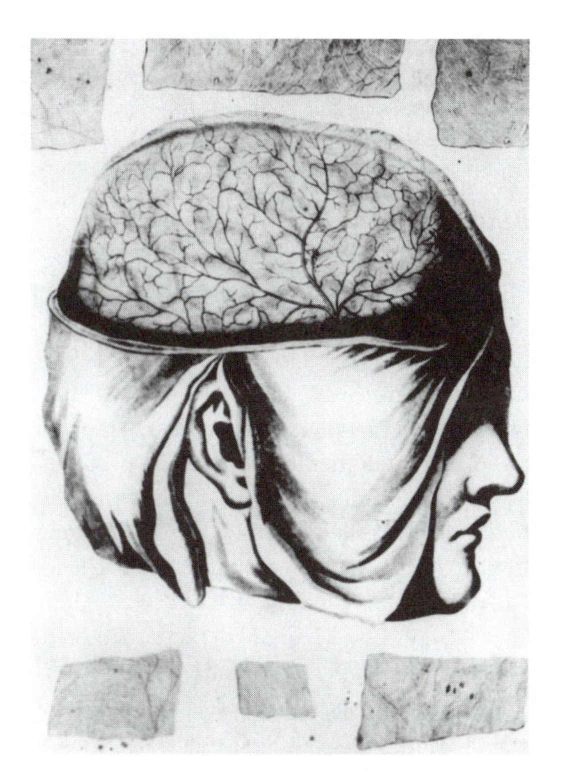

This illustration of a cutaway brain is from Félix Vicq d'Azyr's Treatise on Anatomy and Physiology *(1786). Azyr, founder of the Royal Society of Medicine, was also a pioneering brain anatomist. (National Library of Medicine)*

work on the anatomy of the heart was the Pavia professor Antonio Scarpa (1752–1832), author of *Neurological Tables* (1794), which for the first time clearly delineated the nerves of the heart and also described the condition later known as arteriosclerosis. Scarpa's contemporary, the Siena professor Paolo Mascagni (1752–1815), was the author of *Lymphatic Vessels of the Human Body* (1787), one of a number of important books on the lymphatic system published in the period.

A perpetual problem was getting corpses for dissection. Most places relied on the bodies of executed criminals, and people resisted handing over bodies for dissection, viewing the process with horror. (One reason for the bad reputation of medical students was the belief, sometimes justified, that they robbed graves or paid others to do it to get bodies for dissection.) One way of partially compensating for the shortage of bodies was the establishment of an anatomical collection of normal and abnormal body parts. Many great collections were built in the eighteenth century. Both William Hunter and John Hunter had large collections, as did Albrecht von Haller at the University of Göttingen. Another German collector was Johann Friedrich Meckel (1781–1833).

Another problem with relying on executed criminals was that the vast majority of them were men. Although there was great interest in the anatomy of the female reproductive system, general knowledge of the female body lagged behind knowledge of the male body. The German anatomist Samuel Thomas von Soemmerring (1755–1830), best known for his classification of the cranial nerves and his work on sense organs, attempted to remedy this situation in his *Table of Women's Bones with Description* (1797). Soemmerring attempted to establish the difference between men's and women's bodies by a painstaking examination and analysis of a female skeleton. Like other late-eighteenth- and early-nineteenth-century anatomists, notably the Dutchman Pieter Camper (1722–1789), Soemmerring was also interested in finding anatomical bases for racial divisions.

See also Hunter Family; Medicine; Physiology.

References

Persaud, T. N. *A History of Anatomy: The Post-Vesalian Era.* Springfield, IL: Charles C. Thomas, 1997.

Porter, Roy. *The Greatest Benefit to Mankind: A Medical History of Humanity.* New York: W. W. Norton, 1998.

Schiebinger, Londa. *Nature's Body: Gender in the Making of Modern Science.* Boston: Beacon, 1993.

Arcueil, Society of

The Society of Arcueil, which existed from 1807 to 1813, brought together many of Napoleonic France's most brilliant physicists and chemists in an informal setting. Its leaders were Claude-Louis Berthollet and Pierre-Simon de Laplace, who owned adjoining houses in the village of Arcueil, outside Paris. Other leading members included Alexander von Humboldt, Joseph-Louis Gay-Lussac, Louis-Jacques Thénard (1777–1857), and Pierre-Louis Dulong (1785–1838). The friendship and intellectual alliance between the chemist Berthollet and the mathematical physicist Laplace greatly strengthened the society, as its members approached problems in a way that integrated mathematics, physics, and chemistry. Most members were much younger than the two leaders, and Berthollet and Laplace acted as patrons to the younger men. Although scientific meetings had been going on for several years at Arcueil, the founding of the society is dated to 1807, when the first volume of *Memoirs of Physics and Chemistry of the Society of Arcueil* appeared, most prominently featuring a collaborative paper by Humboldt and Gay-Lussac on terrestrial magnetism. (Subsequent volumes were published in 1809 and 1817.) Meetings were irregular and devoted to conversation, the performance of experiments (Berthollet's house included an excellent laboratory), and the reading of memoirs. After some inactive years from 1808 to 1810, when members were busy with other projects, the society's activity reached its height in the period from 1811 to 1813. It apparently collapsed as an organized group in the chaos of the last years of Napoléon's empire, but Arcueil remained a center of scientific activity in the Paris region.

Important scientific advances associated with the society include Gay-Lussac's discovery of the law of combining volumes of gases and the discovery of the polarization of light by Étienne-Louis Malus (1775–1812). Outside physics and chemistry, the *Memoirs* published important work by Humboldt on isothermal mapping and by Augustin-Pyrame de Candolle (1778–1841) on plant geography. Several younger members of the society went on to be leaders of French science in the first half of the nineteenth century.

See also Academies and Scientific Societies; Berthollet, Claude-Louis; Gay-Lussac, Joseph-Louis; Humboldt, Alexander von; Laplace, Pierre-Simon de; Napoleonic Science.

Reference

Crosland, Maurice. *The Society of Arcueil: A View of French Science at the Time of Napoleon I.* Cambridge: Harvard University Press, 1967.

Astronomy

Although astronomy did not undergo the revolutionary changes in the eighteenth century that it had earlier, such as the triumph of heliocentrism or the introduction of the telescope, it developed on a number of fronts. Newton's understanding of the motions of the planets was further refined and perfected by a group of mostly French scholars, whose most distinguished members were Pierre-Louis Moreau de Maupertuis, Alexis-Claude Clairaut (1713–1765), Jean Le Rond d'Alembert, Joseph-Louis Lagrange, Pierre-Simon de Laplace, and the Swiss Leonhard Euler. More powerful telescopes made possible the discovery of many more celestial objects, including the first new planet ever to be discovered, and, in the early nineteenth century, the first asteroids. Painstaking observations led to new discoveries about the motion of the stars, and of the solar system itself.

Purely in terms of the resources devoted to it, astronomy was a booming science in the eighteenth century. Rulers and governments funded a number of new observatories, and the practical applications of astronomy in navigation and cartography were apparent to all. Expeditions and coordinated observations such as those connected with the transits of Venus in 1761 and 1769 fostered cooperation between national scientific groups and made astronomy a worldwide science. At home, astronomy employed more people than any other science save medicine. Beyond the community of astronomers, astronomical knowledge was spread to large numbers of people through textbooks, lectures, and the sale of astronomical equipment.

The struggles over Newtonianism and its eventual triumph led scientists working on celestial mechanics to solve the remaining problems of planetary motion. The Moon presented a number of questions. The rotation of the Moon was a special case of what became known as the "three-body problem," the determination of the mutual effects of the gravitation of three bodies, in this case the Moon, the Sun, and Earth. Determining the perturbations of the orbit of one planet by the gravitational force of another is also a three-body problem. The three-body problem did not admit of a general solution, but in 1753 Euler demonstrated a way of solving it practically by successive approximations in his *Theory of the Moon's Motion, Showing All Its Inequalities*. Euler also contributed to the debate over the other major lunar problem, the "secular acceleration"—the fact, discovered by Edmond Halley (1656–1742), that the Moon's motion in its orbit of Earth seemed to be accelerating over long periods of time. Euler suspected that the acceleration could not be explained by Newtonian forces and suggested it was due to the resistance of an ethereal fluid. Lagrange also failed to explain the acceleration in Newtonian terms. The solution of the problem was left for Laplace, the greatest celestial mechanist, who announced in 1787 that the acceleration was caused by changes in the orbit of Earth around the Sun, and that after 24,000 years the Moon would decelerate. Laplace's other discoveries, such as the explanation for the changes in speed between Saturn and Jupiter, tended in the direction of establishing the stability of the solar system, a doctrine enshrined in his five-volume *Treatise on Celestial Mechanics* (1799–1825), the culmination of eighteenth-century Newtonianism in astronomy.

The refinement of telescopes and the spread of observatories enabled astronomers to discover a range of new celestial objects. Comet hunting was one popular activity. The greatest comet hunter of the mid-eighteenth century, the Frenchman Charles Messier (1730–1817), sought to reduce the confusion over what was a comet and what was a bright patch in the sky. His catalog of nebulae, first published in 1774, eventually included 101 items. These brighter nebulae are still referred to by "Messier numbers." Messier's catalog inspired the greatest astronomical observer of the period, William Herschel, who unlike Messier used a reflecting rather than a refracting telescope, a choice that would become the norm for those astronomers interested in new discoveries. Herschel, along with his sister Caroline, found an enormous number of new astronomical phenomena, including thousands of nebulae and several planetary moons. The most spectacular of all was the planet Uranus, the first planet discovered in historic times.

The first new kind of object to be discovered in the solar system was anticipated before its discovery as the result of an irregularity in the distribution of the planets. In 1772 the German astronomer Johann Bode (1747–1826) presented what became known as "Bode's law" (although Bode himself acknowledged its earlier formulation by J. D. Tietz [1729–1796]). Bode's law expresses the distances between the planets in terms of a doubling sequence, three, six, twelve, twenty-four, and so on, added to a base of four. After cheating slightly for Mercury, Bode's formula for the distances between the planetary orbits fitted the existing data well, with one exception, and seemed to receive further confirmation with the discovery of Uranus, which was very close to where Bode's formula would have predicted it. The exception was the gap between Mars and Jupiter, which Bode's formula indicated should have been filled by another planet. In 1801, Giuseppi Piazzi (1746–1826), while working on a star catalog, found a tiny object in the gap, although at first he was unsure whether it was a planet or a comet. Carl Friedrich Gauss, in his first important astronomical work, established the body's orbit. Piazzi gave the object the name Ceres after the Roman goddess of agriculture, establishing the tradition that asteroids (a term coined by Herschel) are named for goddesses. Measure of the object revealed that it was less than 300 kilometers in diameter, far too small to be the missing planet. Subsequent asteroids discovered included Pallas, discovered in

1802 by Heinrich Wilhelm Olbers (1758–1840); Juno, discovered in 1804 by Karl Ludwig Harding (1765–1834); and Vesta, discovered in 1807 by Olbers. Vesta was the last asteroid found until 1845. Some astronomers theorized that the missing planet had at one point fallen into fragments.

Outside the solar system, improved telescopes made possible the discovery and mapping of many faint stars. Accuracy in stellar mapping was facilitated by English astronomer James Bradley's discovery of two sources of distortions, the aberration of starlight and the nutation of Earth. Bradley's work was further refined by the German astronomer Friedrich Wilhelm Bessel (1784–1846), director of the observatory at Königsberg, in the early nineteenth century. This was one aspect of the transfer of leadership in precision astronomy and the making of precision astronomical equipment from England to Germany beginning in the late eighteenth century.

The most dramatic development in stellar astronomy during this period was the discovery of the "proper motion" of the stars. Proper motion means that the stars are not static entities whose apparent motions are all the result of the motion of the observing platform, that is, Earth, but are themselves moving. The idea of proper motion was put forth by Halley in 1718, in a paper comparing ancient Greek observations of the stars to modern ones. The idea received backing from the French astronomer Jacques Cassini in 1738, who demonstrated the idea entirely on the basis of modern observations. Johann Tobias Mayer published a table of stellar motions in 1760, and speculated on how it could be used to establish the overall motion of our own solar system among the stars. Herschel used Mayer's idea that the stars would appear to disperse in the direction in which the Sun and planets were heading, and converge in the opposite direction, to establish the Sun's motion in the direction of a point in the constellation Hercules.

The detection of proper motions helped encourage astronomers and cosmographers to think of the universe as a system in change and development, as opposed to a static one. Thomas Wright (1711–1786), Immanuel Kant, and Johann Heinrich Lambert (1728–1777) all put forth theories of the structure of the universe, including the solar system. Lambert's work inspired Herschel to study the Milky Way and give an approximate description of its shape.

See also Bradley, James; Cassini Family; Comets; Herschel Family; Longitude Problem; Maskelyne, Nevil; Mayer, Johann Tobias; Meteors and Meteorites; Observatories; Telescopes.

Reference

North, John. *The Norton History of Astronomy and Cosmology.* New York: W. W. Norton, 1995.

Atomism

The eighteenth-century atomistic tradition rested on the ideas of Sir Isaac Newton (1642–1727) as set forth in the "Queries" attached to the second edition of his *Opticks* (1706). Newton followed the tradition of the ancient atomists in describing atoms as solid, massy particles; indestructible; and eternal. (The word *atom* is derived from the Greek word for *uncuttable.*) He agreed that atoms existed in a void and added, against ancient atomists but like most Christian atomists, that atoms had been directly created by God. Where Newton innovated was in describing the way that atoms cohered, always a weak point of classical atomism. Rather than use the classical explanations that turned on particle shape (Cartesianism, although not an atomistic philosophy, used similar explanations on how minute particles interacted), Newton theorized that atoms interacted through forces of attraction and repulsion, which operated at very short ranges, just as objects in the visible world interacted by gravity, operating at long ranges. This remained the dominant theory of matter through the early nineteenth century, when it was the basis of the physics of Pierre-Simon de Laplace and his followers, the "Laplacian physicists." Only a few thinkers in the eighteenth century denied the existence of atoms outright, the most notable being the Irish idealist George Berkeley (1685–1753), who denied the existence of matter altogether, and Immanuel Kant, who held matter infinitely divisible because space was infinitely divisible.

Several eighteenth-century thinkers portrayed atoms as more energetic than did Newton. Vitalists such as Denis Diderot suggested that atoms were somehow alive. The most radical approach to atomism was that of Ruggiero Giuseppe Boscovich, who realized that once one had the attractive and repulsive forces, there was no need for the material atom itself. Boscovich's theory of "point atoms," followed by Joseph Priestley, treated the atom entirely as a locus of forces rather than a material body. Atoms were identical; individual entities differed in the arrangement of their atoms.

The most important post-Newtonian development in atomism was the rise of chemical atomism. Chemists' adoption of atomism was partly a reaction to Antoine-Laurent Lavoisier's emphasis on the conservation of matter. John Dalton, whose *A New System of Chemical Philosophy* (1808) was the founding text of chemical atomism, conceived the fixed ratios at which substances combined chemically in terms of individual particles, each one characteristic of one particular substance. What distinguished the atoms of one substance from the atoms of another was primarily weight. It was not clear to many, however, whether atoms, which could not be directly perceived, were merely a handy way of understanding chemical combinations, or actually existed—a question chemists debated through the nineteenth century.

See also Avogadro, Amedeo; Boscovich, Ruggiero Giuseppe; Chemistry; Dalton, John; Physics.

References

Pullman, Bernard. *The Atom in the History of Human Thought*. Translated by Axel Reisinger. Oxford: Oxford University Press, 1998.

Toulmin, Stephen, and June Goodfield. *The Architecture of Matter*. Chicago: University of Chicago Press, 1962.

Avogadro, Amedeo (1776–1856)

The physical scientist Amedeo Avogadro was the first to use the concept of a "molecule," but his innovation spread very slowly, partly due to his isolation. From a noble family in the small kingdom of Sardinia in northwestern Italy, Avogadro was educated not as a physician or a scientist, but as a lawyer. He was self-educated in the sciences. The first evidence of Avogadro's scientific activity is two essays on electricity he submitted to the Academy of Turin in 1803 and 1804. This was part of the burst of excitement in European science caused by Alessandro Volta's battery of 1800. Avogadro made a definitive career change in 1806, when he took a teaching position at a boarding school in Turin, the former capital of the kingdom, then under French occupation. He became professor of physics at the College of Vercelli in 1809, and professor of mathematical physics at the University of Turin in 1820.

Avogadro's most important contribution to physical science emerged from his attempt to reconcile Joseph-Louis Gay-Lussac's law of the combining volumes of gases with the atomic hypothesis of John Dalton. Two articles printed in the French *Journal of Physics* in 1811 and 1814 set forth his "molecular hypothesis." Avogadro argued that the equal volumes of any gas, under the same conditions of temperature and pressure, contained an equal number of molecules, a claim that became known as "Avogadro's law." (Avogadro had nothing to do with "Avogadro's number," which was determined only in 1941.) He also claimed that some gases, such as oxygen, were in their natural state composed of double molecules—two simple molecules (Avogadro did not use the term *atom* in this period of his career) combined. Avogadro's dramatic claims rested on the interpretation of already existing data rather than new experimental evidence—he was never a significant experimenter.

The reason for the nearly unbroken silence with which the molecular hypothesis was greeted is one of the classic questions in the history of science. Even when the hypothesis was discussed in the ensuing decades, credit was as likely to be given to the later work of André-Marie Ampère as to Avogadro. Avogadro's isolation from the centers of European science north of the Alps, accentuated when Sardinia regained its independence at the fall of the Napoleonic empire in 1814, was one factor contributing to this, as were his modest and retiring personality and lack of a reputation as an experimenter. Some major chemical works he published in the early 1820s in *Memoirs of the Royal Academy of Sciences of Turin* attracted no international attention although written in French, and Avogadro turned his attention from chemistry to physics. His enormous *Physics of Ponderable Bodies* was published in four volumes from 1837 to 1841 in Italian. It too had limited influence, partly because Avogadro's conservative physics was still basically "Laplacian," in line with the dominant school of the Napoleonic era. For example, Avogadro still accepted the caloric theory of heat, then in decline. When he died in 1856, Avogadro was a prophet without honor in his own country or anywhere else, and his reputation began to climb only with the acceptance of the molecular theory.

See also Atomism; Chemistry; Physics.

Reference

Morselli, Mario. *Amedeo Avogadro: A Scientific Biography.* Dordrecht, Netherlands: D. Reidel, 1984.

B

Ballooning

The most dramatic, if not the most significant, evidence of humanity's increasing power over nature in the Enlightenment was manned flight by balloon. Initial development of flying balloons was due to two paper-manufacturing brothers in the French town of Annonay, Joseph-Michel Montgolfier (1740–1810) and Jacques-Étienne Montgolfier (1745–1799). Both had some scientific education and shared in the movement toward state-supported technical experimentation common in late-eighteenth-century France. Joseph-Michel seems to have originated the idea of flying balloons, influenced by the discovery of different gases, some of which were lighter than air. The early balloons, however, used heated air rather than a specific gas. Hot air expanded to a lesser density than the surrounding atmosphere, although there was some confusion over whether hot air or smoke was the lifting agent. By the late 1770s Joseph-Michel was making calculations and performing small experiments, but the practical Jacques-Étienne was the one in charge of the plant and its resources. Jacques-Étienne at first dismissed his brother's activities, but by late 1782 the two were working together. On 4 June 1783 the brothers demonstrated an unmanned hot-air balloon before the provincial Estates of Vivarais.

Word quickly reached Paris, and the inevitable step of appointing a commission from the Royal Academy of Sciences followed. The group invited Jacques-Étienne Montgolfier to Paris, where he leveraged his newfound celebrity to the advantage of the Montgolfiers and their paper mill. The experimental physicist Jacques-Alexandre-César Charles (1746–1823) on 27 August successfully launched a small balloon whose lift came from hydrogen rather than hot air. (Charles's balloon research ultimately led to the formulation of "Charles's law" of the relation of pressure, temperature, and volume in an enclosed gas.) Montgolfier's 12 September launch of a large hot-air balloon failed due to its destruction by sudden rain. Another launch before the king and court at the Royal Palace of Versailles on 20 September successfully carried a sheep, rooster, and duck. The next step was to carry a human being. The first people to fly were a young man named Jean-François Pilâtre de Rozier (1756–1785), who had actively promoted the idea, and his friend François Laurent, Marquis d'Arlandes. On 21 November they flew for about seventeen minutes at a height of about 1,500 feet. (In 1785, Pilâtre de Rozier became the first balloonist killed, when his combination of a hydrogen and a hot-air balloon blew up in an unsuccessful attempt to cross the English Channel.) Charles continued to work on the hydrogen balloon, which had the advantage of avoiding the cooling problem, and reached a height of 9,000 feet on 1 December. By this time, ballooning was becoming a European craze, with free speculation that humans would eventually reach the Moon and the planets via balloon. The first British balloon ascents took place in 1784, and the first American ascent in 1793. The first woman ascended in 1798. Hydrogen balloons won the struggle with hot-air balloons.

Ballooning, whether hot air or hydrogen, presented a number of technical problems, principally in the area of regulating rise and descent. Leonhard Euler's last calculation, found after his death on his chalkboard, was about the mechanics of the rise and fall of balloons. The one to solve the problem was the French engineer Jean-Baptiste-Marie-Charles Meusnier de la Place (1754–1793). Meusnier suggested that the balloon incorporate a bladder, which could take on or discharge air from the atmosphere, thus providing a way to stabilize the balloon rather than oscillating between rising by throwing off ballast and descending by releasing gas from the balloon itself. His work won him admission to the Royal Academy of Sciences.

The use of the balloon for science reached a peak in 1804. Three attempts to measure the density, humidity, and temperature of the upper atmosphere as well as the performance of magnetic, electrical, and optical equipment took place that year. The first was carried out by the Imperial Academy of Sciences of St. Petersburg, with little success. Two French attempts involved Joseph-Louis Gay-Lussac and the physicist Jean-Baptiste Biot (1774–1862). On the second expedition, carried out by Gay-Lussac alone, he reached a height of 23,000 feet, the highest yet reached. The ascents did reveal new data about the atmospheric and electrical conditions at great height, but interest in scientific ballooning waned thereafter.

See also Popularization; Royal Academy of Sciences; Technology and Engineering.

References

Bacon, John MacKenzie. *The Dominion of the Air: The Story of Aerial Navigation.* Philadelphia: David Mackay, 1903.

Gillispie, Charles Coulston. *Science and Polity in France at the End of the Old Regime.* Princeton: Princeton University Press, 1980.

Banks, Sir Joseph (1743–1820)

Sir Joseph Banks, explorer, collector, and president of the Royal Society for more than four decades, dominated the public face of late-eighteenth- and early-nineteenth-century English science. From a rising family of lawyers and gentry, Banks manifested an early interest in natural history. In 1776 he accompanied a naval expedition to Newfoundland, collecting specimens and making observations. The same year he was elected a fellow of the Royal Society.

Banks made his reputation by accompanying Captain James Cook in the *Endeavour* voyage in the South Pacific from 1768 to 1771 as a natural historian and representative of the Royal Society. Banks was accompanied by an entourage of eight men and his friend, the disciple of Carolus Linnaeus, Daniel Carl Solander (1733–1782). Banks himself was a great admirer of Linnaeus, employing the Linnaean binomial nomenclature and classifying plants by the Linnaean "sexual system." He made enormous collections and kept a voluminous journal of his observations, not published until the twentieth century. (Banks published very little, mostly because of his own reluctance.) On his return to England he became the great celebrity of the voyage. Plans to accompany Cook's next South Sea voyage never came off, and Banks went on a short exploratory trip to Iceland instead.

Banks's friendship with George III (r. 1760–1820) enabled him to make an important institutional contribution to botany, the establishment of Kew Gardens. Banks was appointed director of the gardens at Kew by the king in 1772 and made them one of the great botanical centers of the world. He supervised an enormous network of botanists and natural historians on British ships and throughout the British Empire. (Not all of them were British. Banks had a particularly high regard for graduates of German and Scandinavian universities.) Like many of Banks's projects, Kew combined scientific curiosity and the hope of fostering economic development through finding useful plants for cultivation. Banks was heavily involved in a project to end Britain's unfavorable trade balance with China by acclimating tea in India, a project that succeeded only after his death. Another project that he became involved in later in response to a royal request was one to improve the quality of English wool by importing and acclimating Spanish merino sheep. He received a number of honors from the king and served on many government commissions.

In 1778 Banks was elected president of the Royal Society on the resignation of Sir John Pringle (1707–1782). Banks was a successful but controversial president. His strengths were conscientious devotion to duty and an amazing capacity for work—the letters he wrote and received in his presidency number well over 100,000. His social skills and hospitality also made his house in Soho Square a gathering place for British and European scientists. His diplomatic nature and French respect for him enabled Banks to maintain scientific communication between the two countries during the Revolution and Napoleonic Wars and on occasion to procure the release of British savants held by the French and Frenchmen held by his own government. He served as patron to many struggling naturalists, the most important being his librarian, Robert Brown, Britain's leading botanist.

Banks's weakness, which increased in severity with age and gout, was an autocratic tendency. He was strongly identified with natural history, and some more mathematically inclined fellows believed that their kind of science was not being adequately supported. He successfully fought off one challenge from the mathematicians in 1782 and 1783, and was not seriously challenged again. As his presidency wore on, Banks was also seen as an obstacle by those (including his immediate successor, Sir Humphry Davy) who wanted to reform the society to make it a strictly professional group composed only of scientists. Banks also put up an increasingly hopeless resistance to the foundation of specialized scientific societies in the early nineteenth century, fearing that they would weaken the Royal Society.

See also Academies and Scientific Societies; Botany; Brown, Robert; Cook, James; Exploration, Discovery, and Colonization; Kew Gardens; Royal Society.

References

Miller, David Philip, and Peter Hanns Reill, eds. *Visions of Empire: Voyages, Botany, and Representations of Nature.* Cambridge: Cambridge University Press, 1996.

O'Brian, Patrick. *Joseph Banks: A Life.* Boston: David R. Godine, 1993.

Bartram Family

The extended family centering on John Bartram (1699–1777) was a powerful contributor to the development of natural history in British North America and the early United States. Beginning around 1727, Bartram, a farmer, became interested in botany, a subject in which he was largely self-taught. Through a London correspondent and fellow Quaker, Peter Collinson (1694–1768), Bartram became a participant in the international world of botanical exchange, sending seeds, bulbs, and cuttings of American plants to Europe and receiving European and other foreign specimens in exchange. Bartram's skill became recognized in the learned community. He acquired Benjamin Franklin as a patron and Carolus Linnaeus as an admirer. Bartram contributed specimens to enable Mark Catesby, currently in London, to finish his *Natural History of Carolina, Florida, and the Bahama Islands* (1731–1748). In 1728 he established a botanical garden that attracted many visitors. He was a member of the American Philosophical Society.

The height of Bartram's botanical career

was an expedition to the southeastern British colonies in 1765, when Bartram was appointed by George III (r. 1760–1820) as king's botanist. This appointment shocked some in America's learned community, as Bartram, for all his undoubted skill in collecting, was not a learned botanist—he had little knowledge of Latin and the science of botany as it had developed in the eighteenth century. Regardless of this skepticism, the trip was a success.

John Bartram's work attracted his cousin Humphrey Marshall (1722–1801) to science as both an intellectual activity and a business. He supplemented Bartram's export of botanical specimens by sending small American animals and insects to collectors in Europe. Marshall, whose interests were broader than Bartram's, was also a keen astronomer whose observations of sunspots were published in *Philosophical Transactions.* His catalog of American trees and shrubs, *Arbustrum Americanum* (1785), attracted considerable interest in Europe, with two reprintings.

Two of John Bartram's sons, Isaac (1725–1801) and Moses (1732–1809), were apothecaries, active members of the American Philosophical Society, and dabblers in the sciences, but the most important scientist among the next generation was William Bartram (1739–1823). William Bartram's keen interest in botany and natural history was manifest in his youth, causing his father to despair of finding him a remunerative career. Bartram was a gifted illustrator, and accompanied John on his trip to the southeastern colonies in 1765. William Bartram is best known for his narrative of a series of expeditions in southeastern North America from 1773 to 1776, *Travels through North & South Carolina, East and West Florida, the Cherokee Country, the Extensive Territories of the Muscogulges, or Creek Confederacy, and the Country of the Choctaws, Containing an Account of the Soil and Natural Productions of These Regions, Together with Observations on the Manners of the Indians* (1791). This trip was sponsored by Bartram's patron, the Englishman Lionel Fothergill. William Bartram's interests were broader than his father's; in addition to botany his book has a great deal of information about animal life and both English and Indian society. Its influence was greatest not among natural historians, but among romantic poets, most notably Samuel Taylor Coleridge (1772–1834) and William Wordsworth (1770–1850).

The last of the succession was John Bartram's great-grandson Thomas Say (1787–1834). Conversations with his great-uncle William helped propel Say to a career in natural history. In addition to his explorations of the American interior, Say is known for his three-volume *American Entomology* (1824–1828) and seven-volume *American Conchology* (1830–1834). These were among the first systematic works on American animals. Bartram also influenced Alexander Wilson (1766–1813), a Scottish immigrant whom he admitted to his library and collection to discuss natural history. Wilson's nine-volume *American Ornithology* (1808–1814), the first comprehensive treatment of American birds, featured plates colored by Bartram's niece and pupil Anne Bartram, a professional natural-history illustrator.

See also American Philosophical Society; Botany; Colonial Science; Exploration, Discovery, and Colonization; Sloane, Sir Hans; Zoology.

References

Earnest, Ernest. *John and William Bartram, Botanists and Explorers.* Philadelphia: University of Pennsylvania Press, 1940.

Slaughter, Thomas P. *The Natures of John and William Bartram.* New York: Alfred A. Knopf, 1996.

Bassi, Laure Maria Catarina (1711–1778)

Laure Bassi was the first woman university professor in the sciences, and a leader in the introduction of Newtonianism to Italy. Italy was unique in its tradition of individual women who received university degrees and sometimes appeared at university functions, although usually in a decorative role. Bassi was groomed for this role by her family, who recognized her talent early on. She was educated in the Aristotelian and Cartesian natural philosophy that dominated Italian science in the early eighteenth century and received a degree from the University of Bologna, her hometown, in 1732. The degree came with an appointment at the institute of the university, but Bassi was not expected to function as a teacher and researcher the way her male colleagues were. Instead, she was supposed to lecture and dispute as a marvel on great occasions, such as the visits of dignitaries or the annual "carnival anatomies," where various natural philosophers and scientists gave lectures while a body was dissected in public.

Bassi challenged this role in several ways. She married, which ended the university's attempt to display her as a "learned virgin." Placards actually appeared in the streets begging her not to diminish her glory by marrying. Bassi and her husband and colleague, Giuseppe Veratti (1707–1793), eventually had eight children. She also wanted to lecture on experimental physics, and particularly on Newtonianism, which she had become interested in by the 1730s. In a compromise, she lectured from her home, a not uncommon practice in Bologna. She also built a large cabinet for physical experiments and demonstrations. Bassi benefited from the patronage of Prospero Lambertini (1675–1758), archbishop of Bologna from 1731 to 1740 and pope (as Benedict XIV) from 1740. As Bologna lay in the Papal States, Lambertini's ascent made him even better able to support Bassi. The Senate of Bologna also favored Bassi, who brought glory to their fading city and university, but her male colleagues at the institute were less enamored and tried to discourage her from active participation. Bassi went on regardless and by her death was the highest-paid member of the institute, including the president and secretary. In 1776 she accomplished the extraordinary feat of winning the chair of experimental physics at the institute. Bassi published little, but through her lectures, demonstrations, and correspondence exerted a wide influence over Italian and European intellectual life. Her correspondents included such distinguished physicists as Ruggerio Boscovich and Alessandro Volta. Her most important pupil was her cousin Lazzaro Spallanzani.

See also Bologna Academy of Sciences; Newtonianism; Spallanzani, Lazzaro; Women.

Reference

Findlen, Paula. "Science As a Career in Enlightenment Italy: The Strategies of Laura Bassi." *Isis* 84 (1993): 441–469.

Beddoes, Thomas (1760–1808)

Thomas Beddoes was one of the most scientifically minded physicians of the late eighteenth century. Educated at Oxford University and the medical school of the University of Edinburgh (where he studied under Joseph Black), Beddoes was appointed chemical reader, or lecturer, at Oxford in 1788. His early publications included translations of Lazzaro Spallanzani and Carl Scheele and one of the first historical studies of an individual scientist, the seventeenth-century English physician and experimenter John Mayow (1641–1679). Despite the popularity of his lectures, Beddoes's sympathy for the French Revolution alienated him from conservative Oxford, which he left under a cloud in 1793. He relocated to the village of Clifton outside Bristol, where he devoted himself to reviewing medical and scientific books for the *Monthly Review* (Beddoes was one of the first Englishmen to discuss Immanuel Kant's writings) and his new practice of "pneumatic medicine"—the therapeutic inhalation of the new gases chemists had discovered or separated from the atmosphere. For example, Beddoes claimed oxygen could prevent scurvy. He also experimented on himself, claiming that breathing a mixture of oxygen and nitrogen had made him insensitive to cold and lightened his complexion. As a physician, Beddoes was a Brunonian.

Beddoes's associates in the plan for a pneumatic institute in Clifton included the engineer James Watt (1736–1819), who designed the apparatus for the administration of the

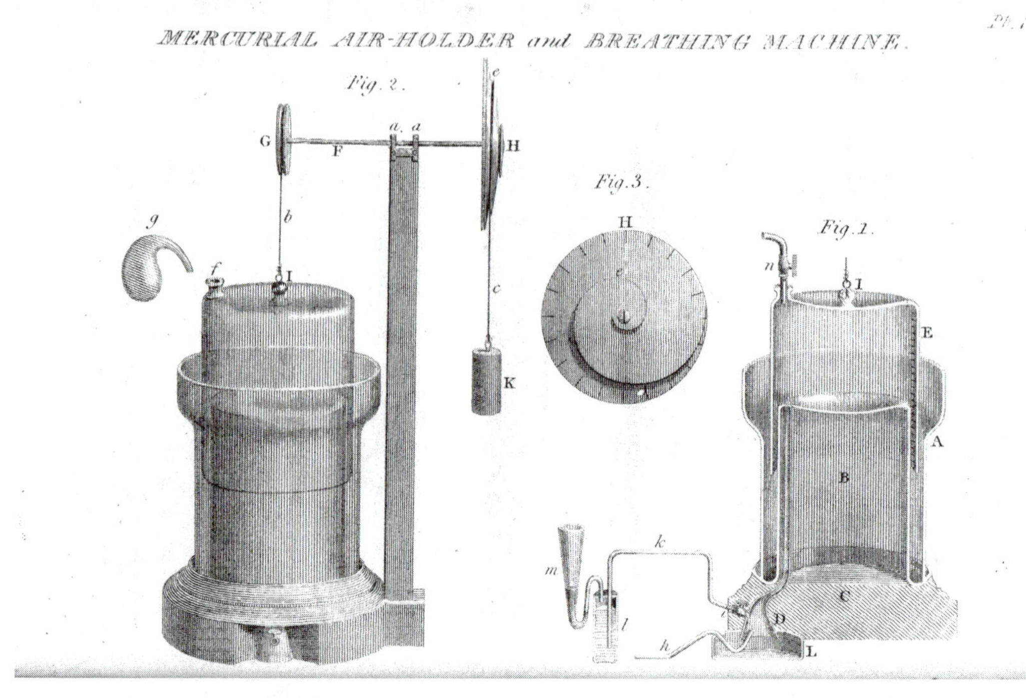

Illustration of a mercurial air holder, a machine for breathing nitrous oxide, as practiced at Thomas Beddoes and Humphry Davy's Pneumatic Institute. (Edgar Fahs Smith Collection, University of Pennsylvania Library)

airs; Erasmus Darwin; and the brilliant young chemist Humphry Davy. The institute, often credited as the first medical research institute, lasted from 1798 to 1802 and became notorious for Davy's experiments with nitrous oxide—"laughing gas." The gas was inhaled at the institute by the romantic poets Samuel Taylor Coleridge (1772–1834) and Robert Southey (1774–1843), among others. Nitrous oxide and Beddoes continued to attract ridicule from conservative writers, who linked both with the excesses of the French Revolution. The institute wound down after Davy's departure for London in 1801, but Beddoes continued an active medical practice with several publications on the subjects of preventive medicine and hygiene, of which he was a pioneer.

See also Brunonianism; Chemistry; Davy, Sir Humphry; Medicine.

References

Golinski, Jan. *Science As Public Culture: Chemistry and Enlightenment in Britain, 1760–1820.* Cambridge: Cambridge University Press, 1992.

Stansfield, Dorothy A. *Thomas Beddoes, M.D., 1760–1808: Chemist, Physician, Democrat.* Dordrecht, Netherlands: D. Reidel, 1984.

Berlin Academy

The first scientific society founded in the capital of Prussia was the Royal Society of Sciences, founded in 1700 by Gottfried Wilhelm Leibniz (1646–1716) under the patronage of the elector of Brandenburg, Frederick III (King Frederick I of Prussia from 1701 to 1713). Frederick hoped to make Prussia a center of learning, an impulse that also led to the founding of the University of Halle and Prussia's shift to the Gregorian calendar. The society received a monopoly on the production, importation, and sale of almanacs and calendars in Prussia. It was the first scientific society to be funded this way, setting an influential precedent. It was divided into four sections: mathematical science, physical science, German language and history, and literature. The new society was slow to get off the ground—its first task, to build an observatory, took eight years, although it did acquire a first-rate astronomer, Gottfried Kirch (1639–1710). (It also acquired the services of his wife, Maria Winkelmann [1670–1720], another fine astronomer. The society's refusal, despite Leibniz's support, to hire Winkelmann as an assistant astronomer after Kirch's death was an important episode in the process by which scientists were defined as masculine.) In 1710, it received permanent quarters and began to publish an irregular series called *Berlin Miscellanies.*

The society was hampered by what was to become a permanent problem in eighteenth-century Berlin science: conflict between its French and German members. The real disaster for the fledgling group, however, was the accession of Frederick William I (r. 1713–1740), the "soldier king," whose attitude to science not of immediate practical use was contemptuous—at one time he had the academicians designated "buffoons of the royal court." Drained of many of its resources, the only accomplishment the society could boast during his reign was the creation of an anatomical theater in 1714, and even that was to be devoted to the training of the medical corps of the Prussian army.

Things improved with the accession of Frederick II "the Great" (r. 1740–1786). The Francophile Frederick wanted a scientific society on the lines of the Royal Academy of Sciences in Paris and lured Leonhard Euler from St. Petersburg to aid in this effort. Euler started the "New Literary Society," which was eventually merged with the Royal Society of Sciences to form the Royal Academy of Sciences and Letters of Prussia in 1744. Frederick also lured Pierre-Louis Moreau de Maupertuis from Paris to be president of the new academy. Maupertuis arrived with a set of statutes modeled on those of the Paris Academy in time for the official launching of the new academy in 1746. Like the old society, the new academy was divided into four sections: mathematical and physical sciences, literature, philology, and speculative philosophy. The society's emphasis on German studies was dropped in accordance with Frederick's French cultural loyalties, and French was the official language.

The new Berlin Academy took its place as Europe's third-leading scientific society, after the Paris Academy and the English Royal Society. It served as an advisory body to the king and acquired a monopoly over the printing, importation, and sale of maps to add to the almanac monopoly it inherited from the old society. It published regular memoirs modeled after those of the Paris Academy and sponsored prestigious prize competitions. It also had a dynamic secretary, Johann Heinrich Samuel Formey (1711–1797). It cooperated in the observations of the transit of Mercury in the early 1750s and adjudicated the dispute between Maupertuis and Samuel Konig (1712–1757), whom Maupertuis charged with having forged a letter from Leibniz that cast doubt on Maupertuis's priority in formulating the principle of least action. This was part of a larger dispute between the Leibnizian and Newtonian supporters in the academy, in which Euler and Maupertuis led the Newtonians. Unsurprisingly, the academy decided in favor of its president.

After Maupertuis's death in 1759, Frederick appointed himself head of the division of speculative philosophy and essentially took over the academy. His closest scientific advisers were not even based in Prussia. The Parisian Jean Le Rond d'Alembert was "shadow president" of the Berlin Academy, with enormous influence over its personnel decisions. This situation contributed to Euler's return to St. Petersburg in 1766. After d'Alembert's death in 1783, another Parisian, the Marquis de Condorcet, succeeded to his position of distant influence. Frederick had little interest in international scientific cooperation, and Berlin played almost no role in the great joint observations of the transits of Venus in 1761 and 1769, despite the urgings of Joseph-Louis Lagrange, who had become head of the mathematical section in 1766. The Berlin Academy also held aloof from the regular exchanges of scientific publications between Europe's leading societies. The academy declined further after Frederick's death in 1786 and suffered during the Napoleonic Wars, when Napoléon carried off much of its collection and equipment in 1806. In the general reform of Prussian institutions in the early nineteenth century, the Berlin Academy was subordinated to the new University of Berlin in 1812.

See also Academies and Scientific Societies; Alembert, Jean Le Rond d'; Condorcet, Marie-Jean-Antoine-Nicolas Caritat, Marquis de; Euler, Leonhard; Klaproth, Martin Heinrich; Lagrange, Joseph-Louis; Marggraf, Andreas Sigismund; Maupertuis, Pierre-Louis Moreau de; Nationalism.

Reference

McClellan, James E., III. *Science Reorganized: Scientific Societies in the Eighteenth Century.* New York: Columbia University Press, 1985.

Bernoulli Family

The greatest dynasty in mathematical history was founded by the Swiss brothers Jakob I (1654–1705) and Johann I (1667–1748) Bernoulli. Their contributions to mathematics extended across a broad intellectual range. The Bernoullis were among the earliest mathematicians to master the Leibnizian calculus, and were responsible for much of its early spread and its defeat of Newtonian calculus on the European continent. Their application of the calculus to mechanical problems was also a foundation of analytical mechanics, the dominant tradition in eighteenth-century theoretical physics. Jakob also gave the first formulation of the law of large numbers in probability theory. Jakob and Johann, two difficult personalities who had bitterly quarreled, successively occupied the chair of mathematics at the University of Basel.

The second generation of Bernoulli mathematicians consisted of the brothers' nephew Nikolaus I (1687–1759), who worked on probability and differentials among other subjects, and the three sons of Johann I, Nikolaus II (1695–1726), Daniel (1700–1782), and Johann II (1710–1790). Daniel was by far the most important mathematician of the group, although he struggled in his early years as his father tried to make him a businessman rather than a mathematician. Eventually, he got a medical degree, although he preferred to work in mathematics rather than medicine. The number of posts for mathematicians in Switzerland was small, and the situation was further complicated by the fact that professorial chairs at Basel were assigned by a lottery among qualified candidates rather than by strict merit. Like many Bernoullis and the mathematicians they trained (the most notable being Johann I's pupil and Daniel's collaborator Leonhard Euler), Daniel and Nikolaus II were forced to find employment elsewhere. Daniel attracted attention when he won a prize from the Paris Academy of Sciences in 1725 for inventing an hourglass that could be used at sea (he would go on to win the prize ten times in all), and he and Nikolaus II were offered chairs at the recently founded Imperial Academy of Sciences of St. Petersburg. Nikolaus II died of a fever shortly after arrival. In Russia, Daniel studied fluid mechanics, which he greatly advanced, and the mathematics of vibrating strings and vibrating systems generally. He demonstrated that the motions of the vibrating string of a musical instrument are composed of an infinite number of vibrations. He also studied probability, originating the famous "St. Petersburg problem," which inspired much fruitful work in the subject.

Like many western Europeans, Daniel did not care for St. Petersburg, and along with Johann II, who had joined him in Russia, returned to Basel in 1734. He supported himself by giving lectures in botany. The next few years of his life were dominated by a savage conflict with his father, Johann I. The two had been in competition for the Paris Academy prize in 1734 and were joint winners. This enraged Johann I, who believed that his son had no right to compete with him. Daniel published his masterpiece, *Hydrodynamica,* in 1738. It contains successful treatments of problems of the pressure and velocity of moving fluids, both liquids and gases. Daniel conceptualized a gas as a collection of moving particles, and deduced several characteristics of the behavior of gases, such as Boyle's law. *Hydrodynamica* is often considered a founding document of the kinetic theory of gases. Johann I published a similar book, *Hydraulica,* drawing on his son's work without crediting it, the next year. Johann I actually backdated *Hydraulica* to 1732, to make it look as if Daniel's book was based on his rather than the other way around. Daniel weathered the conflict to become professor of physics at Basel in 1750 and continued to do prizewinning work, concentrating on the application of hydrodynamics to problems of ships. He also applied the Leibnizian calculus to Newtonian physics, as well as synthesizing Newtonian physics and the Leibnizian mechanical concept of the conservation of the *vis viva.*

Johann II, a far less distinguished mathematician, also had a successful academic career at Basel, succeeding his father on Johann I's death and thus becoming the third Bernoulli to hold the mathematical chair at Basel. He worked mainly on problems involving heat and light.

The third generation of mathematical Bernoullis were the sons of Johann II: Johann III (1744–1807) and Jakob II (1759–1789). Jakob II did important work on mechanics, hydrostatics, and ballistics, but his career was cut short when the fatality of St. Petersburg to the Bernoullis struck again. Jakob II had failed to get an academic post at Basel, so he was forced to follow his uncles' footsteps to Russia, where he drowned swimming in the Neva River.

See also Imperial Academy of Sciences of St. Petersburg; Mathematics; Mechanics; Physics; Probability; Universities.

Reference

History of Mathematics Web site, University of St. Andrews: http://www-groups.dcs.st-andrews.ac.uk/~history/.

Berthollet, Claude-Louis (1748–1822)

Claude-Louis Berthollet was France's leading practical chemist of the late eighteenth and early nineteenth centuries. Born in Piedmont, he obtained a medical degree at the University of Turin in 1768 and then left for Paris. He became interested in chemistry, publishing a paper on tataric acid in 1776 and a major study of ammonia in 1778. The same year he became a naturalized Frenchman, and in 1780 he was admitted to the Royal Academy of Sciences. In 1784 he became director of the government dye works at the Gobelins. The appointment was highly successful. In 1785 Berthollet discovered the power of chlorine (then known as "oxymuriatic acid") to bleach. This was a vast improvement on traditional bleaching practices, which required fabrics to be exposed to sunlight for several weeks. He also discovered potassium chlorate and wrote the standard work on dyeing, *Elements of the Art of Dyeing* (1791). Berthollet was among the very first to adopt Lavoisier's new chemistry, announcing his support in April 1785. His name appears as one of the four authors of *Method of Chemical Nomenclature* (1787), although his interest in nomenclature was small. The same year, one of Berthollet's analyses conflicted with a fundamental tenet of Lavoisier's chemistry. Berthollet demonstrated that prussic acid was a compound of carbon, hydrogen, and nitrogen. Lavoisier's theory required that all acids contain oxygen, so Berthollet avoided directly challenging him by not stating that prussic acid did not contain it. (In 1796, Berthollet demonstrated that hydrogen sulfide had no oxygen yet showed the properties of an acid.)

The French Revolution caused Berthollet some consternation at first, and he temporarily moved to the country in 1789. However, the revolutionaries needed his talents, and he was soon serving on a variety of commissions and teaching a course for gunpowder manufacturers. He also served as a chemistry instructor at the Polytechnic School from 1795 to 1805. The French government sent

Berthollet to Italy during Napoléon's invasion to identify objects of cultural and scientific value to be taken to France, and he became one of Napoléon's leading friends and allies in the scientific community. Napoléon chose Berthollet to head the scientific commission that went on his expedition to Egypt. One Egyptian sight made a particularly strong impression. It was the large deposits of soda found on the shores of salt lakes, arising from the reaction between salt and limestone on the lake bottoms. This was the opposite of the reaction between soda and calcium chloride to form salt and limestone. It led Berthollet to suggest that the reactions of chemical substances were influenced by their concentrations, and the large concentrations of salt had forced the reverse reaction. Berthollet expanded his conclusion to argue, against the usual assumptions made by chemists, that substances did not always combine in fixed proportions. Berthollet's *Essai de statique chimique* (1803), influenced by his growing friendship with Pierre-Simon de Laplace, discussed chemical attractions in a mechanical style, suggesting that the mass of the reactants would affect the nature and products of a chemical reaction. Berthollet's new ideas led to a debate with another French chemist, Joseph-Louis Proust (1754–1826), who argued for the traditional view that chemical substances combined in fixed and definite proportions. Although Proust's views were vindicated by the adoption of chemical atomism, Berthollet is seen as having been the first to point the way to the law of mass action in chemistry.

Berthollet benefited greatly from Napoléon's rise to power. He was appointed to the Napoleonic Senate, and Napoléon gave him money to pay his debts. (Berthollet was very generous and notoriously poor at personal finances.) He also served on the commission to publish the *Description of Egypt,* even decorating his study in an Egyptian style. The principal venue for Berthollet's scientific work during the Napoleonic period was the Society of Arcueil, headquartered at his home outside Paris. The genial Berthollet was very popular both with French scientists and with foreigners visiting the capital of European science. His high position in French science and his interest in fostering the careers of young scientists made him a natural leader of the society. The young men in the Society of Arcueil whose scientific careers Berthollet had encouraged include Joseph-Louis Gay-Lussac and Berthollet's own son Amédée (1780–1810), who committed suicide after the failure of a chemical factory he had helped set up. The suicide plunged Berthollet into depression, and the fall of Napoléon removed his greatest patron. Berthollet continued scientific work after the dissolution of the Society of Arcueil in 1814, but made no further innovations.

See also Arcueil, Society of; Chemical Nomenclature; Chemistry; Egyptian Expedition; French Revolution; Gay-Lussac, Joseph-Louis; Industrialization; Lavoisier, Antoine-Laurent; Napoleonic Science.

References

Brock, William H. *The Norton History of Chemistry.* New York: W. W. Norton, 1993.

Crosland, Maurice. *The Society of Arcueil: A View of French Science at the Time of Napoleon I.* Cambridge: Harvard University Press, 1967.

Berzelius, Jöns Jakob (1779–1848)

Jöns Jakob Berzelius attained a position of unrivaled eminence in both European chemistry and the Swedish scientific world through innovative science, very hard and very skilled work, and a gift for politics. Berzelius studied medicine at the University of Uppsala and was convinced by the new chemical theories of Antoine-Laurent Lavoisier. He was one of the innumerable European scientists enraptured with the possibilities of the voltaic pile in the early nineteenth century. In a series of experiments he reported on in 1803, he separated salts into their constituent acids and bases. Berzelius's observation that some substances were attracted to the positive and others to the negative electrical pole led him to classify chemicals according to their electropositivity or electronegativity. (Berzelius's electropositive and electronegative substances are now known as anions and cations.) He acquired a patron, a mine owner named Wilhelm Hisinger (1766–1852), who invited Berzelius to use his laboratory in Stockholm. Berzelius became independent in 1807, when he received the chair of chemistry and pharmacy at the Caroline Institute, a Stockholm medical school. His two-volume Swedish textbook, *Lessons on Animal Chemistry* (1806–1808), contains original analyses of biological products, such as bone, pointing to his later work in organic chemistry. The introduction to this work contains Berzelius's firm rejection of romantic vitalism. Life is not something added to matter, but a product of material processes. He later modified this position, although his Swedish enemies attacked him as a materialist.

Portrait of Jöns Jakob Berzelius. This portrait of Berzelius emphasizes his social distinction, as signified by his decorations, rather than his career as a scientist. (National Library of Medicine)

Although the Royal Swedish Academy of Science had earned Berzelius's wrath by rejecting several of his early papers, he was admitted to it in 1808 and became its joint secretary in 1818. In 1811 he was made manager of the Swedish Academy of Agriculture, a position he held for the rest of his life. Although he practiced medicine for only a brief period following his graduation from Uppsala, he was also a founder of the Medical Society of Stockholm. He used his position as Sweden's leading scientist to oppose both a purely utilitarian conception of science, a tradition with deep roots in Sweden, and the new German import, *Naturphilosophie,* which he particularly loathed. Berzelius did not reject practical application, but defended the value of pure research. His scorn for *Naturphilosophie* was based on what he saw as its lack of empiricism, and its subordination of evidence in favor of grand theories spun from the heads of *Naturphilosophs.* (He also thought *Naturphilosophie* encouraged sloppy dressing among German university students.)

The 1810s were intellectually very fertile for Berzelius. In 1811 he set forth a new system of chemical classification to bring together chemistry and mineralogy. This was precipitated by his experiences in classifying a mineral collection, during which he learned that existing systems of mineral classification lacked chemical logic. In 1812, after some difficulty caused by the Napoleonic Wars, he obtained a copy of John Dalton's *A New System of Chemical Philosophy* (1808), and enthusiastically adopted the atomic theory and the idea that substances always combine in integral ratios. Berzelius adapted the atomic theory to electrochemistry, and it was his version of atomism, not Dalton's original theory, that became widely accepted. In 1813 Berzelius essentially invented the modern system of chemical notation by using letters or two-letter combinations to denote elements and writing them together to denote compounds, although he subsequently muddled the original clarity of his system by incessantly fiddling with it.

Although Berzelius's theoretical and linguistic contributions to chemistry were immense, he was greatest as a laboratory chemist. Berzelius isolated and identified the elements of thorium, selenium, and silicon, and shared with Hisinger the discovery of cerium. He prepared, purified, and analyzed more than 2,000 compounds. His analytical investigations led to the formation of the first accurate table of atomic weights. In 1836 Berzelius identified catalytic reactions and suggested the term *catalyst.* He raised laboratory work to a higher degree of precision and exactness. Sometimes Berzelius's theoretical commitments held back his perceptions. For a long time he believed that the acids and bases that combined to form salts must both be oxides, and for that reason he refused to accept that chlorine and iodine were elements rather than compounds of an unknown element with oxygen until the 1820s.

Berzelius's domination of European chemistry from about 1820 is an astonishing phenomenon, given the marginality of his base at Stockholm to the centers of European science. He read and wrote English, French, and German, the main languages of international chemistry, fluently. His great *Textbook of Chemistry,* first published in two volumes in 1808 and 1812, went through many editions and was frequently translated, shaping the field for generations of students. Berzelius carried on an extensive chemical correspondence and traveled widely and often despite his frequent bouts of ill health. In 1821 he began publishing a yearbook of physical science for the Swedish Academy. This periodical was soon appearing in French and German translation, and Berzelius's power to select what material was discussed in it helped shape European chemistry.

Although Berzelius did not have many advanced students, he did have some important ones who worked closely with him in the laboratory, notably the Germans Eilhardt Mitscherlich (1794–1863) and Friedrich Wöhler (1800–1882). Mitscherlich in 1818 discovered chemical isomorphism, the doctrine that related chemicals can have the same crystalline form. Berzelius zealously and successfully promoted Mitscherlich's work, which went against the dominant crystallographic doctrine that every substance had a unique crystalline form. Wöhler in 1828 became the first chemist to synthesize an organic compound, urea. Wöhler's discovery initially caused some awkwardness between master and student, as Berzelius had declared in 1814 his adherence to the principle that organic compounds could not be synthesized.

Although the development of organic chemistry by the 1840s was leaving Berzelius behind, he continued vigorous participation in chemical controversy to his death, although more as a critic of the work of others than as a scientific innovator. By the time of his death, he was a member of eighty-four academies and scientific societies.

See also Atomism; Chemical Nomenclature; Chemistry; *Naturphilosophie;* Romanticism; Royal Swedish Academy of Science.

References

Brock, William H. *The Norton History of Chemistry.* New York: W. W. Norton, 1992.

Melhado, Evan M. *Jacob Berzelius: The Emergence of His Chemical System.* Madison: University of Wisconsin Press, 1981.

Melhado, Evan M., and Tore Frangsmyr, eds. *Enlightenment Science in the Romantic Era: The Chemistry of Berzelius and Its Cultural Setting.* Cambridge: Cambridge University Press, 1992.

Bichat, Marie-François-Xavier (1771–1802)

In his short life the French physician Marie-François-Xavier Bichat revolutionized medical theory. Bichat was a child of the French Revolution, whose formal medical studies began in 1791 at Lyons. In 1793 he moved to Paris, where his career was associated with the great hospital the Hôtel-Dieu. There he studied under the great surgeon Pierre-Joseph Desault (1738–1795). Although a fine and beloved teacher who taught many successful doctors, Bichat never attained a teaching post at a major Paris hospital. He taught private courses and performed hundreds of autopsies.

Bichat's central theoretical innovation, set forth in his 1800 *Treatise on Membranes,* was the substitution of the membrane or tissue for the organ as the fundamental unit in the analysis of health and disease. Building on the tradition of pathological anatomy established by Giovanni Battista Morgagni (1682–1771), Bichat insisted that diseases could not be correctly identified as diseases of specific organs, but as diseases of specific tissues. He insisted on the paramount importance of autopsy, but would not use microscopes, which he believed were unreliable. His writings also described a number of experiments he performed, including a series of blood transfusions between dogs to demonstrate the fatal effects of deoxygenated blood on the brain.

Bichat was a vitalist who drew on the tradition of vital materialism associated with the Montpellier medical school, although he had no personal or institutional connection with Montpellier. He distinguished between inorganic matter, governed by mechanism and determinism, and living matter, whose activities were indeterminate and unpredictable. His classification of living tissues further divided them into the two main categories of those that promoted animal life, the voluntary muscles and sense organs, and those that promoted "organic life," the functions that work without the sphere of direct conscious control, such as the lungs, blood vessels, and digestive system. Tissues themselves were divided into twenty-one distinct types, which combined to form organs in a way that Bichat compared to how chemical elements combined to form compounds.

Bichat's last years were spent in a frenzy of publication. *Physiological Studies on Life and Death* (1800) contains his famous definition of life as "the sum of the activities by which death is resisted" and examined the physiological processes of bodily tissues and organs in death. Bichat distinguished between heart and brain death. His most systematic work was *General Anatomy* (1801), which set forth his system in its most elaborate form.

Portrait of Marie-François-Xavier Bichat (Perry-Castaneda Library)

See also French Revolution; Hospitals; Medicine; Physiology; Vitalism.

References

Ackerknecht, Erwin H. *Medicine at the Paris Hospital, 1794–1848.* Baltimore: Johns Hopkins University Press, 1967.

Lesch, John E. *Science and Medicine in France: The Emergence of Experimental Physiology, 1790–1855.* Cambridge: Harvard University Press, 1984.

Porter, Roy. *The Greatest Benefit to Mankind: A Medical History of Humanity.* New York: W. W. Norton, 1998.

Black, Joseph (1728–1799)

The Scottish professor Joseph Black was a leading chemist and theorist of heat. Training as a physician, Black studied with the medical chemist William Cullen (1710–1790) at the University of Glasgow, becoming Cullen's assistant and close friend. In 1752 Black left Glasgow for Scotland's leading university, Edinburgh, where he received an M.D. in 1754. His dissertation on magnesia alba (magnesium carbonate) and other alkaline substances caused a sensation. His interest in the subject began as a medical one, as he hoped to find a superior agent for the dissolution of kidney stones. Careful and rigorous experiments, which he later repeated in front of the Philosophical Society of Edinburgh, showed that magnesia alba when heated expelled an "air" or gas. Black called this substance "fixed air" (a term derived from Stephen Hales, who had used it with a different meaning). Fixed air, now known as carbon dioxide, was the first "air" or gas to be isolated and identified. This dissertation was Black's only major scientific publication in his lifetime, and he founded the British school of "pneumatic chemists" that included Henry Cavendish and Joseph Priestley.

In 1756 Cullen moved to Edinburgh, and Black took his place at Glasgow. There he carried on his studies of heat. These eventually led to the discovery of latent heat, the heat required to change a substance from one state to another (for example, liquid water to steam), and specific heat, the differing amounts of heat required to raise different substances to the same temperature. Black's interest in heat was piqued by Cullen's observations concerning the cold produced by the evaporation of highly volatile substances and by the already well-known fact of the possibility of water cooled below the freezing point remaining liquid, if kept still. His observations led him to distinguish between temperature and quantity of heat, a distinction previous writers had been aware of but none had explored as fully. Black not only established the existence of latent and specific heats, but measured them for particular substances to the degree of accuracy permitted by the instruments available to him. Black favored the fluid theory of heat as opposed to the motion theory, but was more concerned with the phenomena than theoretical explanation. Despite the urgings of his friends, Black never published his work on heat. He believed as a university professor that the appropriate means for disseminating his discoveries were in his lectures to his students, and he also greatly feared plagiarism. Black's aversion to publication, and the fact that his one important publication, the dissertation, was published obscurely in Edinburgh, delayed his work's impact outside Britain.

At Glasgow Black met the engineer James Watt (1736–1819), appointed instrument maker to the university in 1757. It is sometimes claimed that Watt's invention of the condenser for the steam engine was based on Black's discovery of latent heat. This claim, which goes back to Black's student John Robison, who edited his posthumous *Lectures on the Elements of Chemistry* (1803), is not true. Black had revealed the theory of latent heat to Watt earlier to explain a problem in the university's Newcomen steam engine, and Watt did credit Black with assisting his intellectual development and understanding of natural philosophy while denying that Black made any contribution to the condenser specifically. Black's relationship with Watt reflects one persistent aspect of his career, a belief in the importance of the practical applications of science. These pertained to industry as well as medicine, and his social circle included many industrialists, including his brothers James and Alexander, manufacturers in Belfast.

In 1766 Black returned to Edinburgh as a professor of chemistry. Although he produced no more groundbreaking science, Black was immensely influential as a popular teacher at Europe's leading medical school, known for both his penetrating delivery and his smooth hand at performing demonstration experiments. His pupils included Lorenz Florens Friedrich von Crell; Daniel Rutherford (1749–1819), who discovered nitrogen; the leading American physician Benjamin Rush (1745–1813); and John McLean, later the first professor of chemistry at Princeton. Black was one of the first chemists in Britain to accept the new chemistry of Antoine-Laurent Lavoisier, himself a great admirer of Black. In 1790 he wrote Lavoisier, telling him that he had been teaching antiphlogistic chemistry as simpler and more accurate than the traditional system. However, this was not a full endorsement, as Black was suspicious of all systems of chemistry, which he regarded as too young a science to be systematized. With some reservations he also adopted Lavoisier's new chemical nomenclature.

Along with teaching, Black was active in Edinburgh's celebrated intellectual club life. He was a member of the Philosophical Society of Edinburgh, later the Royal Society of Edinburgh, and a founder of a more informal group, the Oyster Club, a weekly dining club that included many luminaries of the Scottish Enlightenment such as Cullen, James Hutton, and Adam Smith (1723–1790).

See also Chemistry; Crell, Lorenz Florens Friedrich von; Heat; Royal Society of Edinburgh; University of Edinburgh.

References

Golinski, Jan. *Science As Public Culture: Chemistry and Enlightenment in Britain, 1760–1820.* Cambridge: Cambridge University Press, 1992.

Guerlac, Henry. *Essays and Papers in the History of Modern Science.* Baltimore: Johns Hopkins University Press, 1977.

Blumenbach, Johann Friedrich (1752–1840)

The German professor and physician Johann Friedrich Blumenbach is best remembered as a founder of physical anthropology and the creator of the most common system of human racial classification. However, he also contributed to a broad range of scientific developments. Blumenbach attended the Universities of Jena and Göttingen. His published Göttingen doctoral dissertation, *On the Natural Varieties of Mankind* (1776), first expounded his theories of human races. In 1778 he was named a full professor of medicine at Göttingen. During his incredibly long career, Blumenbach was a dominant force in German life science. An excellent teacher, his outstanding students included Alexander von Humboldt, Gottfried Reinhold Treviranus (1776–1837), and Carl Friedrich Kielmeyer (1765–1844). Blumenbach rejected the application of mechanical reductionism to the realm of life, while wishing to avoid the sci-

Portrait of Johann Friedrich Blumenbach (National Library of Medicine)

entific use of nonmaterial entities such as the soul. Drawing on the work of his friend Immanuel Kant, Blumenbach's theory of the formative drive ascribed vital power to organized living matter. The theory was taken up by the German romantic movement, although Blumenbach was never a *Naturphilosoph.* Belief in the formative drive led Blumenbach to move from the preformationism of his Göttingen mentor, Albrecht von Haller, to an epigenesist position in embryology. Blumenbach was also a leader in the movement to reduce the chaos of the life sciences by analyzing living beings as variations of a few basic types. He was Germany's leading paleontologist in the late eighteenth and early nineteenth centuries, and articulated many of the ideas later associated with Georges Cuvier's reading of the fossil record, such as extinction and the importance of the geological time sequence.

Blumenbach's physical anthropology can be seen as bringing humans into natural history. His original dissertation divided humans into four races: Caucasian, Mongolian,

Ethiopian, and American. Subsequent editions added a fifth category, the Malay. Blumenbach coined the term *Caucasian* for white people, claiming that the inhabitants of the Caucasus were the most beautiful and perfect of humans. He also broadened the criteria of race beyond skin color, putting particular emphasis on facial structure. Blumenbach's collection of hundreds of human skulls gave him unrivaled authority on the matter. Although both Blumenbach's racial classification system and skull measurement—"craniometry"—were used by nineteenth-century scientific racists, Blumenbach's own position is more complex. He viewed the Caucasian as the norm of the human species, from which other races had deviated, but he attacked those who arranged the races in a hierarchy of worth. (Blumenbach disliked arguments based on a chain of being generally.) To refute those (such as Kant) who denied intellectual capacities to blacks, he wrote a book listing black intellectual attainments. When comparing a human skull to that of an ape, Blumenbach did not use a black person's skull, as was and would remain the custom of European scientists, but deliberately chose a Caucasian skull for the comparison.

See also *Naturphilosophie;* Race; Romanticism; Vitalism.

References

Lenoir, Timothy. *The Strategy of Life: Teleology and Mechanics in Nineteenth-Century German Biology.* Dordrecht, Netherlands: D. Reidel, 1982.

Schiebinger, Londa. *Nature's Body: Gender in the Making of Modern Science.* Boston: Beacon, 1993.

Boerhaave, Hermann (1668–1738)

Hermann Boerhaave was the most powerful and influential person in early-eighteenth-century medicine. The son of a Dutch pastor, Boerhaave was originally intended for the ministry and remained a pious Dutch Calvinist throughout his career. His desired career in the ministry, however, ended before it began when he was accused of being a Spinozist, a follower of the pantheist philosopher Baruch Spinoza (1632–1677). Self-taught in medicine, he was hired as a lecturer in medicine by the University of Leiden in 1701. Boerhaave's rise at Leiden was steady and spectacular. In 1709 he became professor of medicine and botany, which entailed supervision of the university botanical garden. The following year, he married the daughter of a rich Leiden merchant. In 1714 Boerhaave added the chair of clinical medicine, and in 1718 the chair of chemistry. At this point, he was teaching in every medical field Leiden offered except anatomy. In addition to the money he received from the university and from his wife's family, Boerhaave gave private medical and chemical lectures and practiced medicine, building a large fortune. He turned down several offers to become a court physician.

Boerhaave was the leading "iatromechanist," applying a mechanical approach to medical problems as set forth in his 1703 Latin oration, "On the Use of Mechanical Reasoning in Medicine." Boerhaave sought to fill the gap left by the decline of Galenism, Aristotelianism, and Cartesianism with a new comprehensive medical philosophy. Influenced by Newtonianism, Boerhaave viewed many of the processes going on in the human body in terms of fluids under pressure. He distinguished between disorders of fluids, such as the blood, and disorders of the solid parts, such as the muscles. Boerhaave's medical philosophy, set forth in his widely translated textbook *Institutes of Medicine* (1708), was more a systematization of current ideas than an original creation. (Boerhaave's medical mechanism was challenged by the vitalism of his contemporary Georg Ernst Stahl. However, Boerhaave took no notice of Stahl's challenge.) Boerhaave's medicine was not presented as a finished system, but as an open one that encouraged experiment. He upheld the philosophy of Francis Bacon (1561–1626), which endorsed scientific and material progress through observation rather than theorizing. Boerhaave also emphasized bedside learning, reviving clinical instruction at Leiden's Caecilia Hospital on his appointment as professor of clinical medicine. He was also aware of the importance of medical tradition, and edited and published the work of several earlier medical and scientific writers, including Andreas Vesalius (1514–1564), Bartolommeo Eustachio (1520–1574), and Jan Swammerdam (1637–1680). He linked his emphasis on observation to that of the most revered ancient physician, Hippocrates.

Boerhaave's chemistry, set forth in his lectures and his influential two-volume textbook, *Elements of Chemistry* (1732), was based on Newtonian principles. He viewed chemical processes as driven by the short-range interactions of attraction and repulsions between particles. Boerhaave believed that these attractions and repulsions were caused by what he called "fire"—a subtle material permeating the cosmos similar to Newton's idea of the ether. His chemical work stressed quantitative research.

Boerhaave had only two major publications in botany, catalogs of the collections at the Leiden botanical garden. He vastly expanded the garden itself, adding thousands of new species. The Dutch worldwide trading network greatly assisted him in finding and acclimating new plants. Boerhaave's endorsement of the idea that the flowers were the sexual organs of plants helped secure its acceptance.

Boerhaave made Leiden Europe's leading medical school. Among his hundreds of pupils were some of the leading physicians and scientists of the mid-eighteenth century, including Albrecht von Haller; Julien Offroy de La Mettrie, who translated Boerhaave's medical writings into French; and Gerard van Swieten (1700–1772), who founded a medical school in Vienna and became a close adviser of the empress Maria Theresa. Boerhaave's influence was particularly strong at the University of Edinburgh, virtually all of whose medical professors were Leiden graduates. Along with Willem Jakob 's Gravesande (1688–1742), Boerhaave made Leiden the leading center for the dissemination of Newtonianism on the Continent. His leading position in international science was recognized by election as a foreign associate of the Royal Academy of Sciences in 1728 and unanimous election as a fellow of the Royal Society in 1730.

See also Botanical Gardens; Botany; Chemistry; Haller, Albrecht von; La Mettrie, Julien Offroy de; Medicine; University of Leiden.

References

Lindeboom, G. A. *Hermann Boerhaave: The Man and His Work.* London: Methuen, 1968.

Van Berkel, Klaas, Albert Van Helden, and Lodewijk Palm, eds. *A History of Science in the Netherlands: Survey, Themes, and Reference.* Leiden: Brill, 1999.

Bologna Academy of Sciences

Bologna, home of the University of Bologna, was one of the leading centers of natural study in Italy. Although there had been a number of ephemeral academies founded for the study of nature in the late seventeenth century, a formal academy was founded only in 1714. This was part of a project of reform of the university in which the leading spirit was Luigi Ferdinando Marsigli (1658–1730). An institute devoted to the sciences was founded as part of the university, with five salaried professorships for scholars expected to do research and publish. The academy included the five institute professors and other university scholars as the inner group. It was supported by the Senate of Bologna and the pope, as Bologna was part of the Papal States of central Italy. It languished for a few years after its founding, until the appointment of Francesco Zanotti as secretary in 1723. Zanotti encouraged scientific activity and the publication of society-sponsored papers. The necessity to get approval from the Inquisition before publishing anything was a major handicap, and several years often passed between volumes. The academy was expanded and put on a sounder financial footing by Pope Benedict XIV (1740–1758), a native of Bologna and the only eighteenth-century pope to show much interest in science. Desiring to imitate the state-sponsored scientific academies north of the Alps, the pope expanded the class of paid academicians expected to publish and endowed the academy with revenues derived from suppression of a school for the poor.

The society carried on the normal functions of a scientific society, publishing a journal and contributing to the joint observations of planetary transits in 1753, 1761, and 1769. However, despite Zanotti's energy and Benedict's ambition, the Bologna Academy of Sciences had little impact outside Italy. It did not offer prizes, which was one way that provincial societies could attract interest in other parts of Europe. It continued to draw professors from the university, among them Luigi Galvani, who published his famous experiment on frogs' legs in its journal, *Proceedings of the Bologna Academy of Sciences.* Another significant member was the pioneering Italian chemist Jacopo Bartolomeo Beccari (1682–1766), but the society suffered with the rise of new Italian scientific centers in Pavia and Turin.

One way in which the Bologna Academy sought to distinguish itself was its inclusion of women. The most noted was the physics professor Laure Bassi. The academy also looked outside Italy to admit distinguished women who could not hope to be admitted to scientific organizations in their own countries. The French Newtonian Gabrielle-Émilie du Châtelet was admitted to the academy in 1746. Other women invited to join included the mathematician Maria Gaetana Agnesi (1718–1799), although she never actually formally accepted the invitation or came to Bologna. Two more women were admitted in the early nineteenth century, the obstetrician Maria Dalle Donne (1776–1842) in 1800 and the Greek professor Clara Tambaroni in 1802. By this time, the academy was long past the days of its glory. It suffered greatly in the French Revolution and Napoleonic campaigns of the late eighteenth century and closed down for good in 1804.

See also Academies and Scientific Societies; Bassi, Laure Maria Catarina; Galvani, Luigi; Transits; Women.

References

Findlen, Paula. "Science As a Career in Enlightenment Italy: The Strategies of Laura Bassi." *Isis* 84 (1993): 441–469.

McClellan, James E., III. *Science Reorganized: Scientific Societies in the Eighteenth Century.* New York: Columbia University Press, 1985.

Bonnet, Charles (1720–1793)

The Genevan Charles Bonnet enjoyed two scientific careers, the first as a naturalist specializing in very small creatures, and the second, after his eyesight became defective, as a natural philosopher. The young Bonnet's interest in insects (a more loosely defined category then) was inspired by the volume *Spectacle of Nature* by the natural theologian Noel-Antoine Pluche (1688–1761). It was further kindled while he was a student at the Academy of Geneva by a correspondence with the great French naturalist René-Antoine Ferchault de Réaumur (1683–1757). In 1740, following up on an inconclusive experiment of Réaumur's, Bonnet discovered that aphids reproduced parthenogenetically, that is, without a male. Bonnet raised ten generations of aphids to demonstrate that this was their mode of reproduction. Réaumur read Bonnet's discovery to the Royal Academy of Sciences in Paris. Like many others, Bonnet was excited by the discovery of the regeneration of polyps by his cousin Abraham Trembley (1700–1784), and spent much time in 1741 experimenting with freshwater worms, which he discovered shared with polyps the ability to regenerate. Bonnet's discoveries on these and other small animals were set forth in his *Treatise on Insectology* (1745). His career as a microscopic investigator was cut short by deteriorating eyesight (he was already extremely hard of hearing). After a period of despair and inertia, Bonnet turned to natural philosophy in the late 1740s.

As an embryologist, Bonnet was a preformationist, arguing that female organisms (the aphids were a useful example due to their lack of males) carried within them the future generations of their preformed descendants. His *Considerations on Organized Bodies* (1761) established him as a leading preformationist, alongside his friend Albrecht von Haller. Bonnet was also the foremost champion of the ancient idea, recently expounded by the German philosopher Gottfried Wilhelm Leibniz (1646–1716), of the great chain of being. The great chain was a continuous hierarchy of created things with one end reaching to God and the opposite to nonexistence. Thus, Bonnet claimed there were no sharp divisions between plants and animals, the living and the nonliving (asbestos, with its fibrous structure, was the closest mineral to plant life), or the orangutan, which Bonnet believed the most advanced of the apes, and man. Bonnet also believed that the great chain was not static, but that every member advanced together, keeping the hierarchy itself intact. His *Philosophical Palingenesis; or, Ideas on the Past and Present State of Living Beings* (1769) set forth the idea that after great catastrophes, beings further develop—thus, after the next catastrophe (Bonnet appealed to fossil evidence to demonstrate the existence of past catastrophes) stones would become organic, plants would gain the power to move and animals to reason, and humans would become angelic, leaving apes or elephants as the planet's dominant species. Bonnet regarded this idea as an extension of the Christian doctrine of the Resurrection. He also wrote on psychology, memory, and the nature of the soul.

See also Embryology; Haller, Albrecht von; Polyps; Religion; Zoology.

Reference

Anderson, Lorin. *Charles Bonnet and the Order of the Known.* London: Kluwer, 1982.

Boscovich, Ruggiero Giuseppe (1711–1787)

The physicist and natural philosopher Rudjer Josip Boškovic, better known by the Italian version of his name, Ruggiero Giuseppe Boscovich, or the Anglicized version, Roger Joseph Boscovich, was born in the independent republic of Ragusa, modern Dubrovnik. In 1725 Boscovich left Ragusa for the Collegium Romanum in Rome, the pinnacle of the international Jesuit educational system. The pious youth was himself planning to enter the Jesuit order. At the Collegium, Boscovich excelled at mathematics and studied the works of Sir Isaac Newton, which had a deep influence on him. He began writing short studies of mathematical and astronomical problems in Latin, including one on determining the Sun's period of rotation from observations of sunspots. In 1740 Boscovich was appointed chair of mathematics at the Collegium. The same year began the pontificate of Prospero Lambertini, Benedict XIV (1740–1758), the greatest eighteenth-century papal patron of science.

In 1742 the pope appointed Boscovich to a commission to recommend a solution to the cracks appearing in the dome of Saint Peter's. Boscovich's suggestion, that the dome be supported by iron rings, was adopted, and in 1743 he submitted a report on repairing the apse of the Vatican basilica. The following year he passed the examinations to be ordained a priest and became a full member of the Jesuit order. Boscovich continued to perform mathematical work for the pope. From 1750 to 1752 he and another Jesuit, Christopher Maire (1697–1767), surveyed the meridian of the Papal States from Rome to Rimini to make the first accurate map of the pope's territory. In the controversy over the shape of Earth, Boscovich supported and provided evidence for the Newtonian conception of Earth as flattened at the poles.

The Roman circles in which Boscovich moved preserved an association between polished Latin verse and science. Boscovich himself was a member of the poetic academy of the Arcadians, and wrote a long Latin poem on the theory of eclipses. He also published a scientific commentary on another Latin poem by his fellow Ragusan Benedict Stay expounding the Newtonian system. As part of his academic job, Boscovich published a number of Latin dissertations in prose, mostly on celestial and meteorological phenomena such as the transit of Mercury, sunspots, and the aurora borealis. He described a new geometrical method for determining the orbit of a body around the Sun from three observations of its position. As an astronomer, Boscovich accepted the Copernican system and helped persuade the pope to remove the blanket prohibition on Catholics owning or reading heliocentric books in 1757.

Boscovich's greatest work was the 1758 *Theory of Natural Philosophy,* an attempt to reconcile Newton and Leibniz and create a complete system of natural philosophy. The intellectual innovation for which Boscovich is best known is the "point atom," an atom conceived not as a very small body, but as a mathematical point in space that served as the center for both attractive and repulsive forces. Repulsive forces dominate when atoms are very close together, enabling them to maintain their separation. Both the attractive and repulsive forces are different aspects of an underlying universal force. Although the atomic theory of John Dalton, which saw atoms as bodies rather than points, was more influential among chemists in the early nineteenth century, Boscovich's point-atom theory influenced a succession of mostly British chemists and physicists, including Joseph Priestley, Michael Faraday (1791–1867), James Clerk Maxwell (1831–1879), and William Thomson, Lord Kelvin (1824–1907).

In 1759 Boscovich left Rome for a journey through Europe, connected with papal diplomacy and the severe difficulties the Jesuit order was laboring under. He moved in scientific circles in London and Paris. Boscovich was welcomed and honored in both places, but was troubled by the fashionable religious skepticism of Paris, which in some cases expressed itself in hostility to the Jesuit visitor. In London Boscovich was admitted to the Royal Society and met many English scientists and scholars. He hoped to observe the transit of Venus in 1761 from Constantinople, but made it only as far as Venice, where the sky was too cloudy. Boscovich returned to Rome in 1763. He was employed by the pope to study the draining of the Pontine Marshes, and became professor of mathematics at the University of Pavia in 1764. Boscovich was invited to join an expedition to Baja California to observe the 1769 transit of Venus, but the worsening situation of the Jesuit order caused him to decline the offer. That was fortunate, as most of the astronomers who went died of an epidemic.

In 1770 Boscovich was appointed director of the new observatory at Brera near Milan, which he had helped design. He left shortly thereafter after a personality conflict with another astronomer. After political pressure from European Catholic governments forced the dissolution of the Jesuit order by papal decree in 1773, Boscovich relocated to Paris, receiving the position of director of optics for the French navy with the principal assignment of creating achromatic lens telescopes. In 1782 ill health caused him to return to Italy, where he died several years later.

See also Astronomy; Atomism; Observatories; Optics; Physics; Religion; Technology and Engineering; Telescopes.

Reference

Whyte, Lancelot Law, ed. *Roger Joseph Boscovich, S.J., F.R.S., 1711–1787: Studies of His Life and Work on the 250th Anniversary of His Birth.* London: Allen & Unwin, 1961.

Botanical Gardens

Botanical gardens were growing in both number and the range of species included in the eighteenth century. They existed in a number of institutional contexts. The largest and most powerful were associated with royalty and central state authority. Europe's unquestionable leader among botanical gardens was the Royal Botanical Garden of France, and Britain's Kew Gardens was emerging by late in the century. The United States founded a central botanical garden at Washington, D.C., in 1820. Other gardens were associated with universities, usually with medical programs. Several new university gardens were founded in the eighteenth and early nineteenth centuries, such as the Göttingen garden in 1754, Budapest in 1771, Trinity College Dublin in 1806, and Harvard in 1807. The University of Geneva founded a botanical garden in 1817 for the specific purpose of luring the Swiss botanist Augustin-Pyrame de Candolle (1778–1841). Individuals could still found botanical gardens, such as the one founded by the physician John Fothergill in the British town of Upton in 1762.

Whatever its institutional affiliation, the goal of a botanical garden was the collection of plants for purposes of utility and scientific interest. Great gardeners were admired for the skill with which they could acclimate plants coming from vastly different climates and environments. Definitions of utility shifted during the eighteenth century. At its beginning, the dominant use envisioned for botanical garden plants was medical. University botanical gardens were associated with medical programs, and usually administered by medical professors. This tradition retained its vitality and even expanded into new areas—the veterinary school at Lyons, the first ever, acquired a botanical garden in 1763. However, there was a growing tendency to employ botanical gardens to search for new and profitable crops. This was particularly the case for the botanical gardens founded in European colonies. Of these, the most important were the French garden at Pamplemousses on Mauritius, founded in 1735, and three British gardens: St. Vincents in the Caribbean, founded in 1765; Calcutta, founded in 1787; and St. Helena, founded the same year as a way station between the Caribbean and India. The Mauritius garden was associated with one of the great triumphs of acclimatization, the adoption of Brazilian manioc as cheap food for slaves. Robert Kyd (1746–1793), founder of the Calcutta garden, hoped that the cultivation of drought-resistant crops would forestall the repetition of the devastating famine that had hit Bengal in 1770. Both the French and the British incorporated their colonial botanical gardens into a network, run from the Royal Botanical Garden in the French case and by Sir Joseph Banks in the British.

The expansion of scientific functions and the vast numbers of new plants found in the eighteenth century put great pressure on the space allocated to gardens. There were many ways of handling this. John Hope (1725–1786) supervised the transfer of the entire Royal Botanic Garden of Edinburgh to a new location at Leith Walk in 1763, losing only a few plants, while Georges-Louis Leclerc, the Comte de Buffon, intendant and unchallenged master of the French Royal Botanical Garden, in 1782 effectively swindled the monastery of Saint-Victor out of the land between the garden and the Seine. The expansion of botanical gardens also placed severe demands on gardeners, many of whom became botanical experts in their own right. Philip Miller (1691–1771), head gardener of London's Chelsea Physic Garden, wrote *The Gardener's Dictionary* (1731), frequently reprinted into the nineteenth century, while the head gardener at Paris, André Thouin (who inherited the post from his father), was a botanist of great knowledge. The major gardens also published catalogs that acquainted some who could not actually visit with their collections and achievements. All major gardens were also confronted with the question of classification and arrangement, which by the late eighteenth century meant a choice between the Linnaean system and the "natural" system developed at the Royal Botanical Garden.

Unquestionably, the most important garden scientifically was the Royal Botanical Garden at Paris. This institution was more than a garden, including a natural history collection and sponsoring lectures in several fields, including chemistry, mineralogy, and geology as well as botany. This broad range of scientific expertise was recognized after the French Revolution, when the garden was renamed the Museum of Natural History. The garden, and later the museum, was also the institutional home of the illustrious de Jussieu family.

See also Banks, Sir Joseph; Botany; Buffon, Georges-Louis Leclerc de; Colonial Science; Kew Gardens; Museum of Natural History.

References

Fletcher, Harold R., and William H. Brown. *The Royal Botanic Garden Edinburgh, 1670–1970.* Edinburgh: Her Majesty's Stationery Office, 1970.

Gager, Charles Stuart. "Botanic Gardens of the World: Materials for a History." 2d ed. *Brooklyn Botanic Garden Record* 27 (1938): 151–406.

Gillispie, Charles Coulston. *Science and Polity in France at the End of the Old Regime.* Princeton: Princeton University Press, 1980.

Grove, Richard H. *Green Imperialism: Colonial Expansion, Tropical Island Edens, and the Origins of Environmentalism, 1600–1860.* Cambridge: Cambridge University Press, 1995.

Botany

Enlightenment botany was transformed by the acceptance of the idea that flowering plants reproduced sexually, by the constantly swelling volume of information about plants previously unknown to European science, and by the subsequent struggle of classification schemes to order this huge mass of knowledge. The long-standing connection of botany with medicine continued into the eighteenth century. One of the most important botanists of the early part of the century was Hermann Boerhaave, who oversaw the great botanical garden of the University of Leiden and taught botany to hundreds of medical students. However, botany became more of a discipline in its own right during the period, and as the century progressed fewer botanists had medical degrees, practiced as physicians, or taught at medical schools. With the expansion of empire, the usefulness of plants and botanical information grew far beyond the medical field, as the controllers of European empires sought to establish the most profitable crops in their colonial territories. The foundation of Britain's Kew Gardens was associated with the needs of empire, particularly under the directorship of Sir Joseph Banks. The French Royal Botanical Garden disposed of a network of botanists and colonial botanical gardens promoting the transference of crops from the Indian Ocean to France's Caribbean colonies, such as coffee, pepper, cinnamon, and breadfruit.

Eighteenth-century botanists built on the demonstration of plant sexuality by the German Rudolf Jakob Camerarius (1665–1721), although the idea continued to find opponents through the eighteenth century. Sebastien Valliant (1669–1722), a botanist at the Royal Botanical Garden, championed it in a widely published lecture in 1717, and Boerhaave's network of medical students spread it far and wide in the European medical world. Another German, the physician Josef Gottlieb Kölreuter (1733–1806), carried on Camerarius's work in a series of publications on plant sexuality and reproduction between 1761 and 1766, establishing the stability of hybrids and the ubiquity and mechanisms of insect pollination. The Lutheran pastor Christian Konrad Sprengel (1750–1816) published *Revelation of the Secret of Nature in the Construction and Fertilization of the Flower* (1793), which built on Kölreuter's work by explaining in detail the adaptation of nearly 500 different flowers to insect and wind pollination.

Eighteenth-century botanists also inherited a number of schemes for classifying the bewildering variety of plants with which they were confronted, of which the most important were those of John Ray (1627–1705), for whom the key concept was species, and Joseph Pitton de Tournefort (1656–1708), for whom the key concept was genus. The two principal competing systems of botanical classification in the late eighteenth century were those of the Swede Carolus Linnaeus and those associated with French botanists. Linnaeus's system, which built on Tournefort's, was also known as the "sexual system," as it used as the basis of classification only the sexual organs of plants, the stamens and pistils, their number and arrangement. Although in practice Linnaean classifiers did recognize other characteristics, his system suffered from the stigma of being "artificial," isolating a particular characteristic rather than considering the totality of a plant. Linnaeus himself recognized this weakness, but held that artificial systems were much easier to apply in the field. He also introduced the binomial nomenclature, combining a descriptive Latin word for the genus and a "trivial" Latin name for the species, which became the universal scientific method for referring to biological species.

The center of opposition to the sexual system was France's Royal Botanical Garden. Its director during the mid-eighteenth century, Georges-Louis Leclerc, the Comte du Buffon, found Linnaeus's system reductionistic and arbitrary. Although Buffon himself specialized in animals rather than plants, the garden housed a series of botanists of the de Jussieu family who championed natural rather than artificial systems of classification. The first generation of de Jussieus were three brothers, Antoine (1686–1758), Bernard (1699–1777), and Joseph (1704–1779). Antoine, a Montpellier-educated physician, succeeded Tournefort as professor of botany at the Royal Botanical Garden in 1710 and published a work on fungi in 1728. He also brought his brothers to Paris. Joseph accompanied the La Condamine expedition to Peru, where he was stuck for thirty-five years when his money ran out. He did remit specimens back to Paris. Bernard held the modest post of assistant demonstrator of plants at the Royal Botanical Garden and principally influenced the development of botany through his teaching. He was also appointed supervisor of the Royal Garden at the palace of Trianon in 1759. Bernard's teaching emphasized classifying plants according to multiple affinities, rather than the reductionistic approach of the sexual system. The three brothers' nephew, Antoine-Laurent de Jussieu (1748–1836), built on Bernard's teaching to publish *Families of Plants Arranged in Natural Orders* (1789). Another of Bernard's students, Michel Adanson (1727–1806), also published a natural classification system, *Families of Plants* (1763), but despite its theoretical innovations it was cumbersome and had little impact on contemporaries. Adanson's diehard opposition to binomial nomenclature also hindered acceptance.

Antoine-Laurent de Jussieu's system, much more influential than Adanson's, was based on a complete analysis of a plant's characteristics. De Jussieu believed in continuity, holding that there were no sharp divisions between species, but intermediate forms. Even where there seemed to be gaps between groups of plants, there was hope that exploration would reveal intermediate forms lurking in remote areas of the world. De Jussieu's approach to classification was not based on distinguishing between different natural features, an analytic approach, but on grouping together plants that shared natural features, a synthetic one. Jean-Baptiste-Pierre-Antoine de Monet de Lamarck, another Frenchman associated with the Royal Botanical Garden, also put forth a natural system in his *French Flora* (1779) and his work for the *Encyclopédie Methodique.*

In the competition between sexual and natural classification, Linnaeus's system had the advantages of simplicity, ease of learning, and Linnaeus's formidable organizational skill and worldwide network of disciples, the natural systems those of completeness and the backing of the powerful French scientific establishment. Linnaeus's system was more popular among the botanical amateurs who furnished a large proportion of the researchers in the field during the eighteenth and early nineteenth centuries, and its simplicity led to its adoption as a means of identifying plants even at the Royal Botanical Garden itself. The spread of the French approach was also hindered by the plethora of competing natural systems, as opposed to the unity of the Linnaean method, and the turmoil of the French Revolution and Napoleonic periods.

Antoine-Laurent de Jussieu himself survived the Revolution and even prospered in association with the new Museum of Natural History, which succeeded the old Royal Botanical Garden. (His only son, Adrien-Henri-Laurent de Jussieu [1797–1853], last of the botanical de Jussieus, was professor of botany at the museum.) An Afrikaner resident in Paris, Christian Hendrik Persoon (c. 1761–c. 1836) established the first systematic classification of fungi in *Synopsis of the Fungi* (1801). Natural systems of classification, whether de Jussieu's or others', did eventually diffuse from France. In Britain, natural classification was promoted and refined by Robert Brown, author of *Introduction to the Flora of New Holland* (1810). ("New Holland" was Australia.) The other great systematic botanist of the early nineteenth century was the Swiss Augustin-Pyrame de Candolle (1778–1841), author of *Elementary Theory of Botany* (1813). Unlike de Jussieu and, more ambiguously, Brown, Candolle emphasized discontinuities between species and genera rather than natural continuity.

See also Agriculture; Banks, Sir Joseph; Bartram Family; Boerhaave, Hermann; Botanical Gardens; Brown, Robert; Colonial Science; Darwin, Erasmus; Exploration, Discovery, and Colonization; Lamarck, Jean-Baptiste-Pierre-Antoine de Monet de; Linnaeus, Carolus; Linnean Society; Michaux Family; Plant Physiology; Royal Botanical Expedition.

References

Morton, A. G. *History of Botanical Science: An Account of the Development of Botany from Ancient Times to the Present Day.* London: Academic Press, 1981.

Stevens, Peter F. *The Development of Biological Systematics: Antoine-Laurent de Jussieu, Nature, and the Natural System.* New York: Columbia University Press, 1994.

Böttger, Johann Friedrich (1682–1719)

The career of Johann Friedrich Böttger bridged what today seem the vastly disparate realms of alchemical gold making and industrial manufacturing. Something of a child prodigy in chemistry, by 1701 Böttger claimed to have discovered the secret of creating small amounts of gold. Public demonstrations convinced many of his friends and associates. Word of Böttger's feat reached the king of Prussia, Frederick I, who summoned Böttger to his court. Afraid of the consequences of the discovery of his cheating, Böttger fled to Saxony. This provided only a brief respite, as the king sent soldiers after him that forced Böttger to place himself under the protection of the elector of Saxony and king of Poland, Augustus the Strong. An aficionado of alchemy himself, and short of cash due to his extravagant tastes and unsuccessful war with Sweden, Augustus expected Böttger to make gold and plenty of it. He was held prisoner pending success.

While in captivity, Böttger became an alcoholic, and, no longer able to conceal his lack of success in gold making, briefly escaped to Bohemia. Following recapture, Böttger faced execution for his failures, but adroitly shifted to another project of the king, the manufacture of porcelain. Augustus was an obsessed collector of Chinese and Japanese porcelain, a substance Europeans had been unsuccessfully trying to duplicate for decades. The Saxon nobleman Ehrenfried Walter von Tschirnhaus (1658–1708), the leader of Augustus's porcelain-making project, was an admirer of Böttger's chemical knowledge. In September 1705 Böttger was transferred to the Albrechtsburg, the Saxon castle at Meissen, to work on the porcelain problem. Böttger decided that rather than following the usual path of would-be porcelain makers, mixing clay and glass, the most hopeful method would be firing mixtures of clay and rock at extremely high temperatures. Systematic experimentation with firing different mixtures at different temperatures eventually led to success with a blend of alabaster and kaolin clay from a Saxon mine at Colditz in 1708. Success with porcelain did not lead to Böttger's liberation, partly because the setting up of a porcelain factory would require much additional work, and partly because the king still expected Böttger to make gold for him.

Whatever Böttger's gifts, the setting up of a smoothly running factory while maintaining the secret of porcelain manufacture was beyond him. His problems were compounded by the fact that the factory was being established at Meissen while Böttger himself was being held prisoner at the Saxon court at Dresden. However, the factory eventually produced porcelain, and Böttger made progress on refining the process and figuring out how to decorate it properly, which meant emulating Chinese techniques. He was raised to the rank of baron in 1711 and formally liberated in 1714. After years of ill health, Böttger died in 1719, but the Meissen factory dominated European porcelain for decades thereafter.

See also Industrialization; Technology and Engineering.

Reference

Gleeson, Janet. *The Arcanum: The Extraordinary True Story.* New York: Warner, 1999.

Bradley, James (1693–1762)

James Bradley, Savilian Professor of Astronomy at the University of Oxford, made two important astronomical discoveries. The first was the aberration of starlight. Bradley encountered this problem while taking precise observations to detect the stellar parallax predicted by the Copernican theory, the apparent movement of the nearer stars against the background of the farther produced by the motion of Earth around the Sun. Using a zenith sector built by the great instrument maker George Graham (1673–1751), Bradley found a set of stellar motions on an annual cycle that was clearly not the result of parallax, but was not explained by anything else either. Bradley eventually discerned that the motions were caused by the combined effect of Earth's motion and the finite velocity of light. Bradley's announcement of the aberration in a paper given to the Royal Society in 1729 established two key points in addition to the aberration itself. The aberration was the first physical proof of the motion of Earth and the Copernican system, which had little impact because the Copernican system was already nearly universally accepted. Bradley's failure to detect parallax, even with the aberration taken into account, meant that the stars were much farther away than was commonly thought.

Meanwhile, Bradley was working his second great discovery, explaining another mysterious but even tinier apparent motion of the stars. He correctly ascribed this motion to the "nutation," a wobbling in the precessional movement of the axis of Earth caused by the gravitational pull of the Moon on Earth's equatorial protuberance.

Bradley's discoveries were the basis for improvements in observational astronomy in the eighteenth century. They also won him the position of astronomer royal of Great Britain on the death of Edmond Halley (1656–1742). Much of his time was now devoted to the improvement of navigation and the solution of the longitude problem. Bradley was a supporter of the method of lunar distances and spent an enormous amount of time verifying the lunar tables of Johann Tobias Mayer.

See also Astronomy; Longitude Problem.

Reference

North, John. *The Norton History of Astronomy and Cosmology.* New York: W. W. Norton, 1995.

Brown, Robert (1773–1858)

The Scotsman Robert Brown was the greatest botanist of early-nineteenth-century Britain. Failing to graduate from the Edinburgh medical school, he took a position as a military surgeon, botanizing when he could. His big break came in 1800, when Sir Joseph Banks nominated him to accompany a naval expedition to Australia as a botanist. After many delays the *Investigator* departed England in 1801 with the mission of circumnavigating Australia and ascertaining that it was really a continent and not a collection of islands. Brown returned to England in 1805, having collected thousands of specimens, including hundreds of new species. He was appointed secretary to the Linnean Society, a position that gave him plenty of time to catalog his specimens.

As was nearly universal among British botanists, Brown had used the Linnaean or sexual system to identify and classify plants. He began to move toward classification taking a broader range of characteristics into account—the "natural system." Brown's Latin *Introduction to the Flora of New Holland* (1810) employed a natural system following Antoine-Laurent de Jussieu (1748–1836) with some modifications. It introduced 187 new genera. Although Brown's poor Latin excited criticism, his use of it helped the book reach an admiring international audience. Some were disappointed, however, that it was not followed by a catalog of Brown's botanical discoveries with illustrations. The same year as its publication, Brown became Banks's librarian, and the following year he was elected a fellow of the Royal Society. A steady stream of papers and publications followed. Brown made more use of the microscope (preferring the single lens to the compound design) and plant dissection than was common among botanists at the time, and his writings went beyond identification and classification to fundamental matters of plant biology. Brown expanded the range of botanical features to the microscopic level and emphasized the importance of studying the immature as well as the mature plant to place proper weight on those characteristics that disappear in maturity. A paper he read before the Linnean Society in 1825 established the important distinction between angiospermous and gymnospermous reproduction.

Banks's death in 1820 left Brown in control of his library and botanical collection. In 1827, as part of a deal to acquire Banks's legacy, the British Museum appointed him underlibrarian in charge of the botanical library and collections. He actively added to the museum's collections, although, like most people in such positions, he was prone to complain about lack of funds. Brown's acquisitions included some specimens collected by his young acquaintance Charles Darwin (1809–1882), on the voyage of the *Beagle* from 1831 to 1836. As a leading figure in English science and particularly botany, Brown stood for professionalization, being particularly motivated to establish a clear distinction between scientific botanists and the legion of amateur botanists. He was involved in mostly unsuccessful efforts to reform the Royal Society, making it more dominated by working scientists rather than gentlemanly amateurs. Brown's works circulated extensively on the Continent, and in 1833 he received one of the highest honors in European science, selection as one of the eight foreign members of the Academy of Sciences in France.

The same year he was appointed underlibrarian, Brown noted the random movements of grains of pollen suspended in water, visible only under high magnification—"Brownian movement." Brown established that both organic and inorganic particles, if sufficiently small, exhibited such motions. An even more significant microscopic discovery was the nucleus of the cell in 1831. He observed nuclei (a term he is responsible for) first in orchids, a subject in which he was greatly interested. He also cultivated an interest in fossil plants, although he published very little on the subject. (Brown's caution often made him reluctant to publish until he felt absolutely sure.) Brown became involved in a historical controversy over priority in the decomposition of water with the French astronomer François Arago (1786–1853), who maintained that James Watt (1736–1819) had been the first to separate water into hydrogen and oxygen, whereas Brown held the traditional view that it had been Henry Cavendish.

Conflict between the botanical division of the British Museum and Kew Gardens also rendered his final decades contentious.

See also Banks, Sir Joseph; Botany; Exploration, Discovery, and Colonization; Linnean Society; Microscopes; Museums; Royal Society; Surgeons and Surgery.

Reference

Mabberley, D. J. *Jupiter Botanicus: Robert Brown of the British Museum.* Brunswick: Verlag von J. Cramer, 1985.

Brunonianism

Brunonianism was a medical theory named after its originator, John Brown (c. 1735–1788). In his *Elementa Medicinae* (1780) Brown held that all diseases could be reduced to an improper state of excitability in the body. Diseases caused by a deficiency in excitability (asthenia) were more common than diseases caused by overexcitability (sthenia). Brown was influenced by his own gout, for which the usual treatment, abstemiousness, had failed. The proper treatment of asthenic diseases was with stimulants, including rich food, alcohol, and opium. Although traditional physicians sometimes prescribed these things, Brunonians got a reputation for overprescribing, and Brown himself became an opium addict.

Brown, a medical student at Edinburgh, presented his medicine as following the canons of Newtonian and Baconian science, not multiplying entities as did the elaborate disease classification schemes of Edinburgh professors, including Brown's one-time patron William Cullen (1710–1790). Brunonian medicine was taken up by younger Edinburgh physicians, resentful of the university medical faculty and the town's medical elite. It spread to England where it became identified with radical materialism—Thomas Beddoes published an edition of Brown in 1795. On the Continent, Brunonianism was most influential in Italy and Germany. Joseph Frank (1771–1842) introduced it at the Pavia Hospital around 1792, and Adalbert Marcus (1753–1816) at the showcase Bamberg Hospital. Brunonianism also influenced romantic *Naturphilosophie,* particularly the work of Friedrich Schelling (1775–1854). Some early enthusiasts, including Frank, eventually decided that the system was too dogmatic and reductionistic.

See also Beddoes, Thomas; Medicine; Romanticism; University of Edinburgh.

Reference

Bynum, W. F., and Roy Porter, eds. *Brunonianism in Britain and Europe.* Medical History, Supplement 8. London: Wellcome Institute for the History of Medicine, 1988.

Buffon, Georges-Louis Leclerc de (1707–1788)

The Comte de Buffon bestrode Enlightenment natural history alongside his hated rival Carolus Linnaeus. Buffon was a self-made man of bourgeois origin, who used natural history as a vehicle for upward mobility and wealth in the hierarchical society of old regime France. His earliest scientific work was in the field of mathematical probability, and it was on its strength that he was admitted to the Royal Academy of Sciences in 1734. Even more important was his ascension to the post of director, or intendant, of the Royal Botanical Garden in 1739, a position he would hold for nearly fifty years.

Buffon's ambition, enormous capacity for methodical work, and unscrupulousness combined to make him a great scientific administrator, despite spending more than half the year outside Paris at his country estate at Montbard. At the garden, Buffon built the greatest natural history collection in the world, as well as expanding the garden itself by taking over the property between it and the Seine. He also enriched himself—not unusual for an official of the old regime, in which the distinction between public and private property was often hazy. He employed the staff of the garden, which he increased, as assistants to produce the work that made him

Intendant du Jardin du Roi, de l'Académie Française et de celle des Sciences, né à Montbard en Bourgogne en 1707, mort à Paris le 16 Avril 1788.

Buffon, destiné par son Pere à lui succéder dans sa charge de Conseiller au Parlement de Dijon, fut détourné de ce dessein, par le penchant irrésistible qui dirigeoit son Génie vers les Sciences: l'envie de les cultiver exclusivement fut la seule ambition qui l'anima toute sa vie. Sa première étude fut celle de la Géometrie qu'il parvint à apprendre sans Maître. Curieux de voir l'Italie, il visita cette intéressante contrée, parcourut l'Angleterre, et revint en France, où ayant recueilli la succession de sa Mere, il put se livrer tout entier à son penchant pour l'histoire Naturelle qui fut toujours pour lui un objet de prédilection. Animé d'une égale ardeur pour l'étude et pour le plaisir, l'amour de la gloire l'emportoit cependant encore sur toutes ses autres passions, et lorsque son goût pour la société des femmes lui avoit enlevé quelques uns des instans consacrés au travail, il déroboit ceux destinés au someil pour les restituer aux Sciences. Nommé à trente et un Ans Intendant du Jardin du Roi, il exécuta le vaste projet, de réunir dans son enceinte les productions des trois Regnes de la Nature, et cette Collection à peine comencée, devint sous sa direction la plus riche de l'Univers. Ce fut au milieu des soins multipliés que lui demandoit les devoirs de sa place qu'il composa son Histoire de l'Home, *celle des* Oiseaux, *des* Mineraux, *son Ouvrage sur* l'Aimant, *et ses* Epoques de la Nature, *ce chef-d'œuvre de génie, qu'on admire, même en n'adoptant aucun de ses resultats. Cet Homme Immortel, aussi Célèbre par l'élégance et la pureté de son Stile, que par la fécondité de ses idées, et l'exactitude de ses observations, n'arrivoit à cette rare perfection, qu'à force de retoucher ses Ouvrages. Comblé d'honneurs, Buffon qui sentoit sa supériorité, fut fort sensible à la louange; mais s'il lui échappoit quelquefois de faire son éloge lui même, c'étoit toujours d'une manière qui n'avoit rien de choquant pour les autres. La mort de ce Grand Homme laisse un vide bien difficile à remplir dans la carriere des Sciences et des Lettres: Philosophe profond, Génie étonnant, il a dévoilé à nos yeux les sublimes opérations de la Nature: Écrivain éloquent il a su rendre interessa.^ns et clairs, les objets les plus abstraits, et porter dans nos cœurs les douces consolations du Sentiment.*

This engraving of Georges-Louis Leclerc de Buffon, from a series entitled French Worthies, shows the animal and mineral kingdoms paying homage to him. (Edgar Fahs Smith Collection, University of Pennsylvania Library)

the most famous scientist of eighteenth-century France, the forty-four beautifully illustrated volumes of *Natural History.*

The first volume appeared in 1749. It caused controversy, as it contained *Theory of the Earth,* the first of Buffon's two lengthy discussions of the history of Earth. In it, Buffon described Earth history as marked by gradual, cyclic change extending over large periods of time, a picture at variance with Genesis. The faculty of theology at the Sorbonne, guardians of Catholic orthodoxy in Enlightenment France, forced Buffon to print a letter accepting the Genesis account and stating that he had put forth his cyclic theory strictly as a scientific hypothesis. However, the cyclic passages were reprinted unchanged in subsequent editions of *Natural History.* The entire first series comprised fifteen volumes and covered land animals. Buffon was assisted by Louis-Jean-Marie Daubenton (1716–1800), a young man he had hired for the recently created post of curator and demonstrator in the king's cabinet, the area where the garden's natural history collections were held. The second series, appearing between 1770 and 1783, comprised nine volumes on birds. Buffon's major collaborator in this series was the ornithologist Gueneau de Montbelliard (1720–1785), Daubenton having departed for a major government-funded project to improve the quality of French wool.

The third series, seven volumes appearing between 1774 and 1789, was a miscellany that included Buffon's second treatment of the history of Earth, *The Epochs of Nature.* In this work, Buffon shifted from a conception of Earth's past as undifferentiated into a more fully historical conception, dividing the history of Earth into periods linked to Earth's gradual cooling from its initial state as a molten fireball. Each epoch was marked by different forms of life, and Earth's final destiny would be that of a frozen and uninhabitable cinder, although Buffon hoped that humans, by the heat their society generated, could somewhat slow the cooling process. Earth was at least 75,000 years old, and quite possibly much older. Since different regions were affected differently by the cooling, different areas had different types of fauna and flora. This led Buffon into a controversy with Thomas Jefferson (1743–1826), who was insulted by the Frenchman's contention that Old World animals were larger than American animals due to the greater heat of the Old World.

The fourth series included five volumes on minerals, appearing between 1783 and 1788. The last volumes of *Natural History,* covering fish, reptiles, and whales, appeared after Buffon's death and were principally the work of his disciple, the Comte de Lacépède (1756–1825), curator and assistant demonstrator in the king's cabinet.

As a naturalist, Buffon worked mainly from books and from the specimens available in the garden. His interest in animals centered on their appearance and behavior rather than their internal structures, and *Natural History* included little anatomical information, what was present came mostly from Daubenton. Buffon did not arrange his chapters devoted to the animals in a tight system of classification based on natural characteristics, but worked from the more familiar to the less familiar—the first animal discussed is the horse. Buffon's objections to Linnaeus's sexual system of plant classification centered on its artificiality and abstraction. There was no reason, he claimed, why sexual characteristics should be singled out as the basis for classification. Hierarchical classification itself was as questionable as abstracting the complex relationships between species into rigid categories. Buffon's doubts about the imposition of human categories on nature extended to the concept of species itself, fundamentally an imposition. He preferred to define a species in terms of a mutually interfertile population rather than in terms of essential characteristics.

Wealth and honors continued to descend upon Buffon, considered a master of French prose as well as natural history. *Natural History* sold well and was widely abridged, translated, and pirated over the ensuing decades. Buffon was admitted to the French Academy of Literature in 1753 and received the title of count from the king. He quarreled with virtually everyone worth quarreling with in eighteenth-century France, ranging from Jean-Jacques Rousseau (1712–1778) to the abbé Jean-Antoine Nollet. His funeral was the most elaborate of any eighteenth-century scientist's, exceeding even Sir Isaac Newton's in 1727. Fourteen horses draped in black and silver drew his casket, while 20,000 spectators watched the procession.

See also Botanical Gardens; Geology; Linnean Society; Museums; Nollet, Jean-Antoine; Zoology.

References

Gillispie, Charles Coulston. *Science and Polity in France at the End of the Old Regime.* Princeton: Princeton University Press, 1980.

Roger, Jacques. *Buffon: A Life in Natural History.* Translated by Sarah Lucille Bonnefoi. Edited by L. Pearce Williams. Ithaca: Cornell University Press, 1997.

C

Calculus

See Mathematics.

Calorimeters

The calorimeter, a device for measuring heat, was introduced by Antoine-Laurent Lavoisier and Pierre-Simon de Laplace in a paper read before a meeting of the Royal Academy of Sciences in 1783. The initial paper referred to the device simply as a machine, and presented it in a way that was independent of theories of heat's nature. The actual design of the device is Laplace's work, inspired by discussions the two had had over the work on specific heats of an Irish chemist, Adair Crawford (1748–1795), which rested on the concepts of latent and specific heat developed by Joseph Black. The heat of the object placed within the machine is measured by the amount of the ice contained in it that melts. A second layer of ice protects the inner ice from the heat of the room that the calorimeter is in. The melted ice is immediately removed by falling through a funnel into a receptacle in which it is measured. Since the ice is removed from the environment once it melts, all of the heat of the object that is measured goes to melting the ice, overcoming its "latent heat," as opposed to raising the temperature of the resulting water. The idea of using melting frozen water to measure heat was not new—the Swedish experimental physicist Johann Carl Wilcke (1732–1796) had attempted to use melting snow to measure heat in 1781, but his effort had failed. Although Laplace and Lavoisier's machine was an advance on previous measures of heat, it was a finicky device that worked within a relatively narrow range of temperatures—temperatures below freezing meant that the ice would be cooled, some of the object's heat would be absorbed in heating the ice, and temperatures too high would cause warm air to enter the chamber through the funnel.

The calorimeter could measure the specific heat of a substance by measuring the water it melted in cooling from a given temperature to zero. However, this was not a great advance over the measurement of specific heat by mixing a substance with water or another substance at a different temperature—the technique Crawford had used. What Lavoisier and Laplace really hoped to use the device for was the measurement of the heat given off by processes, whether of chemical combination or respiration. This required that air to support the combination or the life of the animal within the calorimeter (small animals could be lowered into the chamber in a wire basket) be introduced, and that the air had to be as close as possible to the temperature of the ice. Such experiments required a second device to introduce the air, and could be made only in the winter. In the winter of 1782–1783 the two experimented with candles, burning coals, and guinea pigs. The evidence pointed to the equivalence of the processes of respiration and combustion. The following winter saw more calorimeter experiments, after which the two went on to other problems. The calorimeter received its name in Lavoisier's 1789 *Elementary Treatise of Chemistry* in which its workings were explained in terms of Lavoisier's theory of heat as a "subtle fluid" called caloric.

Crawford, a defender of the phlogiston theory, wished to use a calorimeter to examine the heat of respiration himself. Crawford claimed that the Laplace-Lavoisier ice calorimeter was not suited for the study of respiration, as animals respire more quickly in a cold environment. The calorimeter Crawford developed, and described in the second edition of his *Experiments and Observations on Animal Heat* (1788), used down for insulation and measured the heat of respiration by the rise in the temperature of water surrounding the central chamber. The data agreed with Lavoisier and Laplace's claim for the equivalence of respiration and combustion. The water calorimeter had the advantage that it could be used in a wide range of temperatures, and eventually became the standard. It was further adapted to the purpose of measuring the heat of combustion by Benjamin Thompson, Count Rumford, in 1812. Ice calorimeters continued to be used for experimental demonstrations, but not for original scientific research.

See also Heat; Instrument Making; Laplace, Pierre-Simon de; Lavoisier, Antoine-Laurent; Royal Academy of Sciences; Rumford, BenjaminThompson, Count.

References

Holmes, Frederic Lawrence. *Lavoisier and the Chemistry of Life: An Exploration of Scientific Creativity.* Madison: University of Wisconsin Press, 1985.

Roberts, Lissa. "A Word and the World: The Significance of Naming the Calorimeter." *Isis* 82 (1991): 199–222.

Cartesianism

The natural philosophy of René Descartes (1596–1650), dominant in many parts of Europe at the opening of the eighteenth century, was defeated and eliminated in the first half of the century by Newtonianism. Cartesianism, which had evolved in the decades since Descartes first expounded it, took a variety of forms. All were mechanical. Cartesians believe that a physical phenomenon could not be understood until it was explained by the mechanical action of bodies in contact. Cartesians believed that the Newtonians' willingness to accept gravity as a quantifiable force without providing a mechanical explanation was intellectually reactionary, a revival of the "occult qualities" of Aristotelianism. They also distrusted the teleological aspects of Newton's physics. Cartesians were vorticists, believing that matter moved in circular paths, and that these circles explained physical phenomena, such as the planets being carried in their orbits. Cartesians put great effort in explaining different kinds of phenomena, such as light, electricity, and magnetism, in terms of different shapes and sizes of particles, or "subtle matters"—explanations sometimes put forth more as evidence of the explicator's ingenuity than as physical theory. Cartesian science was less mathematical than Newtonian, even though some individual Cartesians were excellent mathematicians. Given the increasingly quantified nature of eighteenth-century science, this was an important weakness.

In 1699 declared Cartesians, previously under a religious cloud, were admitted to the Royal Academy of Sciences. The academy, although it never adopted an official dogma, was a Cartesian body in the early eighteenth century, with the Cartesian Bernard Le Bovier de Fontenelle (1657–1757) as secretary. The most important French Cartesian at the time was a priest of the Oratorian order, Nicolas Malebranche (1638–1715). (The Oratorians were the most Cartesian French order in the late seventeenth century, although in the eighteenth century the Jesuit order would also accept Cartesianism.) Malebranche and his followers integrated the Leibnizian calculus (which he was largely responsible for introducing into France) with Cartesian physics. In 1706 even the Jesuits, always publicly opposed to Cartesianism despite some individual members' admiration for it, admitted it as a system that could be taught and defended as a hypothesis.

After the Peace of Utrecht ended the wars of Louis XIV and France became more open to British science, Cartesianism was increasingly opposed by Newtonianism. The Dutch universities, the first in Europe to adopt Cartesianism, were abandoning it for Newtonianism. Fontenelle's eulogy for Newton, much of which turned on a contrast between Newton and Descartes, differentiated the two schools by identifying Newtonianism as empiricist and Cartesianism as based on rational deduction from first principles. The British by the early eighteenth century made anti-Cartesianism a national issue, while some French scientists viewed Cartesianism as a matter of French pride. Some, including surviving members of Malebranche's circle, hoped to save Cartesianism by adopting selected features of Newtonianism. Young and aggressive Frenchmen hoping to make careers, such as Pierre-Louis Moreau de Maupertuis and Voltaire, increasingly identified Cartesianism as an aspect of French backwardness.

What turned out to be the key issue in the conflict between Cartesianism and Newtonianism in France was the shape of Earth. There were two beliefs, one that Earth was a prolate sphere, flattened at the poles, and the other that Earth was an oblate sphere, elongated at the poles. The struggle between these views arose in France independently of the Cartesian-Newtonian struggle—there were Cartesian arguments for both views. However, the prolate view was Newtonian, and the oblate view became identified as Cartesian. The eventual demonstration of the superiority of the prolate theory was a triumph for the Newtonians led by Maupertuis, but by no means the end of Cartesian science. Vortex theory and subtle matters dominated French work on electricity until Benjamin Franklin's system displaced that of Jean-Antoine Nollet. Leonhard Euler presented a simplified Cartesianism in his popularization *Letters to a German Princess* (1768). In France, vorticist works continued to be published into the 1770s, and in Britain the leading opponents of the Newtonians, the Hutchinsonians, combined an essentially Cartesian physical system with biblical literalism.

As a philosophy Cartesianism had a more subtle and long-range impact on French and European science. As Cartesian physics became separated from Cartesian metaphysics in the decades following Descartes's death, Cartesianism, far more than Newtonianism, offered a picture of the world that did not depend on the existence or purposes of a deity. Descartes's interpretation of animals and plants as machines, not animated by any sort of spirit, could easily be extended to man by thinkers who denied that any sort of human soul existed. Cartesianism influenced French materialism, even though materialists rejected specifically Cartesian physical explanations.

See also Hutchinsonianism; Materialism; Newtonianism; Nollet, Jean-Antoine; Physics; Poliniere, Pierre.

References

Greenberg, John L. *The Problem of the Earth's Shape from Newton to Clairaut.* Cambridge: Cambridge University Press, 1995.

Heilbron, J. L. *Electricity in the 17th and 18th Centuries: A Study in Early Modern Physics.* 2d ed. Mineola, NY: Dover, 1999.

Sutton, Geoffrey V. *Science for a Polite Society: Gender, Culture, and the Demonstration of Enlightenment.* Boulder: Westview, 1995.

Vartanian, Aram. *Diderot and Descartes: A Study of Scientific Naturalism in the Enlightenment.* Princeton: Princeton University Press, 1953.

Cassini Family

When the Italian astronomer Gian Domenico Cassini (1625–1712) accepted the offer from Louis XIV (r. 1643–1715) to come to France and head the Paris Observatory, he founded the longest-lived dynasty in the history of astronomy and cartography. His son, Jacques (1677–1756), grandson César-François (1714–1784), and great-grandson Jacques-Dominique (1748–1845) would all head the observatory.

None of his heirs and successors matched the founder's skills as an observational astronomer (among other discoveries, Gian Domenico identified the "Cassini division" in Saturn's rings), but they carried on his work in applying astronomy to the surveying and cartography of the realm of France. Jacques Cassini began assisting his father in an attempt to create a framework of defined points in France to serve as the basis of triangulation around 1700. He published his results after his father's death, in 1718. These results were controversial, as they supported the most common Cartesian theory that Earth was not a perfect sphere, but slightly elongated. This contrasted with the Newtonian theory, which implied that Earth was an oblate sphere, slightly bulging in the middle. The subsequent controversy was resolved in favor of the Newtonian theory by expeditions to measure Earth's curvature at the equator, led by Charles-Marie de La Condamine (1701–1774), and in the far north, led by Pierre-Louis Moreau de Maupertuis, who became Cassini's bitter enemy. All of Cassini's time was not devoted to geodesy. In 1738 he confirmed a shift in the position of Arcturus by comparing a recorded observation of the star in 1672 with a recent one of his own.

Much of the work of the Paris Observatory under the Cassinis was dedicated to accurate surveying, a discipline in which France was the world leader. Beginning with a royal commission in 1739, Jacques and César-François Cassini carried out or arranged for the surveying of France. The most important person in carrying out the fieldwork was the abbé Nicolas-Louis de Lacaille (1713–1762), making use of a recently devised micrometer for sextants and quadrants. The result was a complete triangulation of France, the basis for fixing the location of all geographic features. César-François, more mapmaker than astronomer, coordinated a series of "infilling" efforts to relate specific features to the triangulation. The year his father died, César-François had to deal with a withdrawal of royal funding. He adeptly turned to the private sector. He set up a private company composed of fifty shareholders to finance the completion of the map. The shareholders included many leaders of French science, technology, and society. In 1783 Cassini invited the English government to participate by joining Britain in the same geographic framework. This would be particularly useful to astronomers in establishing the exact geographical relation between England's Royal Observatory and French observatories, but despite the support of the idea in a resolution of the Royal Society, the English government refused. Much of Cassini's map had been completed by his death in 1784, but the full *Carte de Cassini* was not published until 1793, after the French Revolution.

Until 1771 the control the Cassinis exercised over the Paris Observatory had been informal. That year César-François was formally appointed head, and the observatory was withdrawn from the control of the Royal Academy of Sciences. The observatory by this time was far from the center of astronomical research it had been in the days of the first Cassini. The building itself was severely dilapidated, and the instruments were obsolete. France's leading astronomical observers worked elsewhere. Jacques-Dominique, the last Cassini to head the Paris Observatory, was not a particularly talented astronomer, but he did attempt a reform of the observatory. He was handicapped by the government's insistence that the new instruments be of French manufacture rather than English, which were of higher quality. A workshop for the production of instruments was set up on the grounds of the observatory itself, and it was hoped that the effort would spur French instrument makers to equal or surpass the English. This failed, and Cassini ended up having to buy a quadrant from the great English instrument maker Jesse Ramsden (1735–1800).

New regulations for the Paris Observatory promulgated in 1785 provided for paid staff to perform round-the-clock observations and the publication of an ongoing record of the observatory's work. In 1787 Cassini cleared up some unfinished business from earlier in the decade by participating in a linkage of the French and British cartographic systems in conjunction with the British soldier and surveyor William Roy (1726–1790), finally establishing the geographical relation between Paris and Greenwich. The work of the observatory, however, was severely hampered by the outbreak of the French Revolution. Jacques-Dominique himself, despite his connections with many leading politicians of the old regime, survived the Revolution, although he was briefly imprisoned in 1794. The geodetic tradition of the Cassinis' Paris Observatory led to the great measurements that established the metric system. By that time, however, Jacques-Dominique, disgusted with the Revolution, had withdrawn entirely from both astronomy and the observatory.

See also Astronomy; French Revolution; Instrument Making; Metric System; Nationalism; Observatories.

References

Gillispie, Charles Coulston. *Science and Polity in France at the End of the Old Regime.* Princeton: Princeton University Press, 1980.

Hankins, Thomas L. *Science and the Enlightenment.* Cambridge: Cambridge University Press, 1985.

Cavendish, Henry (1731–1810)

Henry Cavendish was England's most profound and versatile experimental scientist of the late eighteenth century, but his impact on science was limited by his reluctance to publish many of his experimental results. Cavendish came from the most exalted social background of any major eighteenth-century scientist. He was a member of the great Cavendish house headed by the duke of Devonshire. His father, Lord Charles Cavendish, had been an active supporter of the sciences, an experimenter, and a leading member of the Royal Society, receiving the Copley Medal in 1757. Henry Cavendish was educated at the Newtonian stronghold of Cambridge University, and Newton always remained for him the model of scientific practice. Cavendish left Cambridge without taking a degree, normal behavior for an eighteenth-century aristocrat.

His father's influence procured Cavendish's election to the Royal Society and to its dining club, the Society of Royal Philosophers, in 1760. He became a faithful attendant at both bodies, and served many times on the council of the Royal Society, being the most eminent physicist to support Sir Joseph Banks in the struggle over the Royal Society presidency in 1783 and 1784. All of Cavendish's scientific papers appeared in the Royal Society periodical, *Philosophical Transactions.*

Cavendish first made his mark in science as a pneumatic chemist—along with Joseph Black and Joseph Priestley, one of the three great British scientists who pioneered the analysis of the air. His first published scientific work appeared in 1766 and was titled "Three Papers Containing Experiments on Factitious Airs." The first paper discusses the production of "inflammable air"—what we now call hydrogen—by experiments on the dissolution of metals in acid. Cavendish's experiments also established the density and properties of "inflammable air," one of the most important being its explosiveness when exposed to flame in the presence of common air, from which inflammable air got its name. A second paper discussed the qualities of "fixed air"—now called carbon dioxide—and the third dealt with airs produced by putrefying and fermenting organic substances. The quality and precision of Cavendish's experiments greatly impressed his readers and hearers, and he won the society's Copley Medal for this work. One of Cavendish's major contributions to chemistry was his quantitative exactness, first seen in his measurements of the density of hydrogen. Cavendish's balance, probably built for him by his instrument maker William Harrison, was the most accurate anywhere at the time. Cavendish's position as an experimenter was immeasurably strengthened by his access to the best and most precise instruments. In addition to his own acuteness in improving instrumental design, this was aided by his wealth and his residence in London, the capital of scientific-instrument manufacturing.

Cavendish's pneumatic chemistry led to the "water controversy," a bitter priority dispute over the synthesis of water from inflammable air and "dephlogisticated air"—now known as oxygen. Another experimenter, John Warltire, had exploded inflammable air and common air with an electrical spark in a glass vessel, noting the generation of heat and light, as well as a dew, which Warltire dismissed as of little significance. Cavendish repeated the experiment in 1781, identifying the dew as pure water. He told Priestley of the analysis, who then informed James Watt (1736–1819), who interpreted the results. Antoine-Laurent Lavoisier also heard of the experiments and repeated them. Cavendish published his results in 1784. Since Lavoisier and Watt had circulated their results and interpretations before Cavendish published, there was question as to priority, and Watt also claimed that Cavendish plagiarized his interpretation. Watt was also motivated by distrust of the aristocrat Cavendish. Unquestionably, though, Cavendish had been the first to identify the substance produced by the experiment as pure water. The conflict did not lead to lasting enmity among the three participants. Another paper by Cavendish in 1785 discussed the production of nitrous acid by the sparking of a mixture of dephlogisticated air and "phlogisticated air"—nitrogen.

Cavendish designed a series of experiments carried out by Thomas Hutchins, the governor of Fort Albany of the Hudson's Bay Company, to determine the freezing point of mercury in the winter of 1781–1782. As a heat theorist Cavendish belonged to the minority that identified heat with motion (a tradition Cavendish derived from Newton) while the majority viewed it as a substance. Although Cavendish's heat theory underlay much of his experimenting, it had little direct influence, as his explicitly theoretical discussions were never published.

Cavendish did publish one major electrical treatise, "An Attempt to Explain Some of the Principal Phaenomena of Electricity, by Means of an Elastic Fluid," in *Philosophical Transactions* in 1771. Building on the work of Franz Maria Ulrich Theodor Hoch Aepinus (1724–1802), Cavendish interpreted electricity in terms of fluid pressure. The charge of a body was determined by whether it was saturated with the electrical fluid (no charge), oversaturated, or undersaturated. Cavendish's difficult and mathematical paper established that the electrical charge on a sphere would be concentrated on its surface. His unpublished work, among many other innovations, contains an experimental demonstration of Coulomb's law, before Coulomb. Cavendish chaired a committee of the Royal Society to examine what kind of lightning rod would best protect the government gunpowder magazines at Purfleet in 1772 and experimented with "torpedoes," or electric eels, to determine if the shock they caused was truly electrical. A Cavendish paper in 1776 suggested the fish's organs were the equivalent of a series of electrical Leiden jars. He even built an artificial torpedo, inviting fellow scientists to be shocked by it. Because so much of Cavendish's work was unpublished, and the mathematics of the 1771 paper was beyond most British electrical researchers, it had relatively little impact on electrical studies.

Cavendish's last great experiment was his "weighing of the world." This idea appears in his correspondence in the early 1780s, but was not carried out until the late 1790s. The method was originated by Cavendish's college friend, the Reverend John Michell (1724–1793). Cavendish used a wire torsion balance to measure the gravitational attraction that lead spheres weighing about 350 pounds exerted on smaller spheres. The attraction exerted by the sphere was then compared to the attraction exerted by Earth to derive a measure of Earth's density. So slight was the attraction of the spheres compared to that of Earth that the experiment, carried out in Cavendish's home, required nearly perfect conditions and meticulous, patient observation—precisely Cavendish's strengths. He estimated the density of Earth to be 5.48 times the density of water, remarkably close to the current estimate, and in line with the thinking of Newton, who had estimated Earth's density as between 5 and 6 times the density of water. The experiments were carried out in 1797 and 1798 and published shortly after their conclusion. The "Cavendish experiment" is often treated by modern physicists as a way of measuring the gravitational constant, but this was not Cavendish's own concern.

Cavendish was a scientific ascetic painfully shy around strangers, who led little social life outside the Royal Society. (He made his library open to visiting scholars and investigators, but they were strictly instructed to say nothing if they encountered him there.) Born in Nice, he never left England as an adult, and in fact seldom left the area of London. His closest scientific associate, the physician Sir Charles Blagden (1748–1820), encouraged him on a series of trips to different parts of England in the later 1780s, where Cavendish observed industrial plants and made notes of the stratigraphy of the country. Unlike his hero Newton, Cavendish was not pious, and his willingness to perform experiments and observations on Sundays would pain his Victorian admirers. After the weighing of Earth, he published no more groundbreaking scientific work, although he devised an instrument and set of procedures for testing the durability of coins containing different alloys of gold for a government coinage commission. He continued to serve the Royal Society and became a manager of the Royal Institution on its founding in 1800. Cavendish's last paper, in 1809, concerned the improvement of astronomical instruments.

See also Chemistry; Electricity; Physics; Royal Society.

References

Brock, William H. *The Norton History of Chemistry.* New York: W. W. Norton, 1993.

Heilbron, J. L. *Electricity in the 17th and 18th Centuries: A Study in Early Modern Physics.* 2d ed. Mineola, NY: Dover, 1999.

Jungnickel, Christina, and Russell McCormmach. *Cavendish: The Experimental Life.* Rev. ed. Lewisburg, PA: Bucknell, 1999.

Châtelet, Gabrielle-Émilie du (1706–1749)

Gabrielle-Émilie du Châtelet was one of the few women to participate in mid-eighteenth century Newtonian physics. From an aristocratic family, Châtelet was exposed to French science at an early age, as the academician Bernard Le Bovier de Fontenelle (1657– 1757) was a friend of her family. By the early 1730s she was permanently settled in Paris, where she began an affair and intellectual relationship with the early French Newtonian Pierre-Louis Moreau de Maupertuis. She also worked with another French Newtonian mathematician, the child prodigy Alexis-Claude Clairaut (1713–1765). Her most important intellectual and romantic alliance, though, was with Voltaire, whom she met in 1733. Voltaire drafted Châtelet, whose mathematical skills far exceeded his own, into his campaign for the Newtonianization of France. They operated from a base at Cirey, a rural estate belonging to the family of Châtelet's tolerant husband. There the two carried on experiments and worked on their writings. Châtelet's superior grasp of mathematics made her assistance vital to Voltaire in writing his Newtonian textbook, *Elements of the Philosophy of Newton* (1738). She herself, however, was influenced away from pure Newtonianism in a Leibnizian direction by Maupertuis's friend, the German mathematician and physicist Samuel Konig (1712–1757), a follower of the Leibnizian Christian Wolff. Châtelet's first published work, the anonymously published *Institutes of Physics* (1740), combined Newtonian with Leibnizian ideas, such as *vis viva* (living force). (The work's anonymity, considered appropriate for the work of a woman, enabled Konig to spread the rumor that he had written it himself.) In the early 1640s Châtelet also carried on a public controversy with Dortous de Mairan (1678–1771), the secretary of the Royal Academy of Sciences, over the formula for momentum. Her other significant work was a French translation of Newton's *Mathematical Principles of Natural Philosophy,* drawing on a number of editions, commentaries, and popularizations of Newton. It included a commentary incorporating work done on mathematical physics since Newton's time. This was completed in 1749, shortly before Châtelet's death in childbirth, but not published until 1759. It remains the only translation of Newton's masterpiece into French.

See also Maupertuis, Pierre-Louis Moreau de; Mechanics; Newtonianism; Voltaire; Women.

References

Sutton, Geoffrey V. *Science for a Polite Society: Gender, Culture, and the Demonstration of Enlightenment.* Boulder: Westview, 1995.

Zinsser, Judith P. "Translating Newton's Principia: The Marquise du Châtelet's Revisions and Additions for a French Audience." *Notes and Records of the Royal Society of London* 55 (2001): 227–245.

Chemical Nomenclature

The eclectic and unsystematized way in which chemistry had developed as a science to the eighteenth century was reflected in its nomenclature, which was unstandardized and even chaotic. Substances' names, usually expressed in Latin, came from different historical sources and traditions. They referred to colors, uses, consistencies, tastes, and the names of discoverers. Alchemical connections between the planets and the metals also influenced some names, as did connection to a place of origin or manufacture. One substance could have multiple names, often based on two or more ways of preparing it. Nitric acid distilled from saltpeter could be called *spirit of nitre* or *aqua fortis* depending on the reagent used. Substances could also have completely different names in their solid state than when in solution. The use of specific names in a chemical work was usually the choice of the individual chemical author. The most influential and widely circulated chemical books did play some part in shaping chemical nomenclature as a whole.

Although criticisms of individual names and name elements had been made for centuries, the expansion of chemical knowledge in the eighteenth century meant that the entire system of chemical nomenclature was increasingly felt to be a problem. The first effort at a reform of chemical nomenclature with some institutional backing was connected with the edition of the *London Pharmacopeia,* which appeared in 1746. This work had influence beyond England, as it circulated widely in France and Germany. A more thoroughgoing attempt to found nomenclature on an accurate and precise basis was made by Pierre Joseph Macquer, in his widely translated *Dictionary of Chemistry* (1766).

The efforts at reform bore some fruit, but many problems remained, and the plethora of new substances being discovered in the mid-eighteenth century, particularly gases, only made the problem worse. The Linnaean system in botany offered an example of a standardized nomenclature. In Carolus Linnaeus's home country of Sweden, Torbern Olof Bergman (1735–1784) put forth a chemical system similar to Linnaeus's, using a binomial Latin designation. Salts would have one Latin word for the base, followed by another for the acid. Bergman's work influenced another system of nomenclature devised and promoted by the French chemist

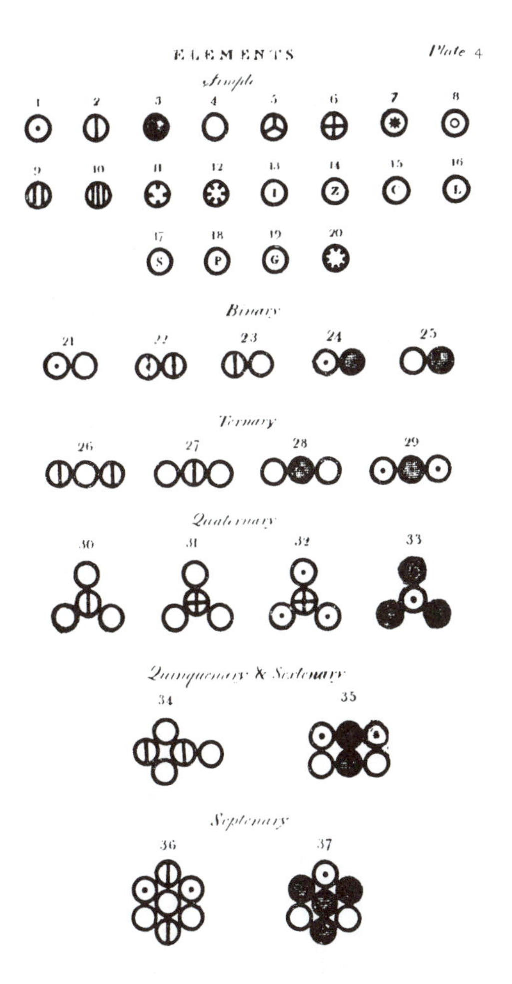

John Dalton's unsuccessful system of representing elemental atoms and their combinations. (Edgar Fahs Smith Collection, University of Pennsylvania Library)

Louis-Bernard Guyton de Morveau (1737–1816). Guyton de Morveau set forth his ideas for naming in a paper in the journal *Observations on Physics* in 1782. He suggested that chemical names be short and reflect the composition of a substance. Although based on classical roots, Guyton de Morveau's terms would have been in French, a language, unlike Swedish, already widely understood in the scientific community.

The culmination of French efforts at reform was *Method of Chemical Nomenclature* (1787). This appeared with the names of four authors, Guyton de Morveau, Antoine-Laurent Lavoisier (who introduced the project with a memoir read before the Academy of Sciences in April), Antoine-François de Fourcroy (1755–1809), and Claude-Louis Berthollet. The *Method of Chemical Nomenclature* set forth a system whereby the names of mixed bodies would be created by combining the names of simple bodies, those that could not be further decomposed. The simple bodies were divided into several classes. Terminations would indicate the proportions—thus, *acid sulfurique* contained more oxygen than *acid sulfureux*. Unlike the previous reform attempts, this was intimately related to a specific chemical theory, that of Lavoisier. Phlogiston was nowhere to be found, and oxygen was so-called from the Greek for "acidifying principle," as Lavoisier believed that oxygen was necessarily a component of all acids. Those doubting Lavoisier's new chemistry looked on the new nomenclature with suspicion. A committee of the Academy of Sciences made a noncommittal report, but criticized the new system severely. Away from France the new system attracted much interest, but was opposed by those who opposed Lavoisier's theory or, like Henry Cavendish, believed that a new system should not be theory-laden. The *Method* was extensively circulated, with several French editions and translations into German, English, Italian, and Spanish. The necessity of knowing the new system to read the works of Lavoisier and the other French chemists of his school meant that even opponents of the new nomenclature had to learn it. The new nomenclature spread with the advance of Lavoisierian chemistry. The manner of its acceptance varied by region, the English mostly simply transliterating the French terms, while the Germans translated many of them, oxygen becoming *Sauerstoff.* In the early nineteenth century the new nomenclature was extended and modified to incorporate new chemical discoveries, and made more quantitative, particularly in the work of Jöns Jakob Berzelius.

See also Berthollet, Claude-Louis; Chemistry; Lavoisier, Antoine-Laurent; Macquer, Pierre Joseph.

Reference

Crosland, Maurice P. *Historical Studies in the Language of Chemistry.* Rev. ed. Mineola, NY: Dover, 1978.

Chemistry

Chemistry underwent the most profound changes of any science during the Enlightenment. Its vocabulary, subject matter, range of available procedures, and fundamental theories were all vastly altered, and its position within the world of science shifted from a marginal one in the physics-dominated early Enlightenment to a central one in the romantic era.

Eighteenth-century chemistry was still in the process of development from the alchemical tradition. As the century wore on, the classical alchemical goal of creating gold came to be the claim of occult charlatans, while some of the procedures and vocabulary associated with alchemy were absorbed into chemistry. The economic rationale of alchemy developed into industrial chemistry, as can be seen in the career of Johann Friedrich Böttger, who went from unsuccessful gold making to successful porcelain making. The chemist was still a rather disreputable figure in some quarters, however.

The connection of chemistry with medicine, going back at least as far as Paracelsus (1493–1541), continued to be vital in the early Enlightenment. In universities chemistry was taught in medical schools, and the two most influential chemists in early-eighteenth-century Europe, Hermann Boerhaave and Georg Ernst Stahl, were physicians and medical professors. Their chemistries, however, differed greatly. Boerhaave was a Newtonian, who sought to reduce chemistry to mechanical principles like those of Newtonian physics, following Newton's suggestion that chemical processes were governed by short-range forces of attraction and repulsion. Stahl was suspicious of the claims of mechanical philosophers, whose theorizing seemed abstract and remote from chemical reality. He inherited the traditions of the alchemists, analyzing chemical processes in terms of "principles" rather than particles and short-range forces. Stahl also thought those who viewed medical problems as basically chemical—the "iatrochemists"—ignored the importance of the distinction between living and nonliving things. The Stahlian view of the fundamental difference of chemistry and mechanical physics became dominant later in the century, and his use of the term *phlogiston* to describe a substance lost in combustion and respiration also became popular.

The advantage of Stahl's chemistry is that it was presented as useful in mining and manufacture. The center of this kind of practical chemistry was Germany, where Stahlian chemistry was taught in the universities and spread by a host of professors and teachers. (Germany and Scotland, with strong university systems, had a heavy contingent of professors among their leading chemists. The situation in England and France was quite different, and few important English and French chemists held university positions in the eighteenth century.) German and Swedish chemists in the mid-eighteenth century, such as Andreas Marggraf and the Swedes Carl Scheele and Torbern Olof Bergman (1735–1784), developed and refined chemical analysis. The introduction of Stahlian chemistry into France by the lectures of Guillame-François Rouelle (1703–1770) at the Royal Botanical Garden beginning in 1742 was motivated by the needs of industry, and was followed by a flood of translations of Stahl's and other German works.

One problem for eighteenth-century chemists was why some substances reacted chemically together, and others did not. This was known as "elective affinity." Many chemists engaged in the identification of affinities and the construction of tables of affinity identified with the short-range forces that Newton and Newtonian chemists such as Boerhaave believed responsible for chemical phenomena. The tables expanded from one in sixteen columns published by Étienne François Geoffroy (1672–1731) in 1718 to one in fifty-nine columns published by Bergman in 1775. Students of affinity hoped to eventually make chemistry as deterministic and quantifiable a science as celestial mechanics—a goal Bergman also advanced by systematically defining the techniques and materials of chemical analysis.

The most important new development in midcentury chemistry was the expansion of chemical analysis to the realm of the air. Traditionally, air had been considered an element, and had not been treated as differentiated into gases. The center of the new research into "pneumatic chemistry" was Great Britain. The movement began with Stephen Hales's research into plants, in which he discovered that plants contained quantities of fixed air, so-called because it was fixed inside the plant rather than being part of the atmosphere. He captured this fixed air in vessels held over water. Hales still thought of the air as undifferentiated, but the technology he used was soon adapted to separate different airs. The first distinct air to be separated and identified was carbon dioxide, which Joseph Black produced in a series of experiments he performed on magnesium carbonate (magnesia alba). He used Hales's term *fixed air* to describe the gas. (The lack of a standard chemical nomenclature led to many problems of this sort.) Henry Cavendish isolated hydrogen (inflammable air) in 1766, and also devised a way of holding gases over liquid mercury rather than water, ending the problem of their dissolution in water. Daniel Rutherford (1749–1819) identified nitrogen in 1772. Joseph Priestley, who came to chemistry from experimental physics, employed and refined the mercury technique and discovered a multitude of airs, the most important being "dephlogisticated air," later named oxygen by Antoine-Laurent Lavoisier.

The name of Lavoisier is associated with the "Chemical Revolution." The best-known aspect of Lavoisier's reshaping of the science of chemistry is his rejection of phlogiston, and its replacement by a theory of combustion as a combination of the burnt substance with oxygen. However, this is only part of Lavoisier's innovations. Along with Cavendish, Lavoisier was responsible for introducing far more precise weight measurement into chemistry. (It is not a coincidence that both Lavoisier and Cavendish were wealthy aristocrats. Their precise balances were very expensive.) Lavoisier also introduced into chemistry the more precise mathematical methods and emphasis on exactitude characteristic of the physical sciences as they had developed in France. He also fundamentally altered the way chemists thought about gases, by introducing the idea that any substance, with enough heat, could become a gas—the theory of the "gaseous state." Lavoisier thought of heat as a chemical substance, caloric, which combined chemically with elements to change their state. The strength of the caloric theory meant that study of heat in the eighteenth century was mainly carried on by chemists, notably Black. Lavoisier also provided a general theory of acidity, claiming that oxygen was the principle of acidity, found in all acids, and introduced several new pieces of equipment, such as the calorimeter.

The vehicles for the transmission of Lavoisier's new chemistry were the *Method of Chemical Nomenclature,* which he and three other French chemists produced in 1787; his 1789 *Elementary Treatise of Chemistry;* and the journal founded the same year, *Annales de Chémie. Elementary Treatise of Chemistry* included an influential list of the simplest "elements," those chemical substances that could not be broken down into other substances, including light and caloric. The notion of a chemical element was associated with Lavoisier's innovations. His new chemistry was widely known as "French chemistry" (although Lavoisier disliked the term, preferring that the glory go to him alone), and its foreign origin complicated its acceptance in the distinct chemical cultures of Great Britain and Germany. In Germany the "French chemistry" became associated with the French Revolution, ironically, considering Lavoisier's fate at the hands of the revolutionists, and was accepted by most German chemists only after it had been demonstrated conclusively that the reduction of the calx (oxide) of mercury produced oxygen. Chemistry also had to overcome revolutionary associations in Britain during the 1790s, although there the "chemical revolutionary" attacked by political and religious conservatives was not Lavoisier, but the arch-phlogistonist Priestley.

Lavoisier's theory had vanquished its rivals by the close of the eighteenth century, and chemistry continued its rapid development. Important developments of the early nineteenth century include the rise of electrochemistry, the introduction of chemical atomism by John Dalton, and the disproof of Lavoisier's theory of oxygen as the cause of acidity. Alessandro Volta's generation of electricity by the juxtaposition of different metals, silver and zinc, in 1800, led to the science of electricity being more closely allied to chemistry than to experimental physics. The foremost champion of electrochemistry was Sir Humphry Davy, who adapted the "voltaic pile" as an instrument of chemical analysis. Davy's first great triumph was the decomposition of water into its components of hydrogen and oxygen in 1801. His demonstration that "oxymuriatic acid" was actually an element, chlorine, dealt the death blow to Lavoisier's theory of acidity, which had been tottering for years. Davy, a master showman, also contributed to making chemistry a fashionable discipline in Great Britain, rendering it safe for the authorities in church and state. Davy was closely associated with romantic writers, and romantics both in England and on the Continent tended to promote chemistry, seen as a science of dynamic and mysterious forces, as compared to the mathematically rigorous and deterministic physics and mechanics.

Two distinguished French chemists, Claude-Louis Berthollet and Joseph-Louis Proust (1754–1826), debated in the first years of the nineteenth century whether chemical elements combined in compounds in fixed or variable proportions. The debate was inconclusive, but the direction of early-nineteenth-century chemistry was clearly toward the traditional idea of fixed proportions. In 1809 Berthollet's disciple Joseph-Louis Gay-Lussac published the law of the combining volumes of gases: that gases combine in whole-number ratios of their volumes. This law stated nothing about the ultimate nature of chemical substances, but John Dalton set forth a theory of chemical atomism in his *A New System of Chemical Philosophy* (1808). The atomic theory was more influential at the time than the molecular hypothesis put forth by Amedeo Avogadro in 1811 and 1814. Chemical atomism was adapted to the needs of chemical analysts by William Hyde Wollaston (1766–1828) and taken up by the dominant force in European chemistry going into the 1820s, Jöns Jakob Berzelius. In 1819 he used atomism to explain the discovery of the isomorphism, or similarity, of the crystals of related metallic salts by his student Eilhardt Mitscherlich (1794–1863).

See also Atomism; Avogadro, Amedeo; Beddoes, Thomas; Berthollet, Claude-Louis; Berzelius, Jöns Jakob; Boerhaave, Hermann; Böttger, Johann Friedrich; Cavendish, Henry; Chemical Nomenclature; Crell, Lorenz Florens Friedrich von; Dalton, John; Davy, Sir Humphry; Gay-Lussac, Joseph-Louis; Klaproth, Martin Heinrich; Lavoisier, Antoine-Laurent; Macquer, Pierre Joseph; Marggraf, Andreas Sigismund; Newtonianism; Phlogiston; Priestley, Joseph; Scheele, Carl Wilhelm; Stahl, Georg Ernst.

References

Brock, William H. *The Norton History of Chemistry.* New York: W. W. Norton, 1993.

Crosland, Maurice. "Chemistry and the Chemical Revolution." In *The Ferment of Knowledge: Studies in the Historiography of Eighteenth-Century Science,* edited by G. S. Rousseau and Roy Porter, 389–416. Cambridge: Cambridge University Press, 1980.

Golinski, Jan. *Science As Public Culture: Chemistry and Enlightenment in Britain, 1760–1820.* Cambridge: Cambridge University Press, 1992.

Hankins, Thomas L. *Science and the Enlightenment.* Cambridge: Cambridge University Press, 1985.

Hufbauer, Karl. *The Formation of the German Chemical Community, 1720–1795.* Berkeley: University of California Press, 1982.

Knight, David. *Ideas in Chemistry: A History of the Science.* New Brunswick: Rutgers University Press, 1992.

Cheyne, George (1671–1743)

George Cheyne was one of the most popular doctors and influential medical thinkers in eighteenth-century England. Of Scottish descent, Cheyne was originally a disciple of the early Newtonian physician Archibald Pitcairn (1652–1713). Cheyne's *New Theory of Fevers* (1701) was an attempt to apply Newtonian categories to disease, treating the body as a collection of vessels and fluids whose behavior could be described mathematically. His one work of pure mathematics, *Inverse Method of Fluxions* (1702), was uninspired and cost him a close alliance with the Newtonians, as he had blundered into the sensitive terrain of the origins of the calculus. Another Newtonian work was the natural-theological *Philosophical Principles of Natural Religion* (1705).

Cheyne's greatest fame as a doctor and writer came after he moved to the resort city of Bath, famous for its supposedly health-giving waters, in 1718. He was a physician to some of the best-known people of his time, such as the novelist Samuel Richardson (1689–1761) and the founder of Methodism, John Wesley (1703–1791). His *Observations on the Nature and the Method of Treating the Gout* (1720), which discussed a condition he suffered from himself, went through several editions. The immensely popular *Essay of Health and Long Life* (1724) went through eight editions in his lifetime and was translated into Latin, French, Dutch, and German. Cheyne recommended maintaining health through exercise and moderation in eating and drinking. This caused some ridicule, as Cheyne weighed 450 pounds before he stabilized his weight at 300 pounds, but he was a renowned wit who gave as good as he got. He was an early advocate of complete vegetarianism, or, if that was not possible, of giving up red meat, and of abstention from alcohol. He linked healthful behavior with morality. Cheyne's *English Malady* (1733) discussed hysteria and hypochondria. It included several case histories, including his own. Cheyne's medical philosophy grew progressively less mechanical but remained broadly Newtonian.

See also Medicine; Newtonianism.

Reference

Guerrini, Anita. *Obesity and Depression in the Enlightenment: The Life and Times of George Cheyne.* Norman: University of Oklahoma Press, 2000.

Colonial Science

Although European colonies had long been sites of scientific inquiry, in the eighteenth century they developed their own scientific communities and institutions, which increasingly claimed independence from their European peers. This process accompanied the spread of colonial institutions, including scientific institutions, to larger portions of the world as European colonial empires grew.

The classic "colonial science" in the early modern era was natural history, frequently seen in the context of identifying colonial resources for exploitation. The immediate fieldwork of natural history could be carried out only on the spot, either by colonial residents or by visitors from the metropolis. Colonial residents joined networks of correspondence and exchange, but usually as subordinates. Some of this work was the absorption or annexation of indigenous knowledge rephrased in European terms. Medicine was another local science drawing on both European and indigenous sources. Colonial medicine was particularly important in the tropics, where diseases existed that had no European precedents. The wealthy French slave colony of Saint Domingue (now Haiti) was the home of several medical writers, including Jean-Barthelemy Dazille, author of

This is the combined laboratory and lecture room of the University of Pennsylvania Medical School in the early nineteenth century. Scientific facilities outside Europe, such as this one, modernized more slowly than those in Europe. (Edgar Fahs Smith Collection, University of Pennsylvania Library)

Observations on the Sicknesses of Blacks (1776) and other works.

The eighteenth century, with the expansion of astronomy and cartography, also saw the spread of observatories and programs of observation outside Europe since it was often necessary or beneficial that astronomical phenomena, such as the transits of Venus, be viewed from many places on the globe. The necessity of exact establishment of the location of isolated ports and islands spurred scientific endeavor. Prominent colonial astronomers included Harvard's John Winthrop (1714–1799) and New Granada's (now Colombia) polymath Francisco José de Caldas (1768–1816). However, colonial astronomy and surveying still filled the traditional colonial role of providing data for compilation and analysis in Europe.

One reason for the impact of Benjamin Franklin was that he was the first colonial scientist, born and resident in a colony, to make an important original contribution to a science other than natural history, namely, experimental physics. This led to his acceptance as a peer by Europe's leading scientists. Franklin was fortunate to be working in a relatively young field where there was not a great deal of previous literature or mathematized theory to master. The disadvantage of colonial isolation for physicists can be seen in the work of his New York contemporary Cadwallader Colden (1688–1776). Colden's *Explication of the First Causes of Action in Matter; and the Cause of Gravitation* (1745) showed little awareness of work in mechanics and physics since Newton and was harshly reviewed when it appeared in Europe.

Science in colonies was practiced in institutions and settings resembling Europe's. The scientific-society movement spread to many European colonies in the eighteenth century. British America's first enduring society was the American Philosophical Society founded in 1768. Dutch Indonesia was the home of the Batavian Society of Arts and Sciences founded in 1784 and lasting until 1795. Saint Domingue had a private group with some public support, the Circle of Philadelphes founded in 1784. The group received a royal charter and a promise of lavish funding as the Royal Society of Sciences and Arts of Cap Français in 1789. Although the group produced some publications, mostly on tropical diseases, it was soon swept away by the revolutions of France and Haiti. Outside formal societies, groups of persons interested in science existed in many colonial cities. A feature of the eighteenth century was increasing interaction between colonial groups and individuals, supplementing their relation with scientific groups in Europe.

A powerful motivation for the foundation of colonial scientific institutions was economic development. The French founded a network of botanical gardens in their island colonies of Mauritius, Réunion, Saint Domingue, and Guadeloupe hoping to acclimate economically productive crops, as coffee was acclimated in the French Caribbean colonies. These colonial gardens often employed botanists trained at the Royal Botanical Garden, and reported back to it. The British were somewhat slower to set up a centralized system, but the time of Sir Joseph Banks's headship of the Royal Society and Kew Gardens beginning in 1778 saw something similar. The Spanish in the same period turned to a more aggressive approach to the resources of their empire, sending the Royal Botanical Expedition and the expedition led by Alejandro Malaspina. The leading scientific center in the Spanish Empire was Mexico City, which opened the Royal Botanical Garden in 1788 and the Royal Mining College in 1792. The Royal Mining College became the center from which the new chemistry of Antoine-Laurent Lavoisier was disseminated in Mexico.

In the New World, colonial scientific communities, like many colonial communities in general, began by the late eighteenth century to emphasize their separateness from Europe. The study of local natural history could contribute to the sense of a separate colonial identity, as could scientific disputes with Europeans. Mexican naturalists attacked the Royal Botanical Expedition's attempt to impose the "European" Linnaean system of classification on Mexican plants. Particularly important for New World natural historians was defending the reputation of the Western Hemisphere from European scientists like Georges-Louis Leclerc de Buffon who claimed it inferior to the Old. Buffon pointed to cases where New World animals were smaller than those of the Old World, as pumas were smaller than lions, to argue that the New World, for reasons of climate and humidity, produced generally inferior fauna. This argument was also given a racist twist, as Native Americans were considered inferior to Old World people, Europeans in particular, and some in Europe went further to assert that Europeans inevitably degenerated in the Americas. This aroused the pride of New World scientists. Thomas Jefferson (1743–1826) defended the fauna of the New World from the aspersions of Buffon and others in *Notes on Virginia* (1785). Among the many Latin American scientific writers who rebutted the European claims was the Lima physician and statesman José Hipolito Unanue (1758–1833), author of *Observations on the Climate of Lima and Its Influence on Organic Beings, Particularly Men* (1806). Resentments between colonial and European scientists and the desire to build an independent scientific culture could easily lead to support for political independence. Franklin and Francisco José de Caldas, the latter executed by the Spanish as a rebel, were both founders of national scientific traditions and heroes of their countries' independence struggles.

The situation was different in Old World colonies that did not see large-scale European settlement, such as India or the coastal colonies of Africa. Although there was interest in science, particularly as related to economic development, scientists and scientific communities did not define themselves against Europe, from which they had come and to which they hoped to return. One question that had to be faced, particularly with the expansion of European direct rule, was if and to what degree native populations should be educated in European science, increasingly seen as evidence of European cultural, or racial, superiority. In India the East India Company opposed education in European science for Indians, whereas schools run by missionaries were much more interested in teaching it. They believed that knowledge of science would overcome Indian backwardness and superstition, a position that eventually won by 1813, when the company began to set aside money for improvement of the sciences. An important reason for this change of policy was the demand for scientific education by the Indian elites themselves. In the French West African colony of Senegal, Jean Dard, founder of the School of Saint Louis, began teaching theoretical and mathematical science to Africans in 1817. Despite his praise for his pupils' acuity, Dard was defeated by colonial advocates of strictly practical and vocational education for Africans.

See also American Philosophical Society; Banks, Sir Joseph; Bartram Family; Botanical Gardens; Exploration, Discovery, and Colonization; Franklin, Benjamin; Nationalism; Race; Royal Botanical Expedition; Sloane, Sir Hans; Transits.

References

Gerbi, Antonello. *The Dispute of the New World: The History of a Polemic, 1750–1900.* Translated by Jeremy Moyle. Rev. ed. Pittsburgh: University of Pittsburgh Press, 1973.

MacLeod, Roy, ed. *Nature and Empire: Science and the Colonial Enterprise. Osiris,* 2d ser., 15 (2000).

McClellan, James E., III. *Science Reorganized: Scientific Societies in the Eighteenth Century.* New York: Columbia University Press, 1985.

Meade, Teresa, and Mark Walker, eds. *Science, Medicine, and Cultural Imperialism.* New York: St. Martin's, 1991.

Petitjean, Patrick, Catherine Jami, and Anne Marie Moulin, eds. *Science and Empires: Historical Studies about Scientific Development and European Expansion.* Boston Studies in the Philosophy of Science, 136. Dordrecht, Netherlands: Kluwer, 1992.

Stearns, Raymond Phineas. *Science in the British Colonies of America.* Urbana: University of Illinois Press, 1970.

Comets

The establishment of the periodicity of some comets was one of the great triumphs of science during the Enlightenment. Edmond Halley (1656–1742), the first to systematically apply Newtonian dynamics to cometary orbits, made the first prediction of a comet's return in 1705, predicting that the comet that had appeared in 1682, now known as Halley's comet, would return late in 1758 or early in 1759. In subsequent publications Halley would grow more tentative in his predictions, as he became increasingly aware of the importance of the gravitational perturbations of the comet's orbit by the heavy planets Jupiter and Saturn and the difficulty of calculating them. As the comet approached in the 1750s, that task was taken up by the French mathematician Alexis-Claude Clairaut (1713–1765), working with the astronomer Joseph-Jérôme Le Français de Lalande (1732–1807) and the experienced astronomical calculator Madame Nicole-Riene Étable de la Brière Lepaute (1723–1788). The group worked for more than a year under the pressure of having to make a prediction about the comet's orbit before it actually appeared. Clairaut announced his predictions in a paper before the Royal Academy of Sciences on 14 November 1758, about a month before the comet arrived.

The predicted reappearance of the comet was widely publicized throughout Europe, with many published guides for finding it. It was first observed on Christmas Day not by any of Europe's famous astronomers but by a Saxon amateur, Johann Georg Palitzsch (1723–1788), who did not realize at first that the comet he observed was Halley's. By January astronomers working in Europe's major scientific centers had found it, and the remarkable accuracy of Clairaut's predictions had been verified, providing additional confirmation of Newtonian dynamics. Establishing cometary orbits continued to be one of the most complex problems in celestial mechanics. Some of the most important mathematicians of the period worked on improved methods for finding orbits from a few observations. As well as Clairaut, they included Leonhard Euler, Pierre-Simon de Laplace, Heinrich Wilhelm Matthäus Olbers (1758–1840), and Carl Friedrich Gauss.

The excitement around comets generated on the part of some astronomers an obsession with finding them. The foremost late-eighteenth-century comet hunter was the Frenchman Charles Messier (1730–1817), who discovered thirteen in the period 1758–1798 and made the first French observations of Halley's comet. Comet hunting placed a premium on observational rather than mathematical skills, making it popular with astronomers like Messier who were not mathematically trained. Messier took particular delight when his observations disproved the predictions of mathematicians. His famous catalog of stellar objects was produced for the purpose of avoiding their confusion with comets. Other successful comet hunters included Caroline Herschel and the most productive of all, Jean-Louis Pons (1761–1831). A Frenchman who spent most of his career in Italian observatories, Pons discovered thirty-seven comets. One comet discovered by both Herschel and Pons was the first periodic comet discovered after Halley's, Encke's comet, named after Gauss's student Johann Franz Encke (1791–1865). In 1819 Encke established that the comet returned every 3.3 years.

Although the traditional idea of comets as providential warnings from God was vanishing among scientists during the eighteenth century (its last major upholder among scientists was William Whiston), it vigorously continued among many Europeans. The founder of Methodism, John Wesley (1703–1791), preached a sermon in 1755 warning that God might punish the Earth for its wickedness by means of a comet. Popular misunderstanding of a Royal Academy of Sciences paper by Lalande in 1773 precipitated a panic in France, as people feared a comet would destroy Earth.

See also Astronomy; Euler, Leonhard; Gauss, Carl Friedrich; Herschel Family; Laplace, Pierre-Simon de.

References

Taton, René, and Curtis Wilson, eds. *Planetary Astronomy from the Renaissance to the Rise of Astrophysics, Part B: The Eighteenth and Nineteenth Centuries.* Cambridge: Cambridge University Press, 1995.

Yeomans, Donald K. *Comets: A Chronological History of Observations, Science, Myth, and Folklore.* New York: John Wiley and Sons, 1991.

Condillac, Étienne Bonnot de (1715–1780)

The abbé Étienne Bonnot de Condillac was both an intermediary between seventeenth-century English philosophical and scientific empiricism and eighteenth-century France, and an important philosopher himself. A priest who made little reckoning of religion (it was claimed that he said Mass only once in his life), Condillac was a habitué of Parisian salons and a friend and colleague of leading Enlightenment thinkers, including Denis Diderot and Jean-Jacques Rousseau (1712–1778). Condillac's program was to bring the empiricism he associated with John Locke (1632–1704) and Isaac Newton (1642–1727) to bear on a broad range of questions, notably the nature of the human mind. For this goal to be achieved, he believed that the "metaphysical systems" of the great seventeenth-century philosophers—René Descartes (1596–1650), Nicolas Malebranche (1638–1715), Baruch Spinoza (1632–1677), and Gottfried Wilhelm Leibniz (1646–1716)—had to be revealed as lacking relation to the observed world. This was the mission of Condillac's *Treatise on System* (1749). As an alternative to the construction of rationalistic systems built on logical operations performed on general principles, Condillac put forth what he and other Enlightenment thinkers considered the Newtonian method, based on the gathering and systematization of observations.

Like his philosophical inspiration, Locke, Condillac believed that knowledge originated in sensory perceptions. He went beyond this to explore how knowledge was formulated and communicated in language. The paradigm for clear communication was the unambiguous language of mathematical symbols. Condillac's emphasis on clarity and distinctness in the use of language had a tremendous influence on the work of French scientists. The reform of chemical nomenclature carried out by Antoine-Laurent Lavoisier, an admirer of Condillac, was partially inspired by his work. However, Condillac's philosophical reputation dimmed by the early nineteenth century.

See also Enlightenment; Newtonianism.

References

Gillispie, Charles Coulston. *The Edge of Objectivity: An Essay in the History of Scientific Ideas.* Princeton: Princeton University Press, 1960.

Hine, Ellen McNiven. *A Critical Study of Condillac's "Traite des Systems."* The Hague: Martinus Nijhoff, 1979.

Portrait of eighteenth-century French philosopher Marquis de Condorcet from The Salon, a Study of French Society and Personalities in the Eighteenth Century *(1907) by Helen Clergue. (University of Colorado at Boulder)*

Condorcet, Marie-Jean-Antoine-Nicolas Caritat, Marquis de (1743–1794)

The Marquis de Condorcet, Marie-Jean-Antoine-Nicolas de Caritat, was a prophet of the application of what he considered scientific principles to politics. From a family of poor nobles, Condorcet attracted through his mathematical talents the patronage of Jean Le Rond d'Alembert, through which (in a rather questionable election) Condorcet became secretary of the Royal Academy of Sciences in 1773. Although his early work on the integral calculus attracted the interest and admiration of leading mathematicians, Condorcet was of only minor importance as a mathematician, often failing to adequately communicate his ideas. His most interesting work applied mathematics to political problems, such as his *Essay on the Application of the Analysis of Probability to Decisions Made on a Plurality of Votes* (1785), the first mathematical treatment of voting. It is famous for "Condorcet's paradox," about how a purely majoritarian system of voting can fail to represent people's true choices.

Condorcet was a tireless propagandist for applying science to politics. Unlike many of his predecessors, going back to Francis Bacon (1561–1626), Condorcet promoted scientific politics in a basically liberal rather than authoritarian style. Informed by science, the average citizen—even including female citizens—could make the right political decisions. Condorcet's eulogies of dead scientists, written in his capacity as secretary to the Royal Academy of Sciences, ceaselessly emphasized the usefulness of science in improving the quality of life of ordinary people, although he sometimes endorsed pure research as well. Condorcet conspicuously failed as a practical politician, however, in meeting the challenges posed to himself and to the academy by the French Revolution. A supporter of the Revolution in its early stages, and a leader in the moderate-liberal Society of 1789 (which attracted a number of other academicians), Condorcet tried to secure the role of the academy, and by extension the scientific community, in the new French society being built. On 20 April 1792 he presented a new scheme for scientific education worked out by a commission of which he had been a member and leading spirit, and based on a superacademy called the "National Society," which was to subject French science and education to tight hierarchical control. Condorcet had the extraordinary bad luck to be presenting his plan to the Legislative Assembly the same day France declared war on Austria, but it never attracted strong political backing and was finally rejected in December as elitist and as insufficiently concerned with the most important task of education, the indoctrination of the young in Republican virtue. Despite its rejection, Condorcet's plan influenced the creation of the National Institute.

On 3 October 1793 the victorious Jacobin party that had overcome Condorcet's own faction, the Girondins, issued a warrant for his arrest. He spent several months in hiding, writing his most famous work, *Sketch for a History of the Progress of the Human Mind* (1795), a classic statement of the belief in science and scientific progress as the motivating forces in the progress of humanity. Attempting to flee Paris in March, he was arrested and put in prison. Rather than being executed, Condorcet killed himself.

See also Enlightenment; French Revolution; Mathematics; Royal Academy of Sciences.

References
Baker, Keith M. *Condorcet: From Natural Philosophy to Social Mathematics.* Chicago: University of Chicago Press, 1975.
Gillispie, Charles Coulston. *Science and Polity in France at the End of the Old Regime.* Princeton: Princeton University Press, 1980.
Hahn, Roger. *The Anatomy of a Scientific Institution: The Paris Academy of Sciences, 1666–1803.* Berkeley: University of California Press, 1971.

Cook, James (1728–1779)

James Cook was the most scientifically minded of eighteenth-century sea captains, and his voyages of exploration and discovery in the Pacific were planned to contribute to scientific knowledge as well as British wealth and power. Cook's achievement is particularly impressive as he came from a poor background and received little formal education. He was largely self-taught in scientific navigation and surveying, in which he became skilled enough to be appointed by the British Admiralty to produce the first exact survey of the coast of Newfoundland from 1763 to 1767. Cook's work on this survey was so impressive that he received command of the *Endeavour,* a North Sea coal hauler refitted by the Admiralty to observe the transit of Venus in 1769 from the Pacific. The Royal Society arranged for two natural historians, Sir Joseph Banks and a Swedish disciple of Carolus Linnaeus, Daniel Carl Solander (1733–1782), to accompany the voyage along with support staff.

The *Endeavour* voyage from 1768 to 1771 circled the world. Not only did Cook observe the transit from the recently discovered island of Tahiti, but under secret orders from the Admiralty, he also explored the South Pacific for the long-conjectured "great southern continent." The *Endeavour* did not find the continent, but demonstrated that it must be smaller than surmised, and by circumnavigating New Zealand (which Cook charted) demonstrated that these islands were not part of it. Cook also charted the eastern coast of Australia. Banks, who received most of the publicity on the expedition's return to England, and Solander brought back thousands of plant specimens and a wealth of natural-historical and anthropological information about the South Pacific. Cook's second voyage, from 1772 to 1775, on the *Resolution* (accompanied by the *Adventure*) finally put to the idea of a great southern continent, as well as providing accurate charts of the South Pacific islands. Cook employed the new chronometers to ascertain the longitude on this voyage, although he also continued to use the rival astronomical method of Nevil Maskelyne. On his return, he was admitted as a fellow to the Royal Society. Cook's first contribution to *Philosophical Transactions* was about preserving the health of seamen, a problem he had tackled with remarkable success. (He persuaded his reluctant men to eat sauerkraut to prevent scurvy by the simple expedient of first having it served at the captain's table, as a delicacy.) Cook received the society's Copley Medal.

Cook's third and final voyage, on the *Resolution* from 1776 to 1779, took him to the North Pacific, where he charted the coast of the far northwest of America and sought to settle the question of the Northwest Passage connecting the Atlantic and Pacific Oceans. This expedition was the first European one to encounter the Hawaiian Islands. Cook mapped the North Pacific coast, from what is now Oregon to Alaska, but found nothing that seemed a possibility for the Northwest Passage. Planning to winter in Hawaii before one last try to find the passage, Cook was killed by native Hawaiians on 14 February 1779.

See also Banks, Sir Joseph; Exploration, Discovery, and Colonization; Longitude Problem.

Reference

Beaglehole, J. C. *The Life of Captain James Cook.* Stanford: Stanford University Press, 1974.

Coudray, Angelique Marguerite Le Boursier du (c. 1714–1794)

By far the most influential midwife of the eighteenth century, Madame du Coudray was equally skilled in delivering children, instructing other midwives, and maneuvering in the complex politics of the French medical world and the French state. Of obscure background, she first emerges in the public record as a Parisian midwife, one of the elite of French midwives. She published a book on midwifery, the frequently republished *Abridgement of the Art of Delivery,* in 1759. The same year she received a brevet, a decree from King Louis XV (r. 1715–1774) permitting her to travel the country giving midwifery classes. On a series of exhausting tours beginning in 1761 and ending in 1783, Coudray taught classes of young girls (usually selected by their parish priests) to be midwives. She preferred to teach novices rather than experienced midwives as they had no "bad habits" to unlearn and could be treated as blank slates. For teaching purposes, she invented a complex mechanical device for demonstrating delivery, which was widely imitated. Despite her implicit endorsement of a mechanical approach to childbirth, Coudray was suspicious of the use of instruments rather than the midwife's hands for delivery, increasingly popular among male practitioners. Through her career, she trained thousands of midwives.

Coudray's program involved her in conflict with a number of vested interests, from traditional midwives in the country suspicious of the Parisian and her young and inexperienced pupils, to surgeons hoping to displace midwives from the lucrative field of delivery. She herself was frequently the target of denunciation from surgeons attacking midwives, and her work frequently went unmentioned in the medical press. To defend herself and promote her vision of improved midwifery, Coudray allied when she could with government officials, beginning with the king and extending to the intendants and local magistrates who governed French provinces. She draped her mission in an Enlightenment rhetoric of benevolence and also in French patriotism. France in the mid-eighteenth century was thought to be a country suffering from population decline, and Coudray argued that much of this decline could be blamed on poor midwifery and reversed by better. Better midwifery would also supply the king with healthy soldiers and mothers, rather than deformed cripples or people with damaged minds caused by botched deliveries. Her midwifery emphasized the safe delivery of the child over the preservation of the life and health of the mother. Unlike some other women who taught midwifery or wrote midwifery manuals, Coudray eschewed antimale rhetoric and profited from alliances with individual surgeons and physicians, including those who attended her classes and assisted her at her demonstrations. Her royalism eventually proved a political liability, and she died while being harassed by the tax collectors of the French Revolution government. Her niece, Madame Coutançeau, survived the Revolution to become the premier midwife of Napoleonic France.

See also Midwives; Nationalism; Women.

Reference

Gelbart, Nina Rattner. *The King's Midwife: A History and Mystery of Madame du Coudray.* Berkeley: University of California Press, 1998.

Coulomb, Charles-Augustin de (1736–1806)

The French military engineer Charles-Augustin de Coulomb was a leader in the development of a precise, quantitative physics in the late eighteenth century, particularly in the field of electricity. Born into a family of lawyers and officials, Coulomb entered the French Royal Engineering School at Mézières in 1760. Graduating with the rank of first lieutenant in 1761, Coulomb was posted to the Caribbean island of Martinique in 1764. This island had recently been returned to France by the English at the end of the Seven Years' War, and Coulomb and other French engineers were to fortify the island and make it defensible. Coulomb supervised the design and building of the fort, and returned to France in 1772 with the rank of captain and ill health that would plague him the rest of his life. In 1773 he summed up much of his experience in a memoir on architectural problems read at the Royal Academy of Sciences, which made fundamental contributions to the theory of the strength of materials, soil mechanics, and the capacity of arches. Stationed at Cherbourg, Coulomb wrote another essay on the production of magnetic compass needles that won him a half share of a prize offered by the Academy of Sciences in 1777. This showed the superiority of magnetic needles suspended by threads over the standard models supported by pivots and laid the groundwork for Coulomb's subsequent use of the torsion balance for very fine measurements of weak physical forces. Coulomb's interpretation of his magnetic experiments endorsed a magnetic theory based on Newtonian attractive and repulsive forces against a neo-Cartesian one of vortices of tiny particles.

He wrote another academy-prizewinning essay on friction in 1781 and was elected to the academy that year. Of the two main eighteenth-century theories of friction, that of it being caused by surface roughness and that of it being caused by a short-range cohesive force analogous to gravity, Coulomb supported the former. His work was the first systematic treatment of friction to appear, and enormously influenced both the science and the engineering of friction well into the nineteenth century.

Coulomb's most important experimental work was in the area of electricity, where he used a torsion balance of his own design for precise measurement of electrical forces. Coulomb's establishment of the law of torsion, that the force of torsion is directly proportional to the angle of twist, enabled him to create a balance using a fine silk thread in which the force of torsion would be negligible. His first memoir on the subject, submitted to the academy in 1785, demonstrated that the repulsive force between two spheres with the same electrical charge was inversely proportional to the square of the distance between their centers. The second memoir, submitted in 1787, demonstrated the inverse-square laws for attractive as well as repulsive electrical forces, as well as for attractive and repulsive magnetic forces, although Coulomb's experiments were open to criticism and are extremely difficult to replicate. The idea of an inverse-square law of electrical or magnetic forces on the analogy of Newtonian gravity was not new, but Coulomb was the first to provide extensive experimental evidence. As a Newtonian he was more concerned with establishing the mathematical equations governing the actions of electrical forces than with discussing their cause. He used a "two-fluid" rather than a "one-fluid" theory, but stated that this was simply because he found it more convenient.

(However, Coulomb's use of the two-fluid theory did help to establish its dominance in French electrical theory.) Subsequent papers discussed the leakage of electrical charge and the distribution of charge on a charged body, establishing that the charge of a conducting body existed only on its surface.

In 1784 Coulomb received the important post of official in charge of the Paris water supply. His career was not immediately affected by the French Revolution, and he served on the commission to establish the metric system. After the abolition of the academy in 1793 he fled Paris for his country estate. He returned in 1795 as a member of the section for experimental physics in the academy's successor, the newly founded Institute of France. By the end of the century his health was failing, but he remained active. Coulomb's last service to the French state was as one of the six inspector generals of French education, a position in which he served from 1802 to his death in 1806.

See also Education; Electricity; French Revolution; Royal Academy of Sciences; Technology and Engineering.

References

Gilmor, C. Stewart. *Coulomb and the Evolution of Physics and Engineering in Eighteenth-Century France.* Princeton: Princeton University Press, 1971.

Heilbron, J. L. *Electricity in the 17th and 18th Centuries: A Study in Early Modern Physics.* 2d ed. Mineola, NY: Dover, 1999.

Crell, Lorenz Florens Friedrich von (1744–1816)

Although not an original scientist himself, Lorenz Florens Friedrich von Crell was in many ways the leader of late-eighteenth-century German chemistry. The son of Johann Friedrich Crell (1707–1747), a medical professor and anatomist, Crell was educated at the undistinguished medical school of the University of Helmstedt. He took chemistry from Joseph Black at the University of Edinburgh in the winter of 1769–1770. Crell joined the Helmstedt faculty in 1774, teaching there until the university was suppressed in the Napoleonic reorganization of the German universities in 1810. Crell's power in German chemistry, however, came not from his teaching position but from his editorship of a series of German-language chemical journals, beginning with the first German chemical periodical, the annual *Chemical Journal for the Friends of Natural Science, Medicine, Domestic Economy, and Manufacturing* in 1778. Crell's endeavor took its final form in 1784 as a monthly, *Chemical Annals.* Crell was a Freemason, and interested in natural theology as well as chemistry.

Crell's program for his journals was explicitly nationalist, to maintain what he claimed to be the historical superiority of German chemistry. Although he occasionally published material from foreign chemists, most of the chemists who appeared in Crell's journals were German, including such leaders of the field as Martin Heinrich Klaproth. About 80 percent of the journal's 400 subscribers were Germans, mostly middle-class, urban Protestant men. Crell's loyalty to the German chemical tradition made him a leader in the resistance to the Frenchman Antoine-Laurent Lavoisier's innovations, both in chemical nomenclature and in antiphlogiston theory. Crell's journal was increasingly perceived as old-fashioned, and beginning in the 1790s it was eclipsed on the German scene by the *Journal of Chemistry* of Alexander Nicholas Scherer (1771–1824). The *Journal of Chemistry* took over Crell's journal in 1804, and Crell finished his career as a professor at the University of Göttingen from 1810 to his death in 1816.

See also Chemistry; Nationalism; Periodicals.

Reference

Hufbauer, Karl. *The Formation of the German Chemical Community, 1720–1795.* Berkeley: University of California Press, 1982.

Cuvier, Georges (1769–1832)

In the early nineteenth century Georges Cuvier dominated French and European zoology. He was a native of Montbéliard, an unusual town in that it was French-speaking but Lutheran (Cuvier's family had produced several Lutheran pastors) and subject to the German duke of Württemberg. Interested in natural history from childhood, Cuvier attended the duke's school at Stuttgart, where he founded a natural history society. On graduation he found no opportunities in the duke's civil service and was forced to try his luck in France, although he retained strong connections to the Germanic world throughout his career. Cuvier was one of the many clever young men who took advantage of the social and intellectual fluidity of revolutionary Paris after the death or exile of many leading scientists and the abolition of centralized scientific institutions such as the Royal Academy of Sciences. He became a successful, popular lecturer at both the Museum of Natural History and the Collège of France.

A politician and scientist, Georges Cuvier recieved many honors from the French state. (Perry-Castaneda Library)

Cuvier turned down an offer to accompany Napoléon's Egyptian expedition, preferring to consolidate his position at the museum and associated Parisian institutions. Cuvier was never very interested in the observation of living animals—all of his work involved the analysis of dead ones. (His brother Frederic Cuvier [1773–1838], by contrast, oversaw and did research at the menagerie attached to the Museum of Natural History and helped found the science of animal behavior.) He was steadily amassing the most remarkable collection of animal specimens in Europe. Cuvier was a brilliant comparative anatomist (one of the first important anatomists with no medical training) and classifier of animals who thought of animals in terms of functioning wholes, each of whose parts was subordinated to an end serving the whole. The structure of an animal or any part of it was determined by its function. Cuvier was an admirer of the biological works of Aristotle (384–322 B.C.E.), adopting his teleological view of zoology. His functionalism led him to deny the possibility of a species changing, as any significant variation would destroy the ability of the animal to function. Cuvier's functionalism, unlike that of British natural theologians who were his contemporaries, was not integrated into an argument from design or any other religious context, but the natural theologians eagerly seized on his work, adapting it to the British context. So strong was Cuvier's belief in the subordination of the part to the whole and the functional nature of the animal body that he boasted (incorrectly) that he could deduce the entire structure of an animal from one bone.

Possibly Cuvier's single most important scientific contribution was his extension of the techniques of comparative anatomy to the fossil record. Cuvier's studies of fossils, collected in the four-volume *Researches on the Fossil Bones of Quadrupeds* (1812), founded paleontology. Cuvier's research led him to support the concept of extinction and also to integrate the fossil record with the geological history of Earth. His geology, first written up in his study of the area around Paris undertaken with Alexandre Broignart (1770– 1847), *Essay on the Mineral Geography of the Paris Region* (1808), was heavily influenced by the neptunist school founded by Abraham Gottlob Werner. The *Preliminary Discourse to Researches,* frequently reprinted separately, set forth the influential theory of "catastrophism," that catastrophes in Earth's past, affecting large portions of its surface, had accounted for the extinction of many species. Although the Flood of Noah was the last of these catastrophes, Cuvier did not set forth his theory as a support to religion and was generally wary of mixing religion and science. The story in the Book of Exodus he treated as a historical account, similar to that found in other Near Eastern documents, rather than as divine revelation. Cuvier's picture of an Earth millions of years old and the history of the rise and fall of its numerous species led the novelist Honoré de Balzac (1799–1850), usually an admirer and intellectual supporter of Cuvier's rival Étienne Geoffroy Saint-Hilaire, to refer to Cuvier as the greatest poet of the nineteenth century. Cuvier's other great work was the five-volume *The Animal Kingdom* (1817), setting forth a new classification scheme, dominant through most of the nineteenth century, which divided animals into four large groups, or *embranchements,* based on the differing structures of their nervous systems. The groups were the vertebrates, marked by a spinal cord; the mollusks, marked by separate neural centers; the *articulata,* or jointed creatures consisting of invertebrates, including insects, spiders, and crustaceans, with a nervous system containing two ventral chords; and the *radiata,* a somewhat catchall group consisting of animals with rudimentary or no nervous systems and defined by radial organization, like starfish. The gaps among these four groups were in Cuvier's mind absolute. He strongly opposed theories linking them, including the chain of being that arranged all animals and other natural entities in a single hierarchy, Geoffroy Saint-Hilaire's attempt to view all animals as variations on a single master plan, and Jean-Baptiste-Pierre Antoine de Monet de Lamarck's evolutionary scheme to derive one species from another. The Cuvier tradition in France would contribute to French reluctance to accept Darwinian evolution, and Cuvier's antievolutionary argument that the fossil record shows no evidence of transitional forms from one species to another continues to be popular among creationists.

The French consulate and empire saw Cuvier go from strength to strength. He produced dozens of papers as well as several multivolume studies, usually prepared with the help of assistants. He also attained an exalted position in the empire as inspector general of the Imperial University, which required extensive travel throughout Europe. Even more important in French scientific life, he acquired one of the two permanent secretaryships to the Institute of France, the successor to the Academy of Sciences, in 1803, a position he would hold for the rest of his life. The responsibility to craft an annual report on French science as well as writing eulogies for recently deceased French and foreign scientists made this position a potential source of power that Cuvier exploited to the fullest. Cuvier's experiences in the French Revolution had made him a strong supporter of authority, transcending his allegiance to any particular regime. He had no difficulty adjusting to the Restoration of the Bourbons in 1815, and served the new government as a liaison to the French Protestant community. His administrative and political tasks caused him to give up most of his teaching from around 1814, but he continued to be an active scientist, publishing along with an assistant eight volumes on the natural history of fish. Cuvier acquired an undeserved reputation as a reactionary, and his contemporaries, in and out of the scientific community, saw him as a master of patronage and as a would-be dictator of science. The resentment built up over the years was expressed in the famous controversy between Cuvier and Geoffroy Saint-Hilaire at the Academy of Sciences in 1830, in which Cuvier successfully defended his four-part scheme of zoology against Geoffroy Saint-Hilaire's single model. Outside the scientific community, however, he was ridiculed as a tyrant and pedant. Cuvier adapted to the French Revolution of 1830 as easily as he had to the Bourbon Restoration, and was made a peer of France in 1831, an extremely rare honor for a Protestant.

See also Fossils; French Revolution; Geoffroy Saint Hilaire, Étienne; Lamarck, Jean-Baptiste-Pierre-Antoine de Monet de; Museum of Natural History; Napoleonic Science; Zoology.

References

Appel, Toby A. *The Cuvier-Geoffroy Debate: French Biology in the Decades before Darwin.* New York: Oxford University Press, 1987.

Coleman, William. *Georges Cuvier, Zoologist: A Study in the History of Evolution.* Cambridge: Harvard University Press, 1964.

Outram, Dorinda. *Georges Cuvier: Vocation, Science, and Authority in Post-Revolutionary France.* Manchester: Manchester University Press, 1984.

D

Dalton, John (1766–1844)

John Dalton was the greatest example of the scientist working in the "provinces" of early-nineteenth-century England, outside London and the universities. He was also the first scientist to establish a quantitative chemical system based on an atomic theory. Ironically, he was not a chemist by training, nor was his principal motivation in first developing his atomic theory solving specific chemical problems. Unlike most English scientists, who were Anglicans and associated with London, Dalton was a Quaker from England's industrial north, who spent most of his career in the industrial city of Manchester. His formal education was extremely limited, and as a Quaker he was barred from attending the English universities. However, there was an active scientific culture where he lived and worked, including several Quakers who took an interest in him, and he picked up a great deal of knowledge through discussion and study. The Quaker school where he was appointed as an assistant in 1781, Kendal Friends School, possessed an adequate scientific library and collection of apparatuses, and he had access to much more when he moved to Manchester in 1792, opening his own academy there in 1800.

Dalton's first scientific interest was meteorology (he kept a weather diary beginning in 1787 and ending the day he died), and his first published book was *Meteorological Observations and Essays* (1793). His intellectual path from meteorology to chemistry began with the question of the composition of the atmosphere. Dalton was interested in how water vapor, and by extension other gases, existed in the mixed atmosphere. During the eighteenth century, chemists such as Joseph Priestley and Antoine-Laurent Lavoisier had divided atmospheric air into its component gases, such as oxygen and nitrogen. The questions on the agenda for meteorologists were how these gases actually mixed in the air and why the atmosphere did not separate into layers, with the heavier gases sinking to the bottom. Many chemists believed that the air was a chemical compound, but Dalton theorized that the particles of the separate gases repelled each other and only each other, so that the gases were independently suspended in the overall mixture. Dalton's exposition of his theory of mixed gases in three papers delivered to the Manchester Literary and Philosophical Society in 1801 sent shock waves through the European chemical world.

In the first decade of the nineteenth century, Dalton developed the idea of these small particles into one that asserted that all basic elements were composed of these particles, or atoms, and that they combined to form compounds in fixed ratios. Elemental atoms were both indestructible and impossible to create, and they could not be transmuted into another element. Atoms of different elements had different weights, the key to the whole system. The atoms of elements combined to form what Dalton called "compound atoms" and what we refer to now as molecules. Since particles of the same element repulsed each other, the most common form of a compound would be one combining one atom of one element with one atom of another element. Because the most common compound of hydrogen and oxygen was water, Dalton reasoned, it was probably composed of one atom of each. Since it took eight times the weight of oxygen to combine with a given weight of hydrogen to form water, the ratio of their particle or atomic weights would be eight to one. Chemical atomism was set forth in a series of lectures Dalton gave in Edinburgh in the spring of 1807 and then in the first volume of his *A New System of Chemical Philosophy* (1808). Subsequent volumes appeared in 1810 and 1827. Although some chemists, such as Sir Humphry Davy, were skeptical of Dalton's system, others, such as Thomas Thomson (1773–1852) and William Hyde Wollaston (1766–1828), adopted it eagerly. Thomson and Wollaston published chemical papers using Dalton's ideas in *Philosophical Transactions* the same year *A New System of Chemical Philosophy* was published.

Atomic theory had been around since ancient Greece, and many of Dalton's contemporary chemists thought in terms of ultimate particles of differing weights. However, Dalton was the first to use chemical atomism as the quantitative basis for chemistry. Dalton was a far more extreme atomist than were most of the chemists who accepted his system. Many found Dalton's system useful for describing chemical phenomena, while reserving their opinion on the subject of its ultimate reality. Dalton believed that the chemical atoms actually existed and that the arrangement of elemental atoms into compound atoms could be divined. Dalton's system of chemical notation, in which different kinds of circles representing elemental atoms were grouped together into compound atoms, was rejected by the chemical community in favor of the Swede Jöns Jakob Berzelius's system of letters (still in use today) because chemists did not want to have to speculate on how the elemental atoms were arranged in a compound atom, and Dalton's notation would have forced them to. Berzelius's system also had the advantage of using existing symbols, whereas Dalton's would have required printers to create new type.

Besides meteorology and chemistry, Dalton's other scientific interests included color blindness, of which he (along with other males in his family) was a victim. In 1794 in the first of 117 papers he would deliver to the Manchester Literary and Philosophical Society, he argued that his own inability to see the color red was the result of the blueness of his eye's aqueous medium. The "Lit and Phil" that admitted Dalton in 1794 was his intellectual and social base, and he served as its president from 1817 to his death.

As Dalton became a prominent scientist after the publication of *A New System of Chemical Philosophy,* he supplemented the income he derived from his school by giving lectures all over Britain, mainly in the north of England. The numerous honors that came to Dalton late in his life, including honorary degrees from the Universities of Oxford and Edinburgh and a government pension awarded in 1833, were connected to the establishment in England of the new role of the professional scientist, as opposed to the amateur virtuoso. Dalton, not a physician, a clergyman, or a gentleman-amateur, was considered a model professional scientist. The cultural gap between Dalton and the London-based and more amateur scientific establishment can be seen in his late admission to the Royal Society in 1822. Dalton had never shown much interest in becoming a fellow of the Royal Society, declining admission when Davy had offered to sponsor him in 1810, and his ultimate admission was done without his knowledge. Even after admission, the amateur-dominated Royal Society played only a minor role in his career despite awarding him a Royal Medal in 1826. He submitted only four papers to it, and did not present himself to be enrolled until 1834. Dalton was far more enthusiastic about his membership in the French Academy of Sciences, a more professionally oriented body that elected him a corresponding member in 1816 and a foreign associate, the highest honor a non-French scientist could achieve, in 1830. The British Association for the Advancement of Science, which Dalton helped found in 1831, was specifically meant as a challenge to the Royal Society in the name of provincial and commercially oriented professional science. He was a very active member of the British Association, serving on a number of committees, although his participation was greatly curtailed by two severe strokes in 1837. Dalton's funeral in 1844 was a great state occasion in Manchester, as more than 40,000 people filed by his coffin over four days. On the actual day of the funeral, shops and offices were closed. His will directed that his eyes be dissected to prove his theories of color blindness, but the dissection disproved them instead.

See also Atomism; Chemistry; Manchester Literary and Philosophical Society.

References

Brock, William H. *The Norton History of Chemistry.* New York: W. W. Norton, 1992.

Thackray, Arnold. *John Dalton: Critical Assessments of His Life and Science.* Cambridge: Harvard University Press, 1972.

Darwin, Erasmus (1731–1802)

The physician Erasmus Darwin was one of the most active promoters and popularizers of science in the north of England. An Edinburgh M.D., Darwin settled in the cathedral town of Lichfield in the English Midlands. He showed an active interest in a broad range of technological and scientific problems and was admitted to the Royal Society at the age of twenty-nine. The sociable Darwin also became part of an active circle of physicians, manufacturers, and engineers in the Midlands, notably centering on the Lunar Society of Birmingham, of which he was a founder. Darwin's move to Derby in 1782 was followed by his founding the Derby Philosophical Society the following year.

Darwin's most important scientific work was in natural history, particularly botany. He was a leader in introducing Carolus Linnaeus's system of plant classification to England, publishing two large translations of Linnaeus's work, *A System of Vegetables* (1783) and *The Families of Plants* (1787). His most famous work, *The Loves of the Plants* (1789), was a verse treatment of plant sexuality inspired by the Linnaean system. Combined with a subsequent work, *The Economy of Vegetation,* it formed *The Botanic Garden.* The notes to *The Botanic Garden* cover a broad range of scientific topics, including many of Darwin's own speculations. Darwin was a fine poet whose work was admired by such poets as William Cowper (1731–1800) and Percy Bysshe Shelley (1792–1822), but the idea of a scientific epic did inspire some ridicule and satire. Darwin's political and religious radicalism—he was a Deist, an open supporter of the French and American Revolutions, and a fierce opponent of the slave trade—also made him open to attack as the climate of Britain grew more reactionary in the 1790s.

Darwin was also a highly regarded physician (although the facility with which he prescribed opium caused serious problems for some of his patients) and published two volumes of *Zoonomia,* in 1794 and 1796. This massive prose treatise attempts to give a theory of living things to establish the causes and classifications of diseases. The discussion of evolution in the book has attracted particular attention due to its similarities to the theories of Darwin's grandson Charles Darwin (1809–1882), who read and admired *Zoonomia* as a young man. Age did not slow Erasmus Darwin's productivity. He published another solid treatise, *Phytologia; or, The Philosophy of*

Agriculture and Gardening in 1800. This work includes a number of new ideas, including discussion of the necessity of niter and phosphorus as plant nutrients. Another scientific poem, *The Temple of Nature* (1803), appeared posthumously.

See also Botany; Derby Philosophical Society; Literature; Lunar Society of Birmingham; Popularization.

Reference

King-Hele, Desmond. *Doctor of Revolution: The Life and Genius of Erasmus Darwin.* London: Faber and Faber, 1977.

Davy, Sir Humphry (1778–1829)

The life of Sir Humphry Davy, the founder of electrochemistry, demonstrates the increasing power of science as a path of upward social mobility in early-nineteenth-century England. From a poor but respectable Cornish family, Davy received a grammar school education and was apprenticed to a local surgeon-apothecary at the age of sixteen. Intellectually, he was a self-made man with little formal training in chemistry or other sciences. His motivation was high, and he seems to have viewed himself as a second Newton from early in his studies, but his lack of formal, systematic training in the sciences would be a handicap in his later career. Davy's interests throughout his career were not limited to the sciences, and he was also fascinated by poetry and literature. The early romantic movement was in full swing, and Davy's early poetry, although not particularly distinguished as poetry, is full of romantic language and images, which would also appear in his scientific prose. An early interest in the concept of the genius suggests that Davy may have consciously presented himself as an example of this romantic ideal.

His studies in chemistry began in 1797, and advanced rapidly. The obscure young man was fortunate that winter, when Gregory Watt (1777–1804), the tubercular son of the great engineer James Watt (1736–1819), was sent to Cornwall for the relatively mild climate. He lodged at the Davy household, and the two became close friends, bringing Davy into contact with the Lunar Society of Birmingham. Another patron, a local member of Parliament and scientific amateur named Davis Giddy (1767–1839), introduced Davy to the physician and chemist Thomas Beddoes, who was impressed with the young man's experiments on light and heat attacking Antoine-Laurent Lavoisier's theory of caloric fluid. Beddoes recruited Davy for his Pneumatic Institute. The institute, based in Beddoes's home outside Bristol, would investigate the medical effects of breathing different gases. This position was a stroke of great fortune, as it enabled Davy to devote most of his time to science and scientific research. It was at the Pneumatic Institute that Davy did the first research that brought him a wide reputation, the experiments on nitrous oxide, "laughing gas," and other oxides of nitrogen. His experimental subjects included himself, and it is possible that his experiments with different gases, including carbon monoxide, shortened his life. Through Beddoes and the institute, Davy also became friends with the leading romantic poets Samuel Taylor Coleridge (1772–1834), William Wordsworth (1770–1850), and Robert Southey (1774–1843).

What revolutionized Davy's science and brought him into national repute was Alessandro Volta's discovery of the "voltaic pile," or battery, in 1799. Davy's analysis of the workings of the pile, which demonstrated that the electrical current was generated not by mere contact but by the oxidization of zinc, was first published in 1800, and led to his election as a fellow of the Royal Society in 1803. Even more important, in 1801 he left the provincial stage at Bristol and moved to London, where Benjamin Thompson, Count Rumford, had recruited him for the newly formed Royal Institution. Davy shone at the institution, where his spectacular lecturing style ensured that his demonstrations were always crowded. In line with the utilitarian emphasis of the Royal Institution, Davy worked on such areas as tanning and soil analysis. He did not suggest any dramatic breakthroughs in these areas, but his work did stand as the most thorough scientific treatment of them for decades. His work on tanning was largely responsible for his Copley Medal in 1805.

Davy's most important scientific work was the series of experiments he performed on the electric decomposition of chemical compounds. The use of electricity to decompose water into oxygen and hydrogen was already known, but Davy (along with his contemporary Jöns Jakob Berzelius) broadened the area of the chemical use of electricity immensely. He discovered, named, and classified as metals potassium and sodium, producing these "new elements" by the electrical decomposition of potash and soda in 1807, and also followed Berzelius in isolating calcium, barium, strontium, and magnesium. His next series of chemical triumphs emerged from his successful attempts to disprove Lavoisier's contention that oxygen was the principle of acidity (although he respected Lavoisier's genius, Davy was a good Englishman never happier than when combating French chemistry—at one point he even toyed with the idea of reviving the phlogiston theory in a modified form). In the course of these experiments, Davy isolated and named chlorine as an element.

Davy's move from the Pneumatic Institute of the politically radical Beddoes to the Royal Institution was accompanied by a move to the conservative end of the political spectrum. Davy emphasized the relation of chemistry to natural theology, thus helping purge it of dangerous materialist associations. He was an avid social climber puzzled, if not shocked, by his chemical contemporary John Dalton's stubborn refusal to move to London and chase after honors. Davy wanted the prizes British society offered, and gained many of them. He was knighted on 8 April 1812 and three days later married an eligible London heiress, Jane Apreece (1780–1855). The marriage was not altogether happy. The next year saw an encounter even more fateful for Davy's reputation, if not his day-to-day life, when he hired the young Michael Faraday (1791–1867) as an assistant. Davy, Mrs. Davy, and Faraday set out on a tour of France and Italy, which was rather difficult because France and Britain were at war. In Paris Davy was unimpressed by Napoléon's empire and got involved in a nasty priority dispute with his great French rival, Joseph-Louis Gay-Lussac, over the identification of iodine.

On his return he accomplished something that spread his fame more widely than anything before: the discovery of a lamp for working coal miners. The candles and lamps used by coal miners were setting off explosions of methane, making an already dangerous job even more so. Davy's laboratory experiments with samples of the gas revealed that a lamp could be made safe through the use of very narrow metal ventilation tubes, later changed to metallic gauzes. The lamp, which Davy refused to patent, thereby forfeiting hundreds of thousands of pounds, did not immediately end explosions in coal mines, partly because of improper use and partly because it was used as a tool to force miners to go into deeper and more dangerous shafts. It was successful in bringing Davy even more honors: the Royal Society's Rumford Medal for useful scientific contributions; a hereditary baronetcy, the highest honor the British crown had ever seen fit to bestow on a scientist; and the presidency of the Royal Society in 1820, on the death of his old patron Sir Joseph Banks. Inevitably, the Davy lamp also brought more priority disputes.

Davy's presidency of the Royal Society was not a success. He was sympathetic to those who wanted to reform the society from a gentleman's club to a more professional body on the model of the French Academy of Sciences, but lacked the executive ability and social standing to overcome the society's entrenched conservatism and control the admission of fellows. Relations between Davy and Faraday grew cold when Faraday allowed his name to be put forward for a fellowship (successfully) in 1823 without consulting

Davy beforehand. The limitations of Davy's power can be seen in the fact that while he, following Banks's example, gave weekly scientific salons in his home, the council vetoed his hope to open them to women (as the Royal Institution's lectures were).

Davy did have some achievements. He promoted the endowment of a new prize, the Royal Medal, and gave up Banks's hopeless struggle against the rise of the new societies specializing in a particular science. His effectiveness as president was hindered by the rapid decline of his health, which forced him to resign the position in 1827. The last few years of his life were busied with experiments on animal electricity, his lifelong passion of fishing, and philosophical speculation.

See also Beddoes, Thomas; Chemistry; Electricity; Gay-Lussac, Joseph-Louis; Nationalism; Popularization; Romanticism; Royal Society.

References

Golinski, Jan. *Science As Public Culture: Chemistry and Enlightenment in Britain, 1760–1820.* Cambridge: Cambridge University Press, 1992.

Knight, David. *Humphry Davy: Science and Power.* Oxford: Blackwell, 1992.

Derby Philosophical Society

Erasmus Darwin founded the Derby Philosophical Society in 1783, shortly after he arrived in Derby the previous year. The society began as informal weekly meetings of gentlemen interested in natural philosophy, but was not inaugurated as a formal society until 18 July 1784. Official meetings took place on the first Saturday of the month. Darwin's ambitions for the society to publish its own papers came to nothing, but it sponsored scientific lectures and built a fine library, particularly strong on the transactions of other scientific societies and in chemistry, a subject of great interest to Derby manufacturers. Like other English provincial societies, it brought together natural philosophers, industrialists, local gentry, and physicians. The society was divided by controversy when it sent a message of condolence to Joseph Priestley after a mob had destroyed his house and laboratory in 1791 (Priestley firmly rejected the society's suggestion that he now abandon theology and stick to natural philosophy) and lost some of its vitality after Darwin's death in 1802, but it survived until 1857, when it merged with the Derby Town and County Museum.

See also Academies and Scientific Societies; Darwin, Erasmus.

Reference

Musson, A. E., and Eric Robinson. *Science and Technology in the Industrial Revolution.* Toronto: University of Toronto Press, 1969.

Desaguliers, John Theophilus (1683–1744)

John Theophilus Desaguliers was by descent a French Protestant whose family had been driven into exile by Louis XIV's Revocation of the Edict of Nantes in 1685. After graduating from Oxford, Desaguliers became a minister of the Church of England, although one inattentive to his religious duties. Far more important to him were his roles as the greatest experimental demonstrator in England after the death of Francis Hauksbee in 1713 and as the most active promoter of European Freemasonry. Desaguliers took over from Hauksbee the position of experimental demonstrator to the Royal Society, receiving as Hauksbee had not the title of curator of experiments (the title lapsed after his death). He was one of the closest allies of Isaac Newton (1642–1727) in the society. But the Royal Society, which paid Desaguliers about ten to forty pounds a year depending on how many experiments he provided, was only one of many venues for his demonstrations. He was the star demonstrator in Britain, with a dazzling array of equipment, including the most sophisticated orrery, or mechanical model of the solar system, ever seen. Although Desaguliers did demonstrate the Cartesian theory of matter (if only to refute it), the bulk of his lectures were Newtonian.

He adopted some procedures from the Oxford professor John Keill (1671–1721), and boasted that his lectures and demonstrations did not just strain for remarkable effects, but taught Newtonian natural philosophy. They also had a practical orientation, with much focus on engines and the application of power to perform particular tasks. The early steam engines of Thomas Savery (c. 1650–1715) and Thomas Newcomen (1663–1729) figured in his presentations.

Desaguliers's demonstrations were particularly important because he was a leader in bringing sophisticated natural-philosophical lectures and experiments out of London and into the English provinces. He lectured before one of the earliest societies for natural philosophy to be formed outside London, the Spalding Society, as well as before less formal groups. He also demonstrated in the Dutch Republic, where he was closely associated with early Dutch Newtonians like Willem Jakob 's Gravesande (1688–1742). His disciple and sometimes assistant Isaac Greenwood (1702–1745) brought experimental Newtonianism to America as the first holder of the Hollis Professorship in Mathematics and Natural Philosophy at Harvard founded in 1727. Desaguliers was aided by his French heritage in spreading his ideas, as his continental audiences would be much more likely to understand French than English. He had a hand in drawing up the Masonic constitution of 1723, spread Freemasonry as well as natural-philosophical Enlightenment, and indeed presented them as one and the same.

Although Desaguliers did sometimes expound his own theories, as when an essay of his on electricity won a prize from the Academy of Bourdeaux in 1742 (there were only two other entrants), he was not an original scientist. His two-volume book based on his lectures, *A Course of Experimental Philosophy* (1743–1744), was a very popular exposition of the Newtonian system, quickly translated into French and Dutch. He also authored a poem on what would seem to be the remarkably unpoetic topic of the application of the Newtonian theory of universal gravity to politics, *The Newtonian System of the World, the Best Model of Government, an Allegorical Poem.* The poem supported the British government of his time.

See also Freemasonry; Industrialization; Newtonianism; Popularization; Royal Society.

References

Heilbron, J. L. *Electricity in the 17th and 18th Centuries: A Study in Early Modern Physics.* 2d ed. Mineola, NY: Dover, 1999.

Jacob, Margaret. *Scientific Culture and the Making of the Industrial West.* New York: Oxford University Press, 1997.

Diderot, Denis (1713–1784)

Denis Diderot, one of the most influential thinkers of the French Enlightenment, vigorously promoted science, particularly in his role as an editor and coordinator of the *Encyclopédie.* His philosophy was deeply influenced by scientific developments, which he deployed in an assault on religion in general and Christianity in particular. Diderot's interest in the sciences moved from mathematics, which he believed to be a "finished" discipline with no great discoveries left to make, to chemistry and natural history. His natural philosophy changed from Newtonian deism to atheistic materialism.

Diderot's *Thoughts on the Interpretation of Nature* (1754), a response to Pierre-Louis Moreau de Maupertuis's *System of Nature,* presented his theory of a self-acting, "alive" matter as the basis of a nonmechanistic materialism. Influenced by the recent discovery of the polyp and the work of Albrecht von Haller, he argued that the material world was not the dead, clockwork universe of vulgar Newtonianism. He attacked the overmathematization of eighteenth-century science and called for a return to the empiricism and experimentalism of Francis Bacon (1561–1626). Diderot integrated his vitalistic theory of matter with militant atheism in a work too dangerous to publish, *D'Alembert's Dream,* written in the late 1760s. In the form of an imaginary dialogue with Diderot's friend and collaborator

ENCYCLOPÉDIE,

OU

DICTIONNAIRE RAISONNÉ DES SCIENCES, DES ARTS ET DES MÉTIERS,

PAR UNE SOCIÉTÉ DE GENS DE LETTRES.

Mis en ordre & publié par M. *DIDEROT*, de l'Académie Royale des Sciences & des Belles-Lettres de Pruſſe; & quant à la PARTIE MATHÉMATIQUE, par M. *D'ALEMBERT*, de l'Académie Royale des Sciences de Paris, de celle de Pruſſe, & de la Société Royale de Londres.

Tantùm ſeries juncturaque pollet,
Tantùm de medio ſumptis accedit honoris! HORAT.

TOME PREMIER.

A PARIS,

Chez
- BRIASSON, *rue Saint Jacques, à la Science.*
- DAVID l'aîné, *rue Saint Jacques, à la Plume d'or.*
- LE BRETON, Imprimeur ordinaire du Roy, *rue de la Harpe.*
- DURAND, *rue Saint Jacques, à Saint Landry, & au Griffon.*

M. DCC. LI.

AVEC APPROBATION ET PRIVILEGE DU ROY.

Title page of Encyclopédie, *an ambitious project that summed up human knowledge (Library of Congress)* on the *Encyclopédie,* Jean Le Rond d'Alembert, *D'Alembert's Dream* attacked the theistic argument from design by suggesting that matter was not only self-acting, but also self-organizing. The world was one of endless flux and development, of which transformation from one form to another, not stability, was the central characteristic. There was no rigid distinction between living and nonliving matter, any more than between animals and humans. (Diderot endorsed the concept of spontaneous generation.) Species were not permanent either, but rose and disappeared.

Diderot had more acquaintance with medicine than did most of the other philosophes, and his interpretation of human nature had much in common with the "medical materialism" of physicians such as Julien Offroy de La Mettrie, whom Diderot bitterly attacked despite his intellectual debt. Diderot's ethics drew from his science. Humans were material beings, with no "cosmic destiny" or "free will," and as material beings they needed to follow their natural biological and social instincts. Religion, and any other force that restrained human nature, was at best suspect, at worst destructive to natural human happiness.

See also Alembert, Jean Le Rond d'; Encyclopedias; The Enlightenment; Materialism; Polyps; Religion; Vitalism.

References

Crocker, Lester G. *Diderot: The Embattled Philosopher.* New York: Free Press, 1966.

Vartanian, Aram. *Diderot and Descartes: A Study of Scientific Naturalism in the Enlightenment.* Princeton: Princeton University Press, 1953.

Earthquakes

Although scientific explanations of earthquakes did not change much in the eighteenth century, they were central subjects of cultural interest. Among the most dramatic and destructive were the Lima earthquake of 1746, the great Lisbon earthquake of 1 November 1755, and the series of earthquakes that hit the Calabria region of the Kingdom of Naples in 1783. Each produced observer accounts of varying degrees of scientific awareness. For most Europeans at the beginning of the Enlightenment, earthquakes were primarily religious phenomena, and every major earthquake was followed by sermons and tracts warning the survivors of God's wrath and urging repentance. Although religious interpretations did not disappear in the Enlightenment, they were increasingly on the defensive against secular and scientific ones. The main scientific theory of earthquakes traced its roots to the ancient Greek philosopher Aristotle (384–322 B.C.E.). His theory was that earthquakes were the result of the explosions or the rapid expansion of pent-up gases. If these gases broke their way to the surface, the result was a volcano; if they remained trapped underground, the result was an earthquake.

Interest in earthquakes increased after the Lima earthquake, an account of which incorporated a theoretical treatment of earthquakes that was widely translated and published (Benjamin Franklin was the printer of the Philadelphia edition in 1749). A series of dramatic although not very damaging earthquakes in England in 1750 produced a flood of accounts directed to the Royal Society journal, *Philosophical Transactions.* What really made earthquakes culturally important, however, was the extremely destructive Lisbon earthquake, which Voltaire used as conclusive evidence against philosophical optimism. The Lisbon earthquake was followed by two substantive works: *History and Philosophy of Earthquakes* (1757) by the English physician and astronomer John Bevis (1693–1771) and *Historical and Physical Memoirs of Earthquakes* (1757) by the Swiss naturalist and geologist Elie Bertrand (1712–c. 1790). Neither broke new ground theoretically, but by gathering together accounts of different earthquakes they provided sources for the most important eighteenth-century treatment of the subject, a series of papers read before the Royal Society in 1760 and published in *Philosophical Transactions* the next year by the English clergyman John Michell (1724–1793). Michell, an astronomer and Woodwardian Professor of Geology at Cambridge University, also accepted the basic theory of subterranean explosions, but explicated it in greater detail.

He also clearly distinguished between the primary and secondary effects of earthquakes and primary earthquakes and aftershocks. His most innovative claim was to distinguish between vibratory and wavelike motions produced by earthquakes.

The Calabria earthquakes also produced much writing, particularly since the first earthquake was followed by others, allowing scientists to come onto the scene. The most impressive writings on the earthquakes are the two-volume work by the Neapolitan court physician Giovanni Vivenzio, *Description of the Earthquake in Calabria* (1783), and the report of the Royal Academy of Science and Letters of Naples, *History of the Phenomena of the Earthquakes near Calabria and Valdemone in 1783* (1784). Vivenzio endorsed the idea of an electrical cause for earthquakes, but this idea, part of the late-eighteenth-century fashion for electrical explanations, had little long-term impact on seismology. Other observers included Sir William Hamilton (1730–1803) and the French geologist Déodat de Gratet de Dolomieu (1750–1801).

Pendulum seismoscopes to detect and measure earthquakes came into widespread use in the eighteenth century. Andrea Bina (1724–1792) refined the device in 1751 to add a recording function by the pendulum making tracks in sand. Another of the many Italians to interest themselves in earthquakes was the physician Domenico Pignataro (1735–1802), a writer on the Calabrian quakes who introduced the first attempt at a scale for earthquake severity, dividing them into the five categories of slight, moderate, strong, very strong, and violent.

Interest in earthquakes diminished in the late eighteenth century, when the French Revolution and its ensuing wars eclipsed any natural disasters, but it picked up again in the second decade of the nineteenth. Michell's work, which had been ignored for some time, was reprinted in *Philosophical Magazine* in 1818, and Karl Ernst Adolf von Hoff (1771–1837) began publishing an annual list of the world's earthquakes in 1826, beginning with those of 1821.

See also Geology.

References

Davison, Charles. *The Founders of Seismology.* 1927. Reprint, New York: Arno, 1978.

Howell, Benjamin F. *An Introduction to Seismological Research: History and Development.* Cambridge: Cambridge University Press, 1990.

Johns, Alessa, ed. *Dreadful Visitations: Confronting Natural Catastrophe in the Age of Enlightenment.* New York: Routledge, 1999.

Education

The eighteenth century was one of fundamental shifts in European education, in which scientific and technical education would assume a more central role. The main process by which this came about was the greater involvement in education by the state as opposed to the church. The basic function of most European schools had been religious indoctrination, the teaching of simple skills, and, for boys, the Latin language. Eighteenth-century educational reformers emphasized utility and practical knowledge, frequently identified with mathematics and science.

Although some reformers, such as the Hungarian Mátyás Bél (1684–1749), succeeded in adding more science to curricula in the first half of the century, in most of the Catholic world the opportunity for encouraging science in the schools emerged most dramatically with the expulsion of the Jesuit order from various countries and its final suppression. This process extended from the order's suppression in Portugal in 1759 to its dissolution by the pope in 1773. (The order was restored by papal decree in 1814.) The vacuum created had to be filled by new teachers. The Jesuits had been identified with outdated science and an emphasis on classical learning, particularly polished Latin. Educational reformers emerging in the wake of the Jesuit collapse advocated more practical education. *Essay on National Education* (1763) by the Breton magistrate Louis-René de Caradeuc de La Chalotais (1701–1785) attracted wide European interest. La Chalotais argued for a basically secular system of education in which religion would play little role and the primary determinant of what was taught would be what was useful to the state. Although the effect was largely limited to the big urban schools, there is evidence that the expulsion of the Jesuits from France in 1764 was followed by an increase in the time devoted to mathematics and experimental physics in the schools they left. The order's expulsion from Portugal, part of a drive toward Enlightenment centralization, was also followed by an increase in technical education. Ironically, many of the Jesuit instructors were able to teach more or less openly in the Protestant world and Orthodox Russia, where the pope's writ did not run.

Eighteenth-century educators believed that different kinds of education were appropriate to different groups in the population. Women's education in the sciences was usually restricted to simple math for doing household accounts and some medicine and "kitchen chemistry." However, some schools for upper-class women taught more advanced science, justified either as a diversion or as a way of demonstrating the wonderful works of God. In addition to more training in advanced mathematics and theoretical physics, there was also an increased need for education in applied technology for workers, particularly as the apprenticeship system was weakening along with the decline of the guilds. For the very prominent in society, the traditional curriculum focusing on Latin remained dominant and even had something of a renaissance in the nineteenth century with the spread of the German gymnasiums.

Scientific education was largely for the middle and professional classes, and in most countries was viewed in light of service to the state. German states built a strong network of technical schools, the most famous in the eighteenth century being the mining academy of Freiberg where Abraham Gottlob Werner taught. In England, an exceptional European country in that neither state nor church took much interest in expanding or reforming education, scientific advance was carried on by Dissenting academies, run by Protestant denominations outside the Church of England, and by a number of for-profit academies founded by individuals. Some of the most important scientists in England, such as the Unitarian Joseph Priestley and the Quaker John Dalton, taught at Dissenting academies. These private academies were also prominent in the British New World colonies. France, by contrast, built an imposing structure of state-sponsored technical and scientific education aimed at producing an elite group that would serve the state, including institutions such as the School of Bridges and Roads, the School of Mines founded in 1783, and the School of Military Engineering at Mézières, founded in 1748, whose instructors and examiners included Jean-Antoine Nollet and Gaspard Monge (1746–1818). These institutions founded under the old regime were joined after the Revolution by the Polytechnic School and the Normal School.

See also Popularization; Universities.

References

Bowen, James. *A History of Western Education.* Vol. 3, *The Modern West: Europe and the New World.* London: Methuen, 1981.

Gillispie, Charles Coulston. *Science and Polity in France at the End of the Old Regime.* Princeton: Princeton University Press, 1980.

Leith, James A., ed. *Facets of Education in the Eighteenth Century.* Studies in Voltaire and the Eighteenth Century, no. 167. Oxford: Voltaire Foundation, 1977.

Egyptian Expedition

The French expedition to conquer and occupy Egypt, led by Napoléon Bonaparte, was accompanied by a large group of scientists who brought many aspects of Egypt into European scientific knowledge. The French occupation, which lasted from 1798 to 1801, also led to the most direct contact between European scientists and Muslim intellectuals of the Enlightenment period. Leading scientists who accompanied Napoléon included the chemist Claude-Louis Berthollet, the zoologist Étienne Geoffroy Saint-Hilaire, the geologist Déodat de Gratet de Dolomieu

(1750–1801), and the mathematicians Gaspard Monge (1746–1818) and Jean-Baptiste-Joseph Fourier (1768–1830).

Napoléon was an enthusiast of science and had hoped for the establishment of a commission of arts and sciences to accompany his army. Monge and Berthollet had been associated with Napoléon during his invasion of Italy in 1796–1797 and were obvious choices to lead the scientific group, although it was formally commanded by a military officer, Brigadier Caffarelli du Falga. During March 1798 a circular called for volunteers to accompany the army, whose destination and purpose remained a military secret. Volunteers were informed that their positions in France would be kept open for them during their absence, and their salaries would continue to be paid to their families. All the leading scientific institutions of revolutionary France—the Institute of France, the Polytechnic School, the Museum of Natural History, and the Paris Observatory—were represented on the commission, as were engineers, surveyors, draftsmen, and students of Oriental languages. The membership of the commission totaled 167 on the eve of departure.

In Egypt many members of the commission worked on projects benefiting the French occupation. The language experts served as interpreters or supervised the printing of Napoléon's proclamations in Arabic, the engineers adapted Egyptian infrastructure to French needs, and the medical men set up facilities and treated the many stricken by heat or disease. These immediate practical tasks were not the sole reason for the presence of so many savants, however. Napoléon promoted the setting up of the Institute of Egypt, on the model of the Institute of France. It was divided into four sections: mathematics, physics (including chemistry, medicine, and zoology), political economy, and literature and art. Each section was supposed to have twelve members, but was in actuality smaller. The first meeting of the Institute took place on 23 August 1798. Napoléon, a member of the mathematical section, proposed that the Institute concentrate on problems such as improving Egyptian baking and purifying the waters of the Nile, as well as improving the Egyptian legal and educational systems. Papers read at the Institute in its early days included Berthollet's studies of the formation of ammoniac salts and the manufacture of indigo. The Institute's proceedings and the discoveries of its scholars filled the pages of the *Decade Egyptienne,* a periodical appearing every ten days.

It was hoped that the Institute and the work of the French scholars would help enlighten the native Egyptian scholars, making them more supportive of the French government. Demonstrations in experimental physics and chemistry, and even a hot-air balloon ascension, were put forward to impress the Egyptians. However, although the Egyptians were impressed with some of these demonstrations, as well as French printed books and accurate astronomical instruments, they felt neither a desire to study European science nor a greater willingness to accept French rule.

The scholars of the expedition performed extensive fieldwork, of which the most notable product was the discovery of the Rosetta stone that eventually led to the deciphering of ancient Egyptian hieroglyphics. Egypt was surveyed according to European techniques and accurate maps produced. Natural historians collected a huge variety of fossils, plants, birds, fish, and other animals unknown to European science. Particularly interesting were the mummified remains of animals from ancient Egypt. The fact that the mummies seemed no different from current animals supported those, such as Geoffroy Saint-Hilaire's rival Georges Cuvier, who argued against evolution and for the fixity of species.

Many of the savants, after the first excitement of discovery had faded, longed to return to France. This was accomplished when English control of the Mediterranean made the French position untenable. Geoffroy Saint-Hilaire became a hero of French science by preserving the specimens collected by the expedition from the English as the French were forced to leave. The savants' work was commemorated in the twenty-three-volume *Description of Egypt,* begun with French government sponsorship in 1803 and published from 1809 to 1826. This work would be the textual basis of the Western study of Egypt through the nineteenth century.

See also Ballooning; Berthollet, Claude-Louis; Exploration, Discovery, and Colonization; Geoffroy Saint-Hilaire, Étienne; Napoleonic Science; Nationalism; War.

Reference

Beaucour, Fernand, Yves Laissus, and Chantal Orgogozo. *The Discovery of Egypt.* Paris: Flammarion, 1990.

Electricity

Although electrical phenomena had received some attention from sixteenth- and seventeenth-century scientists—most notably William Gilbert (1544–1603), who gave it its name; Niccolo Cabeo (1596–1650); and Otto von Guericke (1602–1686)—the science of electricity was largely an eighteenth-century creation. In the beginning it was less a matter for theoretically inclined natural philosophers than for demonstrators like Francis Hauksbee and Pierre Poliniere, who used spectacular electrical effects to amaze their audiences. Hauksbee's generator, which created an electrical charge by rapidly rotating an evacuated glass globe, increased electricity's prominence and also provided an easy way to generate it. Electrical science would long maintain this association with showmanship and display.

Theoretically inclined electricians in the early eighteenth century included the English tradesman Stephen Gray (1666–1736) and the French nobleman Charles-François de Cisternay Du Fay (1698–1739). Gray's experiments revealed that electricity had the power of being conducted across large distances by all kinds of things, most famously the "electrified boy" who hung from silk threads and dispensed electricity from his hands. The "electrified Venus," an electrified woman who kissed men, dispensing a spark from her lips, became a standard part of many electrical demonstrations. Gray conducted electricity across a 650-foot string. Du Fay, a far more systematic investigator, managed to induce electricity in a huge variety of objects. He distinguished between two kinds of electricity, one produced by rubbing glass and another by rubbing amber. Objects electrified with different electricities attracted each other, whereas objects electrified with the same repelled each other. The abbé Jean-Antoine Nollet, a spectacular electrical demonstrator, expanded Du Fay's work to become the most important electrical theoretician. He explained electrical attraction in basically Cartesian terms by an effluvium either departing an electrical body when rubbed or returning to it. Electricity at this point was still discussed almost entirely in qualitative terms, as part of "experimental physics."

The most dramatic innovation in mid-eighteenth-century electrical practice, the primitive condenser known as the Leiden jar, discovered independently in 1745 and 1746, delivered a shock far beyond anything electrical experimenters had previously experienced, and contributed to an outburst of interest in electricity in the mid-eighteenth century. The jar also posed several new problems and forced a rethinking of electrical theory. The most important theory that explained the jar's behavior came from Philadelphia, and has the claim of being the first important scientific theory to arise in North America. This was Benjamin Franklin's one-fluid theory of the electrical atmosphere, which defined electricity as either positive, marked by the presence of the electrical fluid, or negative, marked by its absence. Franklin's famous kite experiment of 1752 also demonstrated that lightning was electrical. Franklin's theory for the first time allowed for the existence of more than one distinct electrical state, rather than a continuum of more or less electrified bodies. Particularly as worked into a coherent system by the Italian professor Giambatista Beccaria (1716–1781), it became quite popular, although not undisputed. (Interestingly, one of the weaknesses of Franklin's theory was its failure to deal with electrical attraction, which until the Leiden jar had been the foremost topic in electrical studies.) Franz Maria Ulrich Theodor Hoch Aepinus (1724–1802) further refined and developed Franklin's theory, ridding it of its last traces of effluvialism and substituting a force acting at a distance, analogous to Newtonian gravity, for Franklin's fluid.

The greatest experimental challenge to Franklin's theory of one electricity was made by the most famous items of clothing in the history of science, the socks of Robert Symmer (c. 1707–1763). In 1758 Symmer, a Scottish government official, fellow of the Royal Society, and experimentalist, noticed that when black and white socks had been electrified on removal from his leg, they ballooned outward as the sock's electrified fabric repelled itself, but when two electrified socks were brought together, they stuck together and lay flat, bulging outward again when drawn apart. The fact that the socks bulged outward *after* they had been brought together indicated that their electrical charges did not equalize on being brought together, but remained separate. This led Symmer to suggest that electricity was two fluids, rather than one, as Franklin and his disciples would have it. Symmer also charged Leiden jars with his socks and demonstrated the astonishing power of electrical cohesion—his socks remained united while bearing a load ninety times their own weight. The debate of "one-fluid" and "two-fluid" theorists continued for the next two decades.

The production of massive electrical sparks by improved equipment, such as the huge electrostatic generator built by the English scientific instrument maker John Cuthberson (1743–1806) for the Teyler Foundation in Haarlem in the Dutch Republic in 1785, seemed to favor the one-fluid Franklin theory. The two-foot-long spark seemed to travel in one direction only, rather than back and forth. In counterpoint to the generation of massive sparks, very weak electrical charges were becoming easier to detect through Alessandro Volta's "condensator," announced in 1782, and subsequent improvements by the Reverend Abraham Bennett (1750–1799) and William Nicholson (1753–1815). This kind of equipment was increasingly required for serious electrical studies, and its expense was driving the amateur investigator of modest resources from the front line.

Aepinus's destruction of effluvialism meant electrical forces could now be quantified as gravity was. Electricity was partially removed from the qualitative domain of experimental philosophy to the mathematical one by the French engineer Charles-Augustin de Coulomb, who used a torsion balance to measure the electrical force and demonstrated what had long been suspected: that electrical attraction operated on the basis of the inverse-square principle. Coulomb was anticipated in these studies by Henry Cavendish and John Robison (1739–1805). He employed the "two-fluid" theory, giving it a further lease of life in France. Coulomb's mathematical approach to electricity culminated in the work of Siméon-Denis Poisson (1781–1840).

The other major development in late-eighteenth-century electrical physics was Volta's development of the "voltaic pile," announced in 1800. The pile, the first battery producing a constant electrical flow, was a stack of alternating zinc and silver disks separated by wet cardboard. The pile emerged from Volta's controversy with Luigi Galvani over "animal electricity," or "galvanism," but its implications were far-reaching. The continuous production of electrical current was immediately seen as useful in chemistry, particularly by Sir Humphry Davy who in 1807 used an electrical current to decompose potash into the elements potassium and sodium. Davy also explained the pile's ability

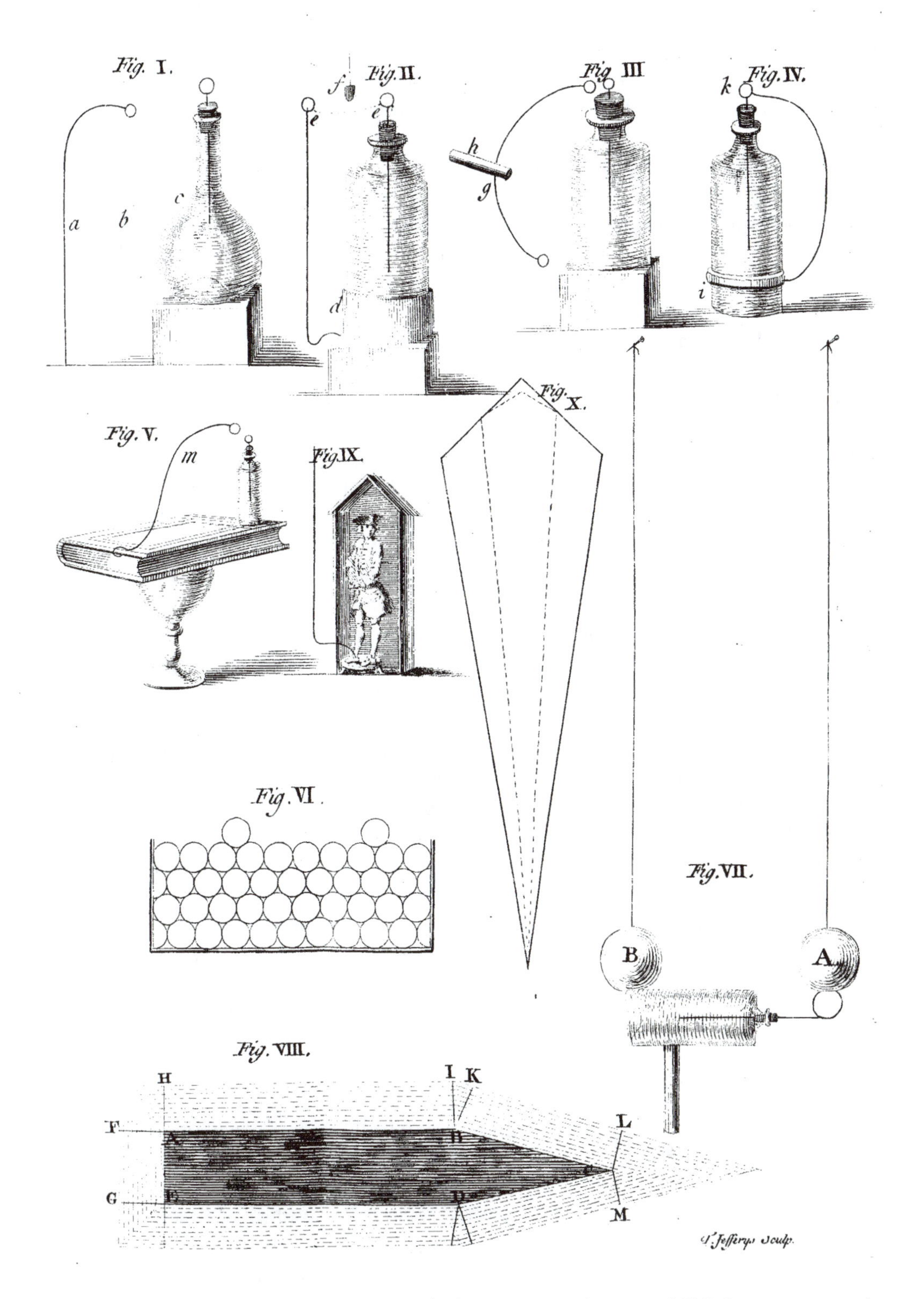

An illustration from Benjamin Franklin's Experiments and Observations on Electricity *(1751) showing experiments he made with Leiden jars and other equipment (Annenberg Rare Book & Manuscript Library, University of Pennsylvania)*

to generate electricity by a chemical process, the oxidation of zinc. Electrical theory began to be incorporated into chemistry, as in Jöns Jakob Berzelius's electrified theory of chemical atomism. The battery itself was further refined by William Cruikshank (1745–1800) who by immersing copper and zinc disks in brine inside a specially prepared box avoided the problem of the cardboard drying out, which had plagued Volta.

In 1820 the Danish romantic, natural philosopher Hans Christian Ørsted published a short Latin paper proclaiming his discovery that passing an electric current through a wire exerted magnetic force on a nearby compass needle. This first solid link between electricity and magnetism was followed almost immediately by the creation of electromagnets by the English soldier William Sturgeon (1783–1850) and the American Joseph Henry (1797–1878). The same year André-Marie Ampère demonstrated attractive and repulsive forces between parallel wires through which a current was flowing, founding the science of electrodynamics.

See also Ampère, André-Marie; Cavendish, Henry; Coulomb, Charles-Augustin de; Davy, Sir Humphry; Desaguliers, John Theophilus; Franklin, Benjamin; Galvani, Luigi; Hauksbee, Francis; Leiden Jars; Lichtenberg, Georg Christoph; Lightning Rods; Nollet, Jean-Antoine; Ørsted, Hans Christian; Physics; Poliniere, Pierre; Volta, Alessandro Giuseppe Antonio Anastasio.

References

Brock, William H. *The Norton History of Chemistry.* New York: W. W. Norton, 1992.

Dibner, Bern. *Alessandro Volta and the Electric Battery.* New York: Franklin Watts, 1964.

Heilbron, J. L. *Electricity in the 17th and 18th Centuries: A Study in Early Modern Physics.* 2d ed. Mineola, NY: Dover, 1999.

Embryology

Embryology, the science of conception and fetal development, was one of the most contentious of all disciplines during the Enlightenment. The basic division was between epigenesists, who believed that the embryo was gradually formed out of preexisting materials, and preformationists, who believed that the embryo was already formed before conception. A further division ran between those preformationists who believed that the fetus was preformed in the male sperm (spermists) and those who believed that the fetus was preformed in the female egg (ovists). Epigenesists also disagreed about the role of the egg and sperm. The sperm itself brought up the question of whether the fertilizing principle resided in the spermatozoa or was located in the liquid semen and the sperm were parasitic worms. Epigenesists, whose position's origins within Aristotelianism made it seem somewhat old-fashioned, had to come up with a mechanism for fetal formation. Preformationists, many of whom argued that their position was more compatible with mechanical philosophy, had to explain how fetuses could be preformed inside fetuses, ad infinitum—a task made easier by the belief that the Earth and living generations had been around for only a few thousand years. Most problems of conception and generation could not be resolved by empirical observation of fertilization and fetal development, as microscopes were not powerful enough. Inadequate microscopes also meant that the feature identified by eighteenth-century scientists as the mammalian egg was not actually the egg, but the graafian follicle.

Preformationism seemed more conceptually elegant for most mechanical philosophers in the early eighteenth century, as it was compatible with an idea of matter as passive rather than self-acting or self-organizing. The passivity of matter in the preformationist theory also had the advantage of preserving a role for God, that of creating all the preformed fetuses at the beginning of creation. Epigenesis was associated with the idea of self-active and self-moving matter, which seemed to deny any role for God. Ovism was generally the majority position among preformationists. It seemed wasteful for each spermatozoon, only one of which would fertilize the egg, to carry a preformed individual.

Several biological discoveries in the early eighteenth century seemed to support one side or the other. Studies of the regeneration of limbs in creatures such as crabs and lizards counted against the preformationists, as there seemed to be no way that both the first and the second limbs could be preformed. Charles Bonnet's experimental demonstration of aphid parthenogenesis, reproduction without males, lent powerful support to ovist preformationism. However, it was quickly followed by the discovery of the regeneration of severed polyps by Abraham Trembley (1700–1787), which seemed to support epigenesism by demonstrating how matter could organize itself to produce two living beings out of one. The claim in 1748 by English Catholic priest John Turberville Needham (1713–1781) for the spontaneous generation of microscopic organisms from the decay of broth supported epigenesis, as it provided an example for organisms created from organic matter without previously existing in it. Needham was an ally of Georges-Louis Leclerc de Buffon, who argued that a particular kind of "organic matter" was both eternally existing and self-organizing. The disproof of Needham's experimental claims for spontaneous generation and epigenesis was the work of Lazzaro Spallanzani, a preformationist like most eighteenth-century biological experimentalists.

Epigenesism received a theoretical boost in 1745, with the appearance of Pierre-Louis Moreau de Maupertuis's anonymous *Venus Physique*. This was presented as the application of a Newtonian view based on attractive forces to the problem of generation. Those particles that would form specific areas of the living being were attracted to each other by forces like gravity in physics or those short-range attractive forces that were part of chemical theory. This theory was further developed in Maupertuis's *System of Nature* (1751), in which he argued that particles arranged themselves properly to form living beings because they were endowed with "memory." The foremost champion of epigenesis in the late eighteenth century was the German Caspar Friedrich Wolff (1738–1794), whose *Theory of Generation* (1759) criticized preformationism as both philosophically incoherent and experimentally dubious based on a series of experiments and observations on the classic experimental subject of embryology, the fetal chicken in the egg. Although Wolff drew on experiment and observation, his approach to science was fundamentally rationalistic and deductive, in the tradition of Gottfried Wilhelm Leibniz (1646–1716) and Wolff's teacher at the University of Halle, Christian Wolff. Caspar Friedrich Wolff also believed that mechanical forces alone were inadequate to explain the workings of living bodies, and added another category of material forces, "vegetative forces."

Wolff's assertions led to a protracted debate with Albrecht von Haller. Haller's own development on embryology shows the range of positions available. He was originally a spermaticist preformationist in the tradition of his teacher Hermann Boerhaave, converted to epigenesism following the polyp discoveries, and ended as an ovist preformationist, influenced by piety, the apparent impossibility of coming up with a mechanical model for epigenesis, and his own experiments on fetal chickens. The decades-spanning debate between Haller and Wolff led to no conclusion, but the issues it raised led Wolff to study monsters, which were difficult to explain by any preformationist theory. The birth of a monstrous or highly deformed animal or human suggested that its preformation was somehow deficient, which would question both the reliability of the process and the perfection of God's handiwork. Wolff's chicken experiments also led to the discovery of the "Wolffian body," an organ that developed and disappeared in the course of fetal development, another phenomenon hard to explain on a preformationist basis.

In the course of the eighteenth century, both preformationists and epigenesists altered their positions to take into account new evidence. Preformationists abandoned the idea of fully formed miniature individuals, and a "Chinese box" picture of nested generations, and asserted merely that the form of the living creature in some way preexisted its conception. Epigenesists abandoned the concept of self-organizing matter. By the end of the century, epigenesism was strongest in the Germanic world, where Wolff's rationalism and opposition to mechanical reproduction were widely shared. Wolff's doctrines were taken up by Johann Friedrich Blumenbach and the subsequent school of *Naturphilosophie.* Drawing on Wolff's suggestion that the embryos of different species are more alike than are the adult forms, *Naturphilosophs* like Carl Friedrich Kielmeyer (1765–1844) tried to relate embryonic development to the hierarchy of the "great chain of being," suggesting that each individual organism develops through vegetable and animal stages until it reaches its species position on the great chain. The relation of embryological development of the individual to evolutionary development of the species was put forth by another *Naturphilosoph,* Johann Friedrich Meckel (1781–1833), in 1811. Romantic interest in monsters was also congenial to epigenesists, and, along with the Frenchman Étienne Geoffroy Saint-Hilaire, many Germans began to experiment with the production and analysis of monstrous creatures.

The Germans, aided by superior microscopes, were also doing the most searching experiments, observations, and dissections on fetal development. They found no trace of a preformed organism, but described how differentiated organs emerge from undifferentiated tissues at the earliest stages following conception. The culmination of this work was the theory of fetal development set forth by Karl Ernst von Baer (1792–1876) who in 1827 distinguished the egg from the graafian follicle and gave the fullest description of the development of different organs and parts of the fetal organism from the earliest beginnings.

See also Haller, Albrecht von; Maupertuis, Pierre-Louis Moreau de; *Naturphilosophie;* Polyps; Spallanzani, Lazzaro.

References

Mason, Stephen F. *A History of Sciences.* Rev. ed. New York: Collier, 1962.

Pinto-Correia, Clara. *The Ovary of Eve: Egg and Sperm and Preformation.* Chicago: University of Chicago Press, 1997.

Roe, Shirley. *Matter, Life, and Generation: Eighteenth-Century Embryology and the Haller-Wolff Debate.* Cambridge: Cambridge University Press, 1981.

Encyclopedias

The eighteenth century saw the creation of the modern encyclopedia, a development intertwined with the expansion of science. The modern encyclopedia drew from the different traditions of the early modern encyclopedia, an ambitious attempt to summarize all knowledge, and the dictionary of arts and sciences, a more modest genre primarily focused on technology and science rather than history, biography, and religion and arranging its contents alphabetically rather than by a grand scheme of organization of the disciplines. Given the rate at which science was developing, dictionaries of arts and sciences were less attempts to create a permanent summary of knowledge than to capture the state of fields at a given time. All successful science-based encyclopedias and dictionaries went through multiple editions, expansions, and supplemental volumes.

The first noteworthy dictionary of the arts and sciences was the French scholar Antoine Furetière's (1619–1688) posthumous three-volume *Universal Dictionary of the Arts and Sciences* (1690). In the early eighteenth century, the most scientifically sophisticated dictionaries were produced by the English. The first was the single-volume *Lexicon Technicum* (1704) of John Harris (c. 1666–1719). Harris, an Anglican clergyman, mathematician, fellow of the Royal Society, and Boyle Lecturer, was a perfect representative of England's Newtonian establishment. His greatest coup was securing a heretofore unpublished paper by Newton on acidity for publication in a supplemental volume to *Lexicon Technicum* in 1710. *Lexicon Technicum* concentrated on the classical disciplines of applied mathematics such as navigation and architecture rather than manufacturing and industry. Like subsequent books in the arts and sciences tradition, it had little coverage of natural history.

Ephraim Chambers (c. 1680–1740), creator of the two-volume folio *Cyclopaedia* (1728), had been an apprentice to John Senex (d. 1740), one of London's leading sellers of scientific books, instruments, and maps. Chambers put his knowledge of both science and the book trade to use, building on Harris's achievement and eventually supplanting his work. The scientific portion of *Cyclopaedia* drew on many leading English scientists from the late seventeenth century as well as the publications of the Royal Society, of which Chambers became a fellow, and the Royal Academy of Sciences. As Newtonian as Harris, Chambers was more combative, attacking the rival philosophies of Cartesians, Leibnizians, and Aristotelians. The *Cyclopaedia* was a great success, going through several expansions and translations to the end of the eighteenth century.

These English works appeared in only a few volumes. Continental Europe in the mid-eighteenth century was the scene of huge projects to contain all knowledge in multivolume encyclopedias drawing on the Renaissance encyclopedic tradition. The German *Universal Lexicon* of Johann Heinrich Zedler (1706–1751) and Carl Gunther Ludovici (1707–1778) was published in sixty-four volumes between 1732 and 1750. A work on this scale required contributions from different writers, whose work was not always coordinated. The principal interest of Zedler's encyclopedia was historical rather than scientific, and in different places it expounded Cartesian, Leibnizian, and Newtonian sciences. Of comparatively modest dimensions was the ten-volume Italian *New Scientific and Curious Dictionary, Sacred and Profane* (1746–1751) of the Bologna Academy member Gianfrancesco Pivati (1689–1764). Although familiar with the most recent developments in science, Pivati was handicapped by the Inquisition. He praised Galileo, but was forced to describe the Sun as circling Earth.

The greatest and most influential midcentury encyclopedia was the French *Encyclopédie* (1751–1780) of Denis Diderot and Jean Le Rond d'Alembert. The first volume was published in 1751 and the last volume of the main A-Z sequence of seventeen volumes in 1765. Volumes of plates, supplemental articles, and indexes appeared until 1780, and the whole work comprised thirty-five volumes. Originating in a project to translate Chambers's *Cyclopaedia* into French, the *Encyclopédie* drew on both the encyclopedic and the arts and sciences traditions. It was large in the encyclopedic tradition. The completed *Encyclopédie* contained thirty-five volumes of text, plates, supplements, and index, and nearly 72,000 entries. It drew on the resources of more than 100 contributors, in addition to plagiarizing articles from Chambers. (Plagiarism was common among eighteenth-century encyclopedists.) The most important contributors in the sciences included Louis-Jean-Marie Daubenton (1716–1800), the main contributor in natural history; Nicolas Desmarest (1725–1815) in technology and science; and Paul-Joseph Barthez (1734–1806), one of many vitalists of the Montpellier school to contribute, in anatomy. D'Alembert contributed a celebrated prefatory discourse, expounding an arrangement of the arts and sciences based on a threefold distinction among sciences of reason, memory, and imagination. D'Alembert drew on the English philosopher Francis Bacon (1561–1626) for this classification scheme. The *Encyclopédie* faced many difficulties, less for its science than for the religious and political radicalism of many of the articles. Once published, it was widely translated and abridged.

The *Encyclopédie* (which popularized the term *encyclopedia*) had followed traditional alphabetical arrangement. Its publisher, Charles Pannekoucke, capitalized on its success by conceiving a more systematically ordered and expanded version, the *Encyclopédie Methodique.* This giant drew on the talents of some of

France's finest scientists, including Jean-Baptiste-Pierre-Antoine de Monet de Lamarck, and appeared in 196 volumes from 1782 to 1832—and even then was incomplete. Another continental giant that included a great deal of science was the German project begun by Johann Georg Krünitz, *Economic-Technical Encyclopedia,* appearing in 242 volumes between 1773 and 1858. Early-nineteenth-century France saw the rise of the gigantic specialized medical encyclopedia, the first example being the sixty-volume *Dictionary of Medical Sciences* (1812–1822). (All of these were dwarfed by a Chinese encyclopedia of 1726 that included 5,044 volumes. However, Chinese encyclopedias included reprints of entire books.)

The first edition of the *Encyclopedia Britannica,* the first encyclopedia to be named for a nation, appeared in 100 weekly installments between 1768 and 1771, and was then collected in three volumes. It was the work of William Smellie (c. 1740–1795), an Edinburgh amateur scientist and like Chambers a Freemason. (Freemasons were involved in several encyclopedic projects.) Smellie made surprisingly little use of either the *Encyclopédie* or the wealth of scientific and literary talent in Edinburgh. The first edition combined long "treatises" over twenty-five pages with shorter entries. It was poorly executed, containing many printing errors and making little use of cross-referencing. The third edition, appearing in 1797 in eighteen volumes, was a departure in soliciting articles from expert contributors. A six-volume supplement to the *Britannica,* appearing from 1815 to 1824, featured a programmatic statement by the Scottish "commonsense" philosopher Dugald Stewart (1753–1828) on the necessity of expert writers, a far cry from the works of Harris, Chambers, and Smellie that were essentially one-man products. The supplement was also the first encyclopedia to prominently feature the names of its expert writers and advertise itself by referring to their authority. Writing encyclopedia articles had become one way for leading scientists to communicate their discoveries to the educated general public.

See also Alembert, Jean Le Rond d'; Diderot, Denis; Popularization.

References

Collison, Robert. *Encyclopedias: Their History throughout the Ages.* New York and London: Harper, 1966.

Kafker, Frank A., ed. *Notable Encyclopedias of the Late Eighteenth Century: Eleven Successours of the "Encyclopédie."* Studies in Voltaire and the Eighteenth Century, no. 315. Oxford: Voltaire Foundation, 1994.

———. *Notable Encyclopedias of the Seventeenth and Eighteenth Centuries: Nine Predecessors of the "Encyclopédie."* Studies in Voltaire and the Eighteenth Century, no. 194. Oxford: Voltaire Foundation, 1981.

Kafker, Frank A., in collaboration with Serena L. Kafker. *The Encyclopedists As Individuals: A Biographical Dictionary of the Authors of the "Encyclopédie."* Studies in Voltaire and the Eighteenth Century, no. 257. Oxford: Voltaire Foundation, 1988.

Yeo, Richard. *Encyclopaedic Visions: Scientific Dictionaries and Enlightenment Culture.* Cambridge: Cambridge University Press, 2001.

Engineering

See Technology and Engineering.

The Enlightenment

The term *Enlightenment* has become standard in discussing the progressive and (mostly) secularly minded intellectuals, or philosophes, of eighteenth-century Europe and North America. However, Enlightened individuals and movements varied in their opinions of many things, including science, and the Enlightenment should not be treated as a set of dogmas. Science unquestionably played a leading role in Enlightenment thought and culture, although most writers preferred the older term *natural philosophy.* Many of the most important thinkers and writers of the Enlightenment came from scientific backgrounds, and the majority took an informed interest in scientific developments. A common career pattern was for a philosophe to establish a reputation in the sciences before turning to social or political questions. This path was followed by Jean Le Rond d'Alembert, Georg Christoph Lichtenberg, and Benjamin Franklin, among others. Even more important, the progress of science furnished a model for the Enlightenment itself. As many Enlightened thinkers saw it, science had made enormous advances in the great age of the seventeenth century by throwing off traditional authorities—Aristotle and the church—and boldly assuming the right to think and speak freely. It was the thinkers of the Enlightenment, and not the scientific revolution itself, who freely applied the term *revolution* to changes in natural philosophy. Even if some scientific ideas became outmoded, the philosophes believed that the empirical and mathematical methods of modern science could be applied to all spheres of thought, and would result in similar intellectual progress. Science was also the one area in which the moderns surpassed not just all previous Western Christian civilization, but also the much admired ancient pagans, and was thus indisputable proof of the truth of progress. Among the moderns, Isaac Newton (1642–1727), in particular, was virtually deified as the greatest man who ever lived. However, many philosophes attacked what Newton himself had seen as the central feature of his system: the alliance between the system of nature and the fundamentally religious truth of the providence of God.

The radical philosophes invoked science as an alternative system of validation to religion and tradition. Many Enlightenment thinkers contrasted an idealized version of a scientific community interested only in truth with a demonized picture of theologians and systematic abstract philosophers motivated by hate, anger, and jealousy. However violent the polemical struggle between Cartesians and Newtonians in the eighteenth century became, not a drop of blood was shed in it, providing a remarkable contrast to the struggle

Portrait of Jean-Jacques Rousseau, Enlightenment thinker. More ambivalent about science than other thinkers of his time, he even suggested that the progress of science had helped corrupt human character. (Library of Congress)

between Catholics and Protestants in the previous two centuries. Whatever the actual reality of scientific practice, the idealized version of science also offered a model of internationalism and interconfessionalism, with cooperation among scientists in different communities. Science was also more socially egalitarian than other aspects of eighteenth-century society, and the privilege of nobility meant little in evaluating a scientist's ideas. (This egalitarianism, like other moves toward greater equality in the eighteenth century, did not extend to gender.)

The spread of true science was necessary for social and intellectual reform, and thus was not a matter for scientists alone. True scientific knowledge had to spread not to the entire population—the philosophes mostly accepted the inevitability of a society divided into classes, with the rural peasantry at the bottom—but to men and women of the literate classes. For this reason Voltaire, not particularly gifted mathematically, took it upon himself to spread the true Newtonian system of the world in France, against the dominant Cartesian school. Voltaire and other philosophes were also capable at times of regretting that the passion for science was driving out that for literature and art, but this was seen at worst as a regrettable necessity, and certainly not a reason to reject science.

Possibly due to the lack of religious freedom and the persistence of only intermittently effective censorship in France, science was a much more radicalizing ideology there than in Britain or Germany. Early Enlightenment French propagandists like Voltaire often linked the discoveries of Newton and other eminent English scientists with the greater degree of personal freedom enjoyed by Englishmen in comparison with the French. The introduction of Newtonian science was part of a greater project of liberalizing French society. The French philosophes struggled long and hard to displace the churchmen and more conservative scholars from the dominating position in French intellectual life, and in the process their criticism of the domination of the Roman Catholic Church and the aristocracy grew more extreme. Such was the radicalization of French Enlightenment science that some of it developed in the direction of materialism. Enlightenment interest in treating human beings as material rather than spiritual creatures was clearly incompatible with traditional religion, and by the late eighteenth century some French philosophes, such as Claude-Adrien Helvétius (1715–1771), were openly calling themselves atheists. This tendency was rejected by other philosophes, and Denis Diderot complained that an overemphasis on the vastness of space should not challenge the centrality of humanity and morality.

A more radical critique of the cultural role of science was made by a somewhat unusual philosophe, Jean-Jacques Rousseau (1712–1778). Rousseau, who prefigured the later romantic reaction against the Enlightenment, stated in his *Discourse Concerning the Arts and Sciences* (1750) that the progress of the sciences had not made people happier, but had made them more corrupt and pushed them further from nature. There was a certain deliberate perversity to this argument, as it was originally made in a submission to a prize competition held by the Academy of Dijon as to whether the arts and sciences had made people happier, and Rousseau was advised that there would be less competition on the negative side of the question. And even Rousseau did not see a return to the primitive state and an abandonment of scientific knowledge and the scientific endeavor as a practical possibility.

The philosophers of the Enlightenment hoped to make philosophy scientific, thus avoiding what they saw as the sterility of Scholastic metaphysics. *Treatise of Human Nature* (1739–1740) by David Hume (1711–1776) was advertised as "An attempt to introduce the experimental method of reasoning into moral subjects." Immanuel Kant's philosophy began as an attempt to provide a solid metaphysical grounding for the natural sciences. Enlightenment social science, one of the major eighteenth-century achievements, began in the claim to apply scientific criteria to thought about society. The hope of building a science of man, both man as an individual and man living in society, was characteristic of the Enlightenment. The "human sciences" could be founded on a scientific and empirical basis, with the study of a variety of human societies in both space and time. Even the established discipline of history could be improved by what were believed to be the methods of the natural sciences. This Enlightenment social science was not morally neutral. The philosophes hoped to found moral philosophy, and thus morality, on a scientific rather than a religious basis.

See also Alembert, Jean Le Rond d'; Condillac, Étienne Bonnot de; Condorcet, Marie-Jean-Antoine-Nicolas Caritat, Marquis de; Franklin, Benjamin; Freemasonry; Kant, Immanuel; La Mettrie, Julien Offroy de; Lichtenberg, Georg Christoph; Maupertuis, Pierre-Louis Moreau de; Newtonianism; Popularization; Religion; Romanticism.

References

Gay, Peter. *The Enlightenment: An Interpretation, Volume II: The Science of Freedom.* New York: Alfred A. Knopf, 1969.

Hankins, Thomas L. *Science and the Enlightenment.* Cambridge: Cambridge University Press, 1985.

Euler, Leonhard (1707–1783)

The Swiss Leonhard Euler outshone all contemporaries in pure and applied mathematics and was the most productive mathematician of all time, contributing to a huge range of fields from number theory to acoustics. The son of a Protestant minister and friend of the Bernoulli family, Euler was originally intended to follow his father into the ministry. His interest and talent for mathematics showed itself early, and he received special instruction from Johann Bernoulli. In 1726 the death of Nikolaus II Bernoulli opened up a slot at the recently founded Imperial Academy of Sciences of St. Petersburg. After losing a competition for a chair of physics at the University of Basel, Euler arrived in St. Petersburg in 1727. The relationship with the academy would last the rest of his life. The same year he inaugurated another long-term relationship by submitting a paper on the mathematically ideal way to arrange masts on a ship for a competition sponsored by the Paris Academy of Sciences. It came in second, but Euler would submit many more papers and win twelve prizes in Paris competitions over the decades.

In 1733 Euler succeeded Daniel Bernoulli as first chair of mathematics at St. Petersburg. He married Katarina Gsell (1707–1783), another Swiss St. Petersburger, the next year. Euler was expected to combine his mathematical research with work for the Russian state in practical areas such as cartography and calendar making, and even creating elementary mathematics textbooks for Russian schools. His cartographic work may have contributed to his growing vision problems—he lost the sight in one eye by 1740. Euler was the most prolific mathematician in history, and a seemingly endless stream of his mathematical papers was published in the *Commentaries* of the St. Petersburg academy. His book *Mechanica* (1736–1737) provided the most mathematically sophisticated treatment of Newtonian and post-Newtonian mechanics to date, putting mechanics in analytic rather than geometric terms. Euler mathematically formulated the concepts of linear momentum and the moment of momentum. He also set forth, in *New Theory of Light and Color* (1746), a theory of light as a vibration, as opposed to the dominant Newtonian theory of light as a particle.

Euler was basically responsible for the subsequent development of mathematics and mathematical physics. This can be seen in the remarkable degree to which Euler's notation became standard. He standardized the use of *e* for the natural base of the logarithms, which appeared for the first time in print in *Mechanica,* after Euler had been using it for several years; *i* for the square root of negative 1 is another Euler innovation. The Greek letter symbol for *pi,* although not originating with Euler, became standard by his use. Euler's *Introduction to the Analysis of Infinites* (1748) established analysis as the central mathematical discipline for the rest of the eighteenth century and also defined the concept of function, for which Euler devised the current expression *f(x).* Infinite processes and infinite series were Euler's bread and butter as a mathematician, and he handled them with breathtaking ease, solving such classic problems as the sums of the reciprocals of the perfect squares (1/1 + 1/4 + 1/9 . . . = *pi* squared/6). In applied mathematics Euler was a leader in celestial mechanics, and his lunar theory, put forth first in *Theory of the Moon's Motions Showing All Its Inequalities* (1753), was the most advanced to date. Euler put forth in this work a method for dealing with the mutual gravitational influences of three large bodies, in this case the Sun, the Moon, and Earth. This was the so-called three-body problem. He returned to lunar problems in 1772 in work—inspired by the speeding up of the Moon's motion over the centuries—so advanced, it had surprisingly little impact.

In 1741 the insecure position of the Imperial Academy led Euler to accept an offer from Frederick the Great of Prussia (r. 1740–1786) to help in the reform of the Berlin Academy. Euler's move to Berlin, where he spent the next twenty-five years, did not result in the severing of his ties to St. Petersburg—he continued to publish in the *Commentaries* of the St. Petersburg academy and was a bridge between the two academies. As in St. Petersburg, Euler was kept busy with practical problems, such as engineering the pipes for the king's new palace of Sans Souci (fluid dynamics was yet another area of Euler's mathematical expertise) and applying his knowledge of probability to the state lottery. (As was typical of Euler, this led to several publications in the theory of lotteries.) Euler got along with the academy's president, Pierre-Louis Moreau de Maupertuis, but his relations with the king himself were rocky. The pious and domestic Euler, who claimed that mathematical discoveries came to him while holding one of his children while others played about his feet, and the worldly, homosexual, atheist Frederick II were a bad fit personally. The situation was not aided by Frederick II's reliance on Euler's mathematical rival Jean Le Rond d'Alembert for advice on the running of the academy after Maupertuis's death in 1759. Although Euler had his doubts about the sufficiency of Newtonian physical theory (his work of scientific popularization, *Letters to a German Princess,* takes a surprisingly Cartesian approach), his basic Newtonianism also led to friction with the academy's Leibnizian faction.

In 1766 Euler returned to St. Petersburg at the invitation of the czarina Catherine the Great (r. 1762–1796), irritating Frederick II greatly. Five years later, he lost the sight of his remaining eye. Although Euler was now totally blind and well past the age of the greatest creativity in most mathematicians, his productivity did not diminish. He did require assistance, which he received from his sons, Johann Albrecht Euler (1734–1800), who followed his father into the St. Petersburg academy, and Christoph Euler. Anders Johann Lexell (1740–1784), who succeeded Euler in the mathematics chair of the academy, and Nicolaus Fuss (1755–1826), Euler's grandson-in-law and another Swiss mathematician who had made his way to St. Petersburg, also assisted him. Euler remained productive until his peaceful death following a brain hemorrhage in 1783.

See also Alembert, Jean Le Rond d'; Berlin Academy; Bernoulli Family; Imperial Academy of Sciences of St. Petersburg; Mathematics; Mechanics.

References

Boyer, Carl B. *A History of Mathematics.* 2d ed. Revised by Uta C. Merzbach. New York: John Wiley and Sons, 1991.

North, John. *The Norton History of Astronomy and Cosmology.* New York: W. W. Norton, 1995.

Exploration, Discovery, and Colonization

The eighteenth century was a great age of exploration, as it became increasingly intertwined with science. Expeditions varied in their relationship to science. A few were undertaken for specifically scientific purposes. Examples of these would be the expeditions to observe the transits of Venus or Mercury, or those to measure Earth or determine its shape. Another category was those expeditions whose purpose was primarily political or economic, but had a major scientific component. Examples of these would be the expeditions of Captain James Cook in the Pacific, or the Lewis and Clark expedition in America. The final category would be expeditions and voyages with no explicitly scientific function, but on which some scientific activity took place. Another division is between oceangoing voyages and expeditions over land, long overland expeditions being a specialty of the nations with extensive, unmapped continents, Russia and the new United States.

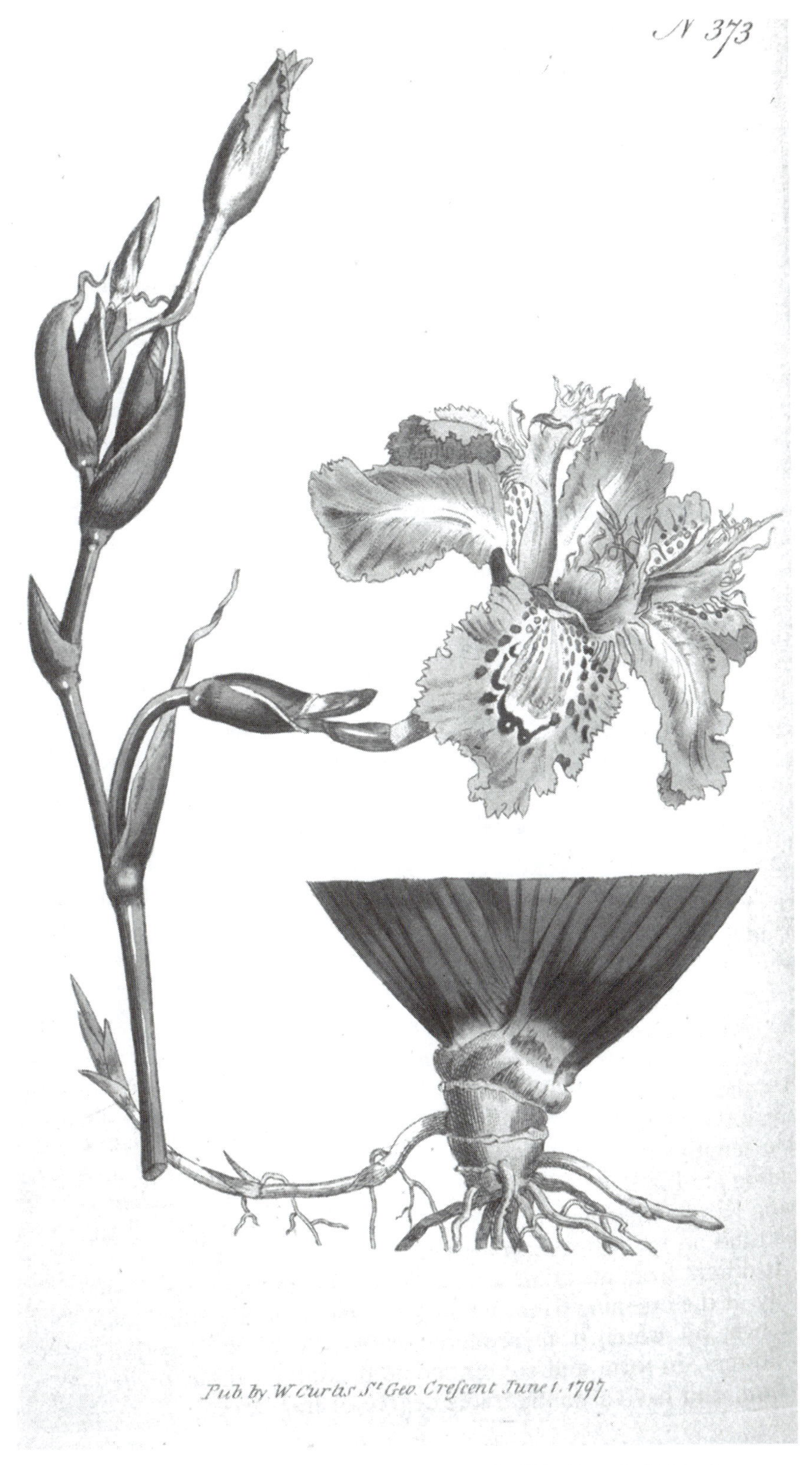

The Chinese iris, illustrated here in the popular late-eighteenth-century periodical The Botanical Magazine, *was introduced to English gardens from China via the East India Company. (Special Collections, National Agricultural Library)*

Exploration led to increases in geographical knowledge, and by the end of the century, the ability to find the longitude made it easier to fix known lands on maps, preventing the same territory from being "discovered" on multiple occasions. The Pacific was the most active site of exploration, particularly as expeditions were driven by imperial competition among Britain, France, the Netherlands, Spain, and Russia. The first half of the eighteenth century saw less activity than the second half. Some significant early-eighteenth-century voyages were the 1721–1722 voyage of Dutch captain Jacob Roggeveen (1659–1729), in which he discovered Easter Island in 1722 (the last Dutch voyage of importance), and the voyages in the Far North in the Russian service in 1728 and 1733–1741 by Danish captain Vitus Johassen Bering (1681–1741).

Activity picked up after 1763, with the end of the Seven Years' War, improvements in navigation, and the publication of the popular *History of the Navigations to the Southern Land* (1756) by French magistrate Charles des Brosses (1709–1777). The most active nations in Pacific exploration were France and Great Britain. The differences between the two are typical of their different scientific styles. French exploratory science was bureaucratized and controlled by the government and the scientific establishment. British exploratory science emerged later and allowed considerably greater scope for private initiative. The British effort also drew on the intellectual resources of Scandinavia and central Europe, the origin of many of the naturalists who sailed on British ships. Much activity was inspired by stories of a great continent in the south that voyagers planned to contact either to colonize or to trade with the inhabitants. The British followed the end of the war with two important circumnavigations, one in 1764–1766 by John Byron (1723–1786) and one in 1766–1768 by Samuel Wallis (1728–1795), his being the first European expedition to land on Tahiti. The Frenchman Louis-Antoine de Bougainville (1729–1811) circumnavigated from 1766 to 1769, arriving on Tahiti eight months after Wallis's departure. Bougainville was a mathematician, and he included an astronomer and natural historian in his crew.

The most important Pacific voyages scientifically and geographically were those of Cook in 1768–1771, 1772–1775, and 1776, which established that there was no great southern continent. The British, who expelled their French rivals from the high seas during the French Revolution and Napoleonic Wars, also sent George Vancouver (1757–1798) to map South Australia, New Zealand, and the northwestern coast of North America from 1791 to 1795. Another British voyage, that of the *Investigator,* circumnavigated Australia and examined its natural history from 1801 to 1805. Opportunities for individual, rather than government-controlled, exploration were shrinking, but there were individual European explorers in the interior of Africa: James Bruce (1730–1794) who followed the course of the Nile, and Mungo Park (1771–1806) who in 1795–1796 followed the course of the Niger. Spain had lagged behind Britain and France, but by the late eighteenth century was attempting both to catch up scientifically and assert its power over its North American colonies with the Royal Botanical Mission and the mission under Alejandro Malaspina (1754–1809).

Accounts of voyages, as they became available in Europe, were collated to form a more complete and exact knowledge of the seas and coasts. The leader in this effort in late-eighteenth-century Britain was Alexander Dalrymple (1737–1808), who on his return to England from the Indian Ocean was responsible for the publication of dozens of travel records and was appointed hydrographer to the East India Company in 1779 and to the Admiralty when the post was founded in 1795. Jean-Baptiste Denis d'Après de Mannevillette (1707–1780) fulfilled a similar function for France, which had founded the Repository of Maps and Plans as part of the Ministry of the Navy (which in France also had responsibility for colonies) in 1720. The

Cook voyages were the beginnings of a tighter, although not totally harmonious, relationship between British science and the Admiralty. On the scientific side, the leader was Sir Joseph Banks who worked with a vast number—well over 100—of collectors and observers, particularly oriented to botany, although they also worked on other disciplines such as oceanography and hydrography. A particular feature of British exploration science in the Banks era was the prominent role played by ships' surgeons, many of Scottish origin. The greatest of these was the *Investigator*'s Robert Brown.

Natural historians often found the flora and fauna of newly explored lands far more intellectually exciting than the well-known animals and plants of Europe, but science on explorations was for utilitarian ends. Even the transit observations were anticipated to provide navigation aids. One concern of European powers was the most efficient and productive exploitation of their colonial empires. Botanists and natural historians were expected to identify resources for exploitation and to evaluate newly acquired territories for their suitability for agricultural use. (The most famous example of this process is the *Bounty* voyage of Captain William Bligh [1754–1817] carrying breadfruit plants from Tahiti to the Caribbean, where it was hoped they would provide cheap food for slaves.) The establishment of a colony was often followed by the establishment of a botanical garden in which different crops could be experimented with.

Some plant gathering closely resembled industrial espionage, as collectors sought tea from China or as a Polish botanist in Banks's service, Anton Hove, sought the seeds of fine cotton from the Indian district of Gujarat in 1788. India, with its political fragmentation and strong British presence, was much more open to this kind of exploitation than China, which managed to keep Westerners confined to port cities in the South, or Japan, which was entirely closed. The few embassies that the Chinese government, headquartered in the northern city of Beijing, deigned to receive were full of natural historians eager to explore the Chinese north. Although the usual plan was to acclimate plants in areas resembling their native habitats, this was not always true. Carolus Linnaeus hoped to render Sweden self-sufficient in tea by acclimating it there, and the crops raised in the British colony in Australia came from many locations.

The desire to make colonies, particularly isolated islands, economically productive raised many scientific questions. The introduction of new species, often in a plantation economy, could have devastating effects on the ecosystem, and islands posed problems of limited resources in their most acute form. Some of the earliest attempts to "scientifically" manage the environment occurred in island colonies, where rulers did not have to deal with the dense thicket of customs and property rights that limited state intervention in Europe itself. Pierre Poivre (1719–1786), a French naturalist and environmentalist, charged that the deforestation of the French possession of Mauritius had led to the island's desiccation. While intendant of Mauritius from 1767 to 1772, Poivre attempted to alleviate this problem with forest-protection policies.

See also Banks, Sir Joseph; Cook, James; Egyptian Expedition; Humboldt, Alexander von; Lewis and Clark Expedition; Longitude Problem; Oceanography; Race; War.

References

Beaglehole, J. C. *The Exploration of the Pacific.* London: Adam and Charles Black, 1947.

Fry, Howard Tyrrell. *Alexander Dalrymple (1737–1808) and the Expansion of British Trade.* London: Cass for the Royal Commonwealth Society, 1970.

Grove, Richard H. *Green Imperialism: Colonial Expansion, Tropical Island Edens, and the Origins of Environmentalism, 1600–1860.* Cambridge: Cambridge University Press, 1995.

MacLeod, Roy, ed. *Nature and Empire: Science and the Colonial Enterprise. Osiris,* 2d ser., 15 (2000).

Miller, David Philip, and Peter Hanns Reill, eds. *Visions of Empire: Voyages, Botany, and Representations of Nature.* Cambridge: Cambridge University Press, 1996.

F

Farming

See Agriculture.

Fossils

The Enlightenment inherited from the seventeenth-century scientific revolution the unresolved question of fossils. Some, including Robert Hooke (1635–1703) and Nicholas Steno (1638–1686), had argued that they were organic remnants, turned to stone by the passage of time. Others, such as the Jesuit Athanasius Kircher (1601–1680), argued that the Earth itself formed these stones by means of an innate power, or "plastic virtue." These were extreme positions, while others claimed that some fossils were organic remnants, and some other fossils formed by the Earth. (The problem was particularly complicated due to the wide range of meanings of "fossil," which could extend to any unusual object found in the Earth.) Champions of the "plastic" theory of fossils pointed out that many seemed to resemble animals that did not actually exist, and that marine fossils, such as fossilized seashells, were often found at the tops of mountains. One solution to the second problem was to invoke the biblical deluge, but the first seemed to require admitting the very controversial idea of extinction.

Growing knowledge of fossils in the early eighteenth century, combined with powerful Flood-based explanations, led to the eventual triumph of the view that fossils were organic remnants. The last major dispute occurred in 1726, when the German professor Johann Beringer (d. 1740) published a book describing fossils in the area of Würzburg, *Würzburg Lithography*. Beringer's fossils included many unusual specimens, including Hebrew letters and astronomical objects, and he submitted that they were proof that a "plastic virtue" formed fossils. As it turned out, Beringer had been maliciously deceived by two colleagues, who had faked the fossils and then distributed them.

As in other areas of natural history, the expansion of European exploration and domination added to the wealth of fossil material to be interpreted. The discovery of the first fossil mastodon among many other fossils at Big Bone Lick on the banks of the Ohio River in 1739 was a puzzler, as the beast combined elephantine characteristics, such as tusks, with nonelephantine teeth. (Some believed that the remains of two different animals had been somehow jumbled together.) Identifying fossils was also difficult because very few entire bodies of large animals had been preserved, and it was often a case of trying to reconstruct an animal from a small part. This led to spectacular mistakes, as in the proud identification by Thomas Jefferson (1743–1826) of a fossil claw as belonging to an American lion larger in size than any lion of the Old World. This would have vindicated the American fauna from Georges-Louis Leclerc de Buffon's charge of New World fauna being smaller than Old World fauna, but the claw turned out to be that of a sloth, a somewhat less inspiring creature. The growing number of fossils that seemed to represent no living creature helped to legitimate the idea of extinction, and Buffon included catastrophes that destroyed species, leaving only their fossil remnants, in his influential theory of the Earth, *The Epochs of Nature*.

The first to fully integrate a theory of extinction into life history based on fossils, though, was Georges Cuvier. From the very late eighteenth century and for the next two decades, Cuvier made a series of spectacular anatomical reconstructions of fossil animals, arguing that many species of large animals had become extinct. (It was Cuvier who correctly identified Jefferson's claw as an extinct sloth's, tactfully naming the beast *Megalonyx Jeffersoni*.) Jean-Baptiste-Pierre-Antoine de Monet de Lamarck, Cuvier's colleague, agreed that the beasts were no longer extant, but argued that instead of becoming extinct they had evolved into other creatures. Cuvier expanded his study of fossils from anatomical reconstruction to using them to identify geographic strata, in a study of the region around Paris undertaken with Alexandre Broignart (1770–1847) and published as *Essay on the Mineral Geography of the Paris Region* (1808). His collected papers and other writings on fossils were published as *Researches on the Fossil Bones of Quadrupeds* (1812). The nineteenth-century discipline of paleontology was founded on Cuvier's work, even though it came to accept, as he had denied, evolution.

See also Cuvier, Georges; Geology.

References

Beringer, Johann Bartholomaus Adam. *The Lying Stones of Dr. Johann Bartholomew Adam Beringer, Being His "Lithographia Wirceburgensis."* Translated and annotated by Melvin E. Jahn and Daniel J. Woolf. Berkeley: University of California Press, 1963.

Rudwick, Martin J. S. *The Meaning of Fossils: Episodes in the History of Palaeontology*. 2d ed. New York: Science History Publications, 1976.

Franklin, Benjamin (1706–1790)

Benjamin Franklin was the first native of the European colonies in the Americas to win European recognition as a natural philosopher. Franklin's interest in science was long-standing, and his retirement from his Philadelphia printing business in 1748 may in part have been motivated by the desire to have more time for electrical experiments. Franklin's introduction to electrical studies is a classic example of eighteenth-century scientific internationalism. The English periodical the *Gentleman's Magazine* had printed a translation of a piece in a French-language journal based in the Dutch Republic, *Bibliotheque Raisonée*, by the Swiss Albrecht von Haller, reporting the work of three German electricians, Georg Matthias Bose (1710–1761), Christian August Hausen (1693–1743), and Johann Heinrich Winkler (1703–1770). Franklin and a group of his friends began to make electrical experiments shortly afterward.

Electrical studies became an all-consuming passion with Franklin, particularly after the Philadelphia group began to make use of the recently discovered Leiden jar. The group devised a number of ingenious experiments, some entertaining, and some directed at practical uses. Franklin was interested in the possibility of using electricity to slaughter chickens and turkeys, claiming that not only was this more humane but also resulted in a more tender bird. More important were Franklin's theoretical innovations. Previous electrical theories had been based on "effluvia"—tiny electrical particles. Although he did not dispense totally with effluvia, Franklin identified electricity as a universal fluid and distinguished between electrified states as "positive" (saturated with the electrical fluid) and "negative" (deficient in it). This theory seemed to explain the foremost puzzle in electricity, the behavior of the Leiden jar, better than anything previously. It was communicated in 1750 in a letter to Peter Collinson (1694–1768), a London Quaker who corresponded with several colonial scientists. Collinson had been receiving letters on electrical experiments from Franklin since 1748 and had been communicating them to the Royal Society, where they attracted only minor interest. They attracted more when they were published as a short book, *Experiments and Observations on Electricity, Made at Philadelphia in America, by Mr. Benjamin Franklin, and Communicated to Mr. P. Collinson of London, F.R.S.* (1751). The work went through several other editions, with additional material, and was translated into French, German, and Italian.

The other electrical innovation associated with Franklin is the demonstration that lightning is electrical. The similarity between, or even the identity of, lightning and electricity was one of the biggest clichés in electrical literature, but Franklin was the one who actually devised a way of testing it. This was the famous "sentry-box" experiment, where a man standing in an enclosed box in a high place during a storm would be electrified by a long, pointed metal rod extending into the clouds. Franklin claimed that the man would undergo little risk, but the experiment he himself used to demonstrate the identity of lightning and electricity was the famous key experiment, carried out in 1752. A kite was flown on a day of storm clouds. Electricity was drawn down the kite string to a metal key, which gave off an electric spark when touched. It is not clear whether it was carried out by Franklin himself or by another of the Philadelphia group, but Franklin was undoubtedly its originator. The idea of drawing electricity from the clouds by means of metal rods—lightning rods—was originated by Franklin, and he was the foremost champion both of rods in general and of pointed as opposed to rounded rods. Franklin's work on lightning won him honorary degrees from

Portrait of Benjamin Franklin with scientific equipment (Perry-Castaneda Library)

Yale and Harvard in 1753 and the Copley Medal of the Royal Society the same year. In 1756 he was admitted as a fellow of the Royal Society. As America's greatest living scientist, he was elected president of the newly founded American Philosophical Society in 1769, and did much to promote its publications in Europe. He remained president of the society until his death.

In France Franklin's work was promoted by the Comte Georges-Louis Leclerc de Buffon, who saw in it a weapon against Jean-Antoine Nollet, whom Buffon disliked. Franklin's sentry-box experiment was actually carried out in France at the king's palace of Marly on 10 May 1752. The Marly experiment brought Franklin's work to the attention of every important electrical researcher in continental Europe, although in France itself Nollet continued to be an obstacle to the success of Franklin's system. Franklin's electrical theory was further refined and developed in Europe by Giambatista Beccaria

(1716–1781) and Franz Maria Ulrich Theodor Hoch Aepinus (1724–1802). Franklin himself maintained an interest in science, but increasingly his time was taken up by political projects, culminating in his role as ambassador of the rebellious colonies to France during the American Revolution.

In Paris Franklin's role as a scientist was a tremendous diplomatic asset to the beleaguered revolutionaries. Nollet was dead, and Franklin was universally recognized not only as a great electrical theorist but also as the benefactor of mankind who had invented the lightning rod. He actively involved himself in the affairs of the Academy of Sciences, faithfully attending its meetings and serving on its committees. (He also became head of the famous Parisian Masonic Lodge of the Nine Sisters, which included many French intellectual leaders such as Voltaire. Franklin had been a Mason since 1731.) He chaired the famous Royal Commission that investigated the claims of Franz Anton Mesmer (1734–1815) in 1784. Mesmer had approached Franklin with the hope of winning him over to the cause, but Franklin was skeptical, and some of the experiments that disproved Mesmer's "animal magnetism" were carried out at Franklin's estate at Passy outside Paris.

See also Colonial Science; Electricity; Freemasonry; Nationalism.

References

Clark, Ronald W. *Benjamin Franklin: A Biography.* London: Weidenfeld and Nicolson, 1983.

Heilbron, J. L. *Electricity in the 17th and 18th Centuries: A Study in Early Modern Physics.* 2d ed. Mineola, NY: Dover, 1999.

Freemasonry

The eighteenth century saw the rise of a secret society known as the Freemasons, a development intertwined with the progress of science. The leader of the Masonic movement in early-eighteenth-century London, the base from which it spread over the Continent and to the Americas, was the Newtonian experimentalist John Theophilus Desaguliers, and much of the early leadership of the Grand Lodge of London was composed of members of the Royal Society. Much of the mythology and rhetoric of early Masonry were congenial to Newtonian natural philosophy, as it emphasized a "great architect" who had created the universe in a mathematical fashion. Freemasons also spoke of society as bound together by the "attraction" of its individual members.

As Masonry spread and developed, it took many different forms and attitudes toward science. In Germany much of the Masonic movement went in the direction of occultism, alchemy, Rosicrucianism, and religious conservatism. Some German Masons renounced not only mechanistic physics but even Copernican astronomy as impious! Far different were the lodges founded as intellectual societies, notably the famous Lodge of the Nine Sisters (the classical Greek Muses) in Paris and the True Harmony Lodge of Vienna. The Nine Sisters was founded in 1776 by the French astronomer and atheist Joseph-Jérôme Le Français de Lalande (1732–1807), who served as its first master. Members included Benjamin Franklin, who also served as master; Lalande's fellow astronomer Jean-Sylvain Bailly (1736–1793); Voltaire, admitted somewhat irregularly only a few weeks before his death; the physician and materialist Pierre-Jean-Georges Cabanis (1757–1808); and the chemist Antoine-François de Fourcroy (1755–1809). The group sponsored educational institutions. It was abolished in 1792, as the French revolutionary government held Masonry to be "aristocratic."

The True Harmony Lodge, founded in 1781, did not have so illustrious a membership, but unlike the Nine Sisters it published a scientific journal, appearing annually and particularly strong in geology and mineralogy. The lodge, along with many other Masonic organizations in the Hapsburg Empire, was dissolved in 1786 in a Hapsburg government crackdown on secret societies. More radically Enlightened, although less intellectually distinguished, than the Nine Sisters or

True Harmony was a short-lived group founded in 1776 by the Ingolstadt professor Adam Weishaupt (1748–1830). The Bavarian Illuminati were inspired by the materialism of radical philosophes like the Baron Paul-Henri-Dietrich d'Holbach (1723–1789). The group was quickly suppressed.

See also Bradley, James; Crell, Lorenz Florens Friedrich von; Desaguliers, John Theophilus; Franklin, Benjamin; Voltaire.

References

Jacob, Margaret C. *Living the Enlightenment: Freemasonry and Politics in Eighteenth-Century Europe.* New York: Oxford University Press, 1991.

McIntosh, Christopher. *The Rose Cross and the Age of Reason: Eighteenth-Century Rosicrucianism in Central Europe and Its Relationship to the Enlightenment.* Leiden: Brill, 1992.

Weisberger, R. William. *Speculative Freemasonry and the Enlightenment: A Study of the Craft in London, Paris, Prague, and Vienna.* Boulder, CO: East European Monographs, 1993.

French Revolution

The great French Revolution, which began in 1789, had effects on French science both creative and destructive. In addition to causing the deaths of several leading scientists, most notably Antoine-Laurent Lavoisier, the Revolution ended the existence of most French scientific institutions, most prominently the Royal Academy of Sciences. However, it gave birth to several others, which would go on to maintain the French predominance in science into the Napoleonic era.

French scientists generally supported the Revolution in its early stages, when it seemed to offer hope for social and political reform along Enlightenment lines. No French scientist of any prominence joined the Royalist emigration, and the Revolution, even in its most violent phases, was a high-water mark in the involvement of scientists and mathematicians in politics. Several—including the Marquis de Condorcet, Marie-Jean-Antoine-Nicolas de Caritat; Lavoisier; Jean-Baptiste-Pierre-Antoine de Monet de Lamarck; and the mathematician Gaspard Monge (1746–1818)—were active in a moderate revolutionary group, the Society of 1789, founded in 1790. (The most scientifically and technologically advanced sections of the French officer corps, the artillerists, also showed a significantly higher degree of loyalty to the new regime than did the infantry or cavalry.) The most prominent scientists to be active politicians in the French Revolution included Condorcet, who committed suicide rather than be brought before the revolutionary tribunal, and the astronomer Jean-Sylvain Bailly (1736–1793), who reached the high positions of president of the National Assembly and mayor of Paris before being guillotined in the Terror. The most illustrious (and the most intellectually active) scientist to die was Lavoisier, who had not engaged himself as deeply in revolutionary politics as Bailly and Condorcet but was executed for his prerevolutionary role as a tax farmer. Not all politically active scientists had careers that ended so disastrously. The mathematician and military engineer Lazare-Nicolas-Marguerite Carnot (1753–1823) sat as one of the twelve-man body that ruled France during the Terror, the Committee of Public Safety. Carnot's administrative skills won him the title "organizer of victory." Monge and the chemists Antoine-François de Fourcroy (1755–1809) and Louis-Bernard Guyton de Morveau (1737–1816) also served at the higher levels of the government.

The Revolution in its later phases also struck at scientific institutions. Some damage was caused simply by France's desperate condition in the revolutionary era. The Royal Academy of Sciences had to abandon a project to build a large reflecting telescope to compete with that of the English astronomer William Herschel, and donate the money collected to the convention. This may have contributed to the French weakness in observational astronomy that persisted into the nineteenth century. Important periodicals, such as the *Journal des Savants* and Lavoisier's *Annales de Chémie,* were also forced to shut down for lack of resources.

The resentment the traditional scientific institutions attracted was less because they were scientific than because they were inextricably associated with the old regime and "aristocratic" culture and society. Sometimes this resentment was personal. Among the most radical of the revolutionaries was Jean-Paul Marat (1743–1793), a physician and minor experimental physicist, specializing in optics, who had been trying to storm the barriers of the Royal Academy of Sciences for years. Marat attacked the academy viciously, and many agreed that it, along with the other Royal Academies, needed reorganization or suppression in the new France. The Royal Academy was suppressed on 8 August 1793. The gap was filled in part by two nonstate scientific associations, the Society of Natural History (a revival of the old Linnean Society) and the Philomathematical Society. Eventually, a reorganized Academy of Science was part of the Institute of France, founded in 1795 under the more conservative rule of the Directory.

The major medical associations, such as the medical faculties of the Universities of Paris and Montpellier, the Royal Society of Medicine, and the Royal Academy of Surgeons, were also eliminated in the Revolution. Medically, the surgeons, with their more pragmatic approach, were the gainers over the physicians, as the distinction between the two groups ceased to exist. The intellectual center of French medicine shifted from the colleges and academies to the hospitals, particularly the Paris Hospital, which the Revolution removed from the control of the church and placed in the hands of the medical profession.

The one major prerevolutionary scientific institution to survive was the Royal Botanical Garden, in a new and expanded form as the Museum of Natural History. The revolutionaries were somewhat more lenient to natural history, seen as less elitist, than to the physical sciences. They did, however, provide much employment for physical scientists in practical concerns, from the vast project for creation of new weights and measures that resulted in the metric system to supervising the manufacture of gunpowder for the revolutionary armies. The new institutions founded by the revolutionaries were mainly for applied science. The most important of these besides the Institute of France and the Museum of Natural History was the Polytechnic School, which under the leadership of Monge became Europe's leading center of mathematics, the foundation of French mathematical supremacy. Another educational institution, the Lycée of Arts, attracted lecturers in chemistry and natural history, including Claude-Louis Berthollet and Antoine-Laurent de Jussieu (1748–1836). The social and intellectual fluidity of the revolutionary era also aided the quick rise of talented young men such as Georges Cuvier and Marie-François-Xavier Bichat.

See also Condorcet, Marie-Jean-Antoine-Nicolas Caritat, Marquis de; Hospitals; Laplace, Pierre-Simon de; Lavoisier, Antoine-Laurent; Linnean Society; Michaux Family; Museum of Natural History; Napoleonic Science; Nationalism; Royal Academy of Sciences; Royal Society of Medicine; War.

References

Alder, Ken. *Engineering the Revolution: Arms and Enlightenment in France, 1763–1815.* Princeton: Princeton University Press, 1997.

Gillispie, Charles Coulston. *Science and Polity in France at the End of the Old Regime.* Princeton: Princeton University Press, 1980.

Guerlac, Henry. *Essays and Papers in the History of Modern Science.* Baltimore: Johns Hopkins University Press, 1977.

G

Galvani, Luigi (1737–1798)

The Italian physician and anatomy professor Luigi Galvani performed one of the most important experiments in Enlightenment physics. Galvani, a professor at the University of Bologna and a member of the Bologna Academy of Sciences, had a long-standing interest in the relation of electricity and pneumatic chemistry to the workings of living things. Influenced by the work of Alessandro Volta, later his greatest intellectual rival, Galvani speculated on the relation of electrical discharge, the extinction of flame in a closed container, and respiration. In 1780 Galvani gave Bologna's "carnival anatomy," a public dissection used as a show of scientific virtuosity; evidence indicates he proclaimed that the "electrical fluid" was the spark of life. Many held this idea, but Galvani spent years trying to demonstrate it experimentally. The principal experimental instrument he used was the legs of frogs, from which the skin had been flayed. Originally, Galvani studied the reaction of the legs when electrically sparked from an external source. His breakthrough, reported in *Proceedings of the Bologna Academy of Sciences* in 1791, was when he established that the frogs' legs jumped when a bimetallic connector joined the crural nerve and the leg muscle. This led Galvani to abandon his idea that the legs somehow drew electricity from the air, and embrace the idea that an "animal electricity," with somewhat different properties than conventional electricity, was generated by living things themselves.

Galvani sent a copy of his report to Volta, who eventually put forth the opposing claim that the frogs' legs were not generating the electricity but simply registering the electricity generated by the contact of the two metals in the bimetallic connector. This led to a controversy, in which the lead part in the defense of animal electricity was taken not by the retiring Galvani himself, but by his nephew Giovanni Aldini (1762–1834), a professor of physics at Bologna. Galvani and his supporters lost the debate, although the real losers were Italian frogs, slaughtered in the thousands for their legs. Galvani's final years were also darkened by the loss of his position at the university in 1797 when he refused to swear allegiance to the new republican government put into place by the French army.

See also Bologna Academy of Sciences; Electricity; Volta, Alessandro Giuseppe Antonio Anastasio.

Reference

Heilbron, J. L. *Electricity in the 17th and 18th Centuries: A Study in Early Modern Physics.* 2d ed. Mineola, NY: Dover, 1999.

Gauss, Carl Friedrich (1777–1855)

Carl Friedrich Gauss, often considered the greatest pure mathematician who ever lived, was also an important contributor to problems of applied mathematics in the sciences, particularly celestial mechanics. Born to a family of laborers in the northern German city of Brunswick, Gauss manifested his genius for calculation at an early age. His elementary school teacher arranged for him to be admitted to a secondary school, where he learned the Latin that was indispensable for any German prospective academic. His record continued to be strong, and he was admitted to the Caroline College in 1792, leaving in 1795 for the University of Göttingen. Around this time Gauss made his first great original mathematical discovery, the method for constructing seventeen-sided polygons with a compass and straightedge. He left Göttingen in 1798, moving back to Brunswick. At the request of his patron, the duke of Brunswick, Gauss submitted a doctoral dissertation in mathematics to the University of Helmstedt in 1799 and received his degree in absentia.

Gauss's first important book was *Disquisitiones arithmeticae* (1801), primarily devoted to number theory. (Gauss was one of the last scientists to write in Latin, although he increasingly switched to German in subsequent decades.) Around the same time, he began to work seriously in astronomy and celestial mechanics. Gauss's serious association with astronomy began in his relationship with the German astronomer Franz Xaver von Zach (1754–1832) and the newly discovered asteroid Ceres, the first asteroid to be discovered. The Italian astronomer Giuseppi Piazzi (1746–1826) had discovered Ceres on 1 January 1801, and Zach had made a few observations before Ceres disappeared behind the Sun. The question that exercised celestial mechanists throughout Europe was the position Ceres would occupy after its expected reappearance at the end of the year. Gauss's predicted position and the orbit he ascribed to Ceres using the method of least squares that he devised (although Adrien-Marie Legendre [1752–1833] was the first to publish it) were quite different from those of most European celestial mechanists. However, observations in December showed Gauss's prediction to be substantially correct. This prediction gained Gauss a considerable reputation, leading to among other offers an invitation from the czar of Russia to head an observatory in St. Petersburg. Gauss's new interest in astronomy extended from the theoretical to the observational side, and he began a series of observations in Brunswick, despite its relatively poor observational facilities. He did manage to acquire a reflecting telescope. He also worked out the complicated mathematics of the effects of Jupiter's gravitational attraction on asteroids. In all of this work, Gauss was aided by his own extraordinary facility for mental calculations.

In 1805 Gauss married his first wife, Johanna Osthoff. The marriage was happy, though short. In 1807 Gauss left Brunswick to take up a position as director of the observatory of Göttingen and professor of the university there. Johanna died two years later. Gauss was heartbroken, but remained productive, and remarried shortly thereafter. His second book, *Theory of the Motion of Celestial Bodies Going around the Sun in Conic Sections* (1809), discussed the theory of solar orbits, and demonstrated how to calculate them and then refine the calculation. Gauss's methods were an improvement on those of Pierre-Simon de Laplace. He also worked on the construction of a new observatory. By this time his fame had spread beyond Germany—in 1810 he received a gold medal from the Institute of France. Gauss became a leader of the international scientific community—in 1824 his intervention to defend a Hungarian astronomer, Janos Pasquich (1754–1829), against charges of falsifying data was decisive in preserving Pasquich's reputation. He also turned down an offer to become permanent secretary of the Berlin Academy once Hanover matched the proffered salary.

Gauss's next major venture in applied mathematics was the survey of the Kingdom of Hanover, which absorbed much of his energy from 1818 to 1832. He traveled around the country taking measurements, and even invented a device, the heliotrope, for reflecting dispersed rays of sunlight. He published two major works on "geodesy," the science of measurement of Earth's surface, the first in 1828 and the second in two parts in 1843 and 1846. These works, aimed at a less rarefied audience than his work in celestial mechanics, were published in German. His work on geodesy also rekindled Gauss's long-standing interest in the possibility of a non-Euclidean geometry, although he did not publish on the subject.

By the late 1820s physics was attracting more of Gauss's attention, an interest quickened by the arrival of Wilhelm Eduard Weber (1804–1891) as Göttingen's new professor of physics in 1831. In an 1829 publication Gauss announced the principle of least restraint in mechanics. He also developed the notion of "potential." In 1832 he began his investigations of terrestrial magnetism. Alexander von Humboldt attempted to enlist Gauss in his plan to establish a series of magnetic observatories across the surface of the Earth to fully measure the variations of the magnetic force. Gauss demonstrated mathematically that there could be only two magnetic poles, a fact that up until that time had had no theoretical explanation. Gauss also invented an absolute measurement of the magnetic force. His contributions to the study of magnetism were not merely theoretical, however. At Göttingen he led the building of the first magnetic observatory. This building presented unique demands in that no iron could be used anywhere in its construction—copper rather than iron nails held it together. Gauss also participated in the founding of a journal of magnetism studies and the publication of an atlas of the Earth's magnetic forces in 1840. Under his leadership Göttingen was the center of magnetic studies, and he corresponded with magnetic observers over a wide area. Two strong personalities, Gauss and Humboldt remained friends while clashing over magnetism, although Gauss's procedures for carrying out magnetic observations were superior. Gauss and Weber also devised an early telegraph, although their collaboration was cut short in early 1838 when the liberal Weber was expelled from Göttingen along with six other professors by the new reactionary king of Hanover, Ernest Augustus (r. 1837–1851).

The last decades of Gauss's life were marked by continued activity, but waning creativity. Most of his attention was devoted to practical problems, such as his work from 1845 to 1851 establishing the soundness of the pension fund for the widows of Göttingen professors. Gauss's involvement with science did not cease with his death—he attained an odd sort of posthumous fame when dissection of his body revealed an unusually large brain, with particularly marked and deep convolutions. His brain, preserved by the University of Göttingen, was frequently invoked in late-nineteenth-century debates about the relationship of the size and formation of the brain to general intelligence.

See also Astronomy; Humboldt, Alexander von; Mathematics; Physics.

Reference

Buhler, Walter Kaufmann. *Gauss: A Biographical Study.* Heidelberg: Springer-Verlag, 1981.

Gay-Lussac, Joseph-Louis (1778–1850)

The chemist Joseph-Louis Gay-Lussac was one of the young men who took advantage of the social mobility of revolutionary and Napoleonic France to build a scientific career. From a legal family that suffered hard times in the Revolution, Gay-Lussac entered the Polytechnic School in 1797 and the School of Bridges and Roads in 1800. The same year Gay-Lussac, already planning a scientific career, met Claude-Louis Berthollet, who adopted the young man as a lab assistant and protégé. Working from Berthollet's home at Arcueil, Gay-Lussac almost immediately made an important discovery: that all gases expand by the same proportion for an equivalent rise in temperature, given constant pressure. This discovery is now referred to as Charles's law, after Jacques-Alexandre-César Charles (1746–1823), whose previous work had been much sketchier and unpublished. Gay-Lussac's approach to chemistry was deeply influenced by the mathematical culture of French Newtonian physics, and unlike many of his contemporaries he defined the primary task of the chemist as the establishment of laws that could be expressed mathematically.

In 1806 Gay-Lussac was elected to the First Class of the Institute of France, the successor to the old Royal Academy of Sciences. He was also a member of the exclusive Society of Arcueil, and was appointed teacher of physics at the Faculty of Science in the newly established University of France in 1808. In 1809 he justified the Institute's faith by announcing another law pertaining to gases, Gay-Lussac's law of combining volumes of gases. This law states that gases that combine chemically with each other always do so in simple ratios. Gay-Lussac viewed the key measurement in the chemical combination of gases to be the volume, rather than the weight, of the gases. This was for a long time a successful rival of the atomic theory of John Dalton, which emphasized weight. Gay-Lussac's position at the center of Parisian chemistry was further enhanced in 1810 when he received the chair of chemistry at the Polytechnic School.

From 1807 to 1815 Gay-Lussac's great rival in European chemistry was the Englishman Sir Humphry Davy. The two were often working on similar projects, and accusations of poaching occasionally flew back and forth across the Channel. (The situation was exacerbated by the poor communications between the two countries, then at war.) Gay-Lussac had the advantage of working in the scientific capital of Paris, but he was somewhat timid and perhaps overly respectful of Berthollet and the memory of Antoine-Laurent Lavoisier. Although Gay-Lussac hinted at the elemental nature of chlorine, then known as oxymuriatic acid, it was Davy who boldly proclaimed it and gave the new element its name. On the other hand, while Davy and Gay-Lussac quarreled over priority in the description of iodine, it was Gay-Lussac's exhaustive memoir on the subject in 1814 that laid the basis for the subsequent development of chemical studies of the element.

Gay-Lussac adjusted easily to the Restoration of the French Bourbon dynasty in 1814. He had never been as thoroughly identified with the Napoleonic regime as were the older men Berthollet and Pierre-Simon de Laplace. In addition to his teaching posts, he also became one of the two coeditors of the scientific journal *Annals of Chemistry and Physics.* (The other was an Arcueil colleague, the astronomer François Arago [1786–1853].) The journal, formerly *Annales de Chémie,* was quite successful both financially and intellectually, publishing not only French chemists but also some of the leading foreign chemists, such as the Swede Jöns Jakob Berzelius.

Much of Gay-Lussac's time in the Restoration was spent as an adviser to the government and manufacturers on technical chemical problems. He made great improvements in French munitions and minting, becoming assay master at the Paris Mint in 1829. For purposes of taxation, Gay-Lussac devised a more accurate way of measuring the alcohol content of a fluid, and the "degrees Gay-Lussac" remain the French standard. His most important industrial work was his invention of volumetric analysis, for which he devised the instruments the pipette and burette in their modern forms. He also served in the French Chamber of Deputies from 1831 to 1839. Elected only after the revolution of 1830, Gay-Lussac was a moderate liberal who frequently spoke for the interests of French manufacturers. His contributions to pure science were modest during this period, but he continued to teach thousands, exchanging his chair at the Polytechnic School for one at the Museum of Natural History in 1840. His most important student was the

German chemist Justus von Liebig (1803–1873), whom Gay-Lussac personally invited to assist him in his laboratory, an unusual practice in France at the time.

See also Arcueil, Society of; Ballooning; Berthollet, Claude-Louis; Chemistry; Davy, Sir Humphry.

Reference

Crosland, Maurice. *Gay-Lussac: Scientist and Bourgeois.* Cambridge: Cambridge University Press, 1978.

Geoffroy Saint-Hilaire, Étienne (1772–1844)

Étienne Geoffroy Saint-Hilaire was one of France's leading natural historians and anatomists in the early nineteenth century. His struggle with the doyen of French biology, his friend and rival Georges Cuvier, attracted enormous interest in France and Europe, in and out of the scientific community.

One of the many bright young men making scientific careers in revolutionary Paris, Geoffroy Saint-Hilaire received a professorship at the Paris Museum of Natural History in 1793, a position he would hold for the rest of his life. His first paper was read before the Society of Natural History in 1794, on the classification of the Madagascan aye-aye, which he successfully argued was in a genus of its own. Geoffroy Saint-Hilaire was partly responsible for welcoming Cuvier, another ambitious young scientist, to Paris in 1795. (Even after he and Cuvier had become bitter rivals, he referred to bringing Cuvier to Paris as one of the great achievements of his life.) The two lived together briefly and collaborated on five natural history papers. They parted in 1798 when Geoffroy Saint-Hilaire accepted, and Cuvier declined, a position in Napoléon's expedition to Egypt. Geoffroy Saint-Hilaire's time in Egypt was very productive. The study of Egyptian fish led him to consider the relation of the skeletons of fish and those of other vertebrates. Belief in the "unity of plan," the idea that all vertebrates, and ultimately all animals, were variations on a single design, marked all of Geoffroy Saint-Hilaire's subsequent scientific work. Intellectually, the source for this idea was the *Natural History* of Georges-Louis Leclerc de Buffon, of which Geoffroy Saint-Hilaire was a great admirer, but it was informed by his work in comparative anatomy. Studying the Egyptian fish, he found many parallels, or "homologies," between fish skeletons and those of mammals, but the question of the bones of the operculum, the gill cover of fish, vexed him. There seemed to be no obvious parallel in the human skull. Geoffroy Saint-Hilaire wished to make anatomy "philosophical," to get beyond the mere piling up of facts about various creatures into a universal system modeled on Newtonian physics. His ambitions were not always limited to the science of living things—his fondness for grand theory and an encounter with some electric fish led him in his last days in Egypt to devise a unified theory of the physical universe in terms of forces and fluids, a theory that he seems to have never entirely abandoned.

On Geoffroy Saint-Hilaire's return to France in 1802, he found French natural history dominated by Cuvier, who was suspicious of both grand theory and a universal plan for all animals. Geoffroy Saint-Hilaire unmistakably set forth his own views, and opposed Cuvier's, in *Anatomical Philosophy* (1818). He asserted that all vertebrates had fundamentally the same skeletal system, that each part had a parallel on all other creatures. He had finally solved the problem of the operculum by paralleling it with the bones of the mammalian ear. In the 1820s Geoffroy Saint-Hilaire expanded this parallelism to invertebrate creatures, finding parallels between the exoskeletons of insects and the skeletons of mammals, for example. This was an unmistakable challenge to Cuvier, who had divided all animals into four completely independent classes. It also led to an alliance between Geoffroy Saint-Hilaire and his followers, the "philosophical anatomists," and the German *Naturphilosophs,* who had arrived at similar conclusions by a somewhat different route. As early as 1817 Geoffroy Saint-Hilaire's work had been published in German *Naturphilosoph* journals. He adopted from the *Naturphilosoph* Lorenz Oken (1779–1851) the idea that the skull is composed of fused vertebrae. Despite the benefits of this alliance in spreading "philosophical anatomy" in Germany, association with what Cuvier and other French scientists saw as *Naturphilosoph* mysticism left Geoffroy Saint-Hilaire vulnerable to attack.

During the 1820s Geoffroy Saint-Hilaire also engaged in teratology—the study of monsters—which he was one of the first to integrate into a general biological theory. He hoped to create a classification scheme for human and animal monsters that would emphasize commonalities in the body parts deformed and the way they were deformed. He actually attempted to create monstrous birds from eggs, either by surgically interfering with the mother or by exposing the eggs to extreme conditions. This was not successful in advancing knowledge. Some of his writings from this period touch on evolution, but unlike his friend Jean-Baptiste-Pierre-Antoine de Monet de Lamarck, he never made it a major feature of his biological theory.

Things came to a head between Geoffroy Saint-Hilaire and Cuvier in 1830. Two obscure provincial naturalists submitted a paper supporting the "philosophical-anatomist" position by arguing that the arrangement of the organs of mollusks was homologous to that of vertebrates. In reporting on the paper to the Academy of Sciences, Geoffroy Saint-Hilaire repeated and ridiculed a passage from one of Cuvier's writings on cephalopods, without mentioning the author's name. Cuvier's reply, expounding on the differences between cephalopod and vertebrate anatomy, was generally considered to be successful by his and Geoffroy Saint-Hilaire's scientific peers. But it was a different story outside the scientific community. Geoffroy Saint-Hilaire himself carried the struggle into the public realm, appealing to a broader audience with an account of the controversy, *Principles of Zoological Philosophy,* published the same year. Geoffroy Saint-Hilaire presented himself and was viewed in romantic terms as a rebel standing against the scientific despotism of Cuvier—a role that made him particularly a hero of supporters of the political Left and anticlericals, distrustful of Cuvier's appeals to biblical authority. Pamphlets and dramas presented him as a true philosopher and friend of man and Cuvier as a pedantic fact-grubber and corrupt politician. Johann Wolfgang von Goethe, who admired Geoffroy Saint-Hilaire, published two articles supporting him and considered the conflict between the two French savants as more significant than the French political revolution the same year. Prominent French romantic writers who supported Geoffroy Saint-Hilaire included George Sand (1804–1876) and Honoré de Balzac (1799–1850).

The immediate conflict ended with Cuvier's death in 1832. Geoffroy Saint-Hilaire made an effort to take his place as the leader of French biology and permanent secretary of the Academy of Sciences, but was unsuccessful due to his age and failing powers. He took to giving long-winded and obscure papers expounding a universal system of natural philosophy based on the idea, which he never fully explained, of "self for self"—self-attraction as a universal principle. He successfully toured Germany, whose scientific culture he extolled over that of France. His funeral in 1844 was a great occasion. His son, Isidore Geoffroy Saint-Hilaire (1805–1861), followed in his footsteps as a natural historian and teratologist and wrote his biography.

See also Cuvier, Georges; Egyptian Expedition; Museum of Natural History; Zoology.

Reference

Appel, Toby A. *The Cuvier-Geoffroy Debate: French Biology in the Decades before Darwin.* New York: Oxford University Press, 1987.

Geology

Geology as a science was formed in the eighteenth century from a mixture of intellectual traditions including mineralogy, the practical knowledge of miners, and cosmogonic theories about Earth. The collection and identification of rocks and fossils were allied with life sciences in the discipline of natural history. The most important intellectual development was the alliance of the study of rocks and minerals to theories of Earth's history. This was broadened in the early nineteenth century with the increased use of fossils to reconstruct the chronology of Earth's development.

The histories of Earth created in the late seventeenth and early eighteenth centuries were dominated by the Bible, which on the one hand limited the span of Earth history to a few thousand years and on the other posed the Flood of Noah as a major event in the terrestrial past. Early writers such as John Ray (1627–1705), Thomas Burnet (c. 1635–1715), and John Woodward (c. 1665–1728) were essentially reconciling the data of natural philosophers with the biblical account. Some, like Woodward, thought the Flood a global event and ascribed a central role in Earth history to it; others, like Ray, thought the Flood a local event in the Middle East and asserted that volcanoes and Earth's internal heat were more important in Earth history. This theory was further developed by Lazzaro Moro (1687–1764) of Venice in 1740, who held that the Flood had no geological significance at all. This debate prefigured the much more secular debate between "neptunists" and "vulcanists" by the second half of the eighteenth century. Moro also distinguished between primary and secondary rocks, those formed originally, which contained veins of metallic ore, and those formed out of the primary rocks, which contained fossils. This distinction emerged in the writings of a number of mid-eighteenth-century cosmologists and mineralogists.

The Enlightenment's challenge to biblical authority was carried into the field of Earth history by Georges-Louis Leclerc de Buffon, who argued in 1749 that Earth was much older than the biblical account would indicate, possibly as old as eighty thousand years. This kind of view still had to be put forward with care and discretion, but was voiced more and more openly as the eighteenth century progressed.

Meanwhile, the study and classification of rocks and minerals had developed its own set of problems. Buffon's French contemporaries Jean-Étienne Guettard (1715–1786) and the government official and encyclopedist Nicolas Desmarest (1725–1815) were pioneers in geological fieldwork, surveying a great portion of France and beginning the field of geological mapping. With the collaboration of Antoine-Laurent Lavoisier, Guettard published a large-scale geological map of France in 1780. Germany, with its rich tradition of mining, was another active center of field geology, with workers such as the physician Georg Christian Füchsel (1722–1773), who explored the famous mining region of the Harz Mountains. Füchsel and the Berlin professor Johann Lehmann (d. 1767) began to develop a theory of stratification based on a distinction between primary, secondary, and Tertiary rocks. Mineralogy, supported on the European continent by a number of institutes and academies founded by governments eager to increase their wealth, also developed a laboratory tradition that increased knowledge of how rocks were formed by precipitation and crystallization, and altered by chemical reactions and the application of heat.

Woodward had attempted to bring his fieldwork knowledge to bear in constructing a theory of Earth, but his prickly personality and isolated intellectual position made his attempt abortive, despite his bid to institutionalize geology in Britain with the founding of the Woodwardian Chair in Geology at Cambridge in 1729. British geology remained highly empirical in its orientation, with close description of different kinds of rocks and their arrangement, and little theoretical speculation. The first widely accepted synthesis of mineralogy, fieldwork, and cosmogonic theories of Earth was produced on the Continent, by a German professor at the Freiberg School of Mining, Abraham Gottlob Werner, who together with his international band of disciples dominated European geology for many decades beginning in the late eighteenth century. Werner was best known as an extreme neptunist, whose theory of Earth traced the laying down of successive strata by chemical precipitation and settling of substances dissolved in an original "world ocean." His most important innovation, however, was the identification of the geological unit of study as the "formation," the body of rock laid down at the same time. This removed geology from the timeless world of the mineralogists, who focused on the chemical composition of each rock, to the world of the history of Earth.

Werner's neptunism, despite its enormous influence, was never unopposed. A group of mostly French and Italian geologists carried on the vulcanist tradition, while the Scotsman James Hutton came up with a new theory emphasizing gradual, or "uniform," development of geological features rather than the dramatic catastrophes that both Wernerian neptunists and vulcanists favored. Hutton's theory was set forth in his *Theory of the Earth,* read before the Royal Society of Edinburgh in 1785, and published as a paper in the *Transactions of the Royal Society of Edinburgh* in 1788 and then in a revised and expanded two-volume edition in 1795. It became known as plutonism, after the Roman god of the underworld. Hutton emphasized the deep interior heat of Earth as a causal agent. Despite attracting a few significant followers, the most notable of whom was the Edinburgh professor of mathematics John Playfair (1748–1819), Hutton's theory had little initial impact either on the Continent, where the Wernerians held the key strategic positions, or in Britain, where many viewed Hutton's uniformitarianism as a threat to the Bible. Hutton's obscure style and use of English also limited his work's impact outside Britain. The principal authority for the Huttonian doctrine in the early nineteenth century, in fact, was not Hutton's own work, but Playfair's far better–organized and stylistically masterful *Illustrations of the Huttonian Theory of the Earth* (1802).

The key issue dividing orthodox Wernerians from their opponents, whether vulcanist or Huttonian, in the late eighteenth and early nineteenth centuries was that of the sedimentary versus the igneous origin of certain rocks, notably basalt. Werner himself had declared that basalt was formed from sediment of the primeval ocean, but many other geologists, including some of Werner's disciples, believed it to be created by great heat. Desmarest's studies of the extinct volcanoes of the Auvergne region of France, which he had been the first to identify, linked basalt to volcanoes. In the early nineteenth century Louis Cordier (1777–1861) performed a series of experiments demonstrating the mineralogical and chemical similarity of basalt and modern lava from known volcanoes. The growing acceptance of the volcanic formation of basalt and other minerals in the early nineteenth century led to the modification of the Wernerian system, but not to its abolition. It continued to dominate geological theorizing through the 1820s.

After Werner the key innovation in dating rocks was the incorporation of the study of fossils as a dating tool, a development encouraged by Werner himself. Interest in fossils also originated in early modern natural history. The Welshman Edward Lhwyd (1660–1709) had recognized the importance of collecting fossils within the discipline of natural history with the publication of his illustrated guide to British fossils, *Lithophylacii Britannici iconographia* (1699), the first field manual for collectors. Woodward built a large and highly respected fossil collection, the first to identify the provenance of each item. Interest in the association of fossils with particular geographical strata blossomed in the late eighteenth century. Georges Cuvier's exploration of the geology of the Paris region in association with Alexandre Broignart (1770–1847), published in 1808, was an early example of the usefulness of fossils for geological ordering. Cuvier was a Wernerian in geology, and his biological catastrophism—his belief in waves of extinction that wiped out hundreds of species in a relatively short interval of time—also supported geological catastrophism, as vast geological catastrophes were the easiest explanations for biological catastrophes. Cuvier's writings on paleontology also contributed to the backdating of the formation of Earth, suggesting an Earth history that had lasted for millions rather than thousands of years.

In Britain interest in geology was fostered by the canal building and coal mining associated with the development of the Industrial Revolution. Intellectually, the empirical British tradition reached its summit in the geological map of Britain produced by William Smith (1769–1839), known as "Strata Smith," in 1815. Smith was a self-taught geologist with little education in geological theory. Although geological mapping had a longer history on the Continent, Smith's map was a remarkable achievement in its care and detail. Growing interest in geology in early-nineteenth-century Britain was manifested in the founding of the Geological Society of London in 1807 and the Wernerian Society in Edinburgh the following year. One of the first specialized scientific societies to flourish in Britain, the Geological Society combined a generally Wernerian theoretical structure with a strong emphasis on empiricism as opposed to theorizing, whereas the Wernerian Society was more dogmatically Wernerian.

See also Earthquakes; Fossils; Hutton, James; Volcanoes; Werner, Abraham Gottlob.

References

Hallam, A. *Great Geological Controversies.* 2d ed. Oxford: Oxford University Press, 1989.

Laudan, Rachel. *From Mineralogy to Geology: The Foundations of a Science, 1650–1830.* Chicago: University of Chicago Press, 1987.

Porter, Roy. *The Making of Geology: Earth Science in Britain, 1660–1815.* Cambridge: Cambridge University Press, 1977.

Germain, Sophie (1776–1831)

Sophie Germain was the most distinguished female mathematician and mathematical physicist of the early nineteenth century. Social attitudes toward women made participation in advanced mathematics very difficult for her, beginning with her childhood, when she had to read the works of Isaac Newton (1642–1727) and Leonhard Euler in secret, under the blankets, as her parents did not approve of girls taking an interest in mathematics. Her parents did eventually relent, and their financial support enabled her to live without marrying. Rather than attending Joseph-Louis Lagrange's courses on mathematical analysis at the École Polytechnique, she was forced to obtain lecture notes. When she emerged on the mathematical scene with a paper submitted to Lagrange, she used the male pseudonym "M. Leblanc" fearing that as a woman she would not be taken seriously, a consideration that also led her to avoid all association with other learned and scientific women throughout her career. Germain also used a male pseudonym in commencing a mathematical correspondence with Carl Friedrich Gauss in 1804, although Gauss eventually learned his correspondent's true gender and identity when the French occupied his hometown of Brunswick in 1806. Germain's most important work in pure mathematics was in number theory, where she defined a number known as a "Sophie Germain Prime." This is an odd prime number, p, with the property that $2p + 1$ is also a prime. Germain used the concept in her work on the last theorem of Pierre de Fermat (1601–1665).

In mathematical physics Germain's most important work was in the theory of elasticity. In 1808 the Institute of France set up a prize competition for a mathematical theory of elasticity consistent with experimental evidence. Germain was the only competitor the first two times entries were judged in 1811 and 1813, but despite her innovations her lack of formal mathematical training prevented her from expressing her theory in a way acceptable to the academy. She was also handicapped by her ignorance of the dominant physical theory of the time, that of Pierre-Simon de Laplace. In 1815 she finally received a medal of one kilo of gold from the Institute.

This was a defeat for Laplace and his disciple Siméon-Denis Poisson (1781–1840), Germain's chief competitor as well as one of the judges of the contest. Despite this triumph for Germain, as the theory of elasticity became further developed by male mathematicians with secure institutional positions, her work was increasingly ignored. Undaunted, Germain continued to work on mathematical problems until her death from breast cancer. She also wrote an essay on the development of the arts and sciences whose account of the history of science anticipated the positivism of Auguste Comte (1798–1857).

See also Gauss, Carl Friedrich; Mathematics; Women.

Reference

Petrovich, Vesna Crnjanski. "Women and the Paris Academy of Sciences." *Eighteenth-Century Studies* 32 (1999): 383–390.

Johann Wolfgang von Goethe was one of Germany's leading writers and an influential natural philosopher. (Perry-Castaneda Library)

Goethe, Johann Wolfgang von (1749–1832)

Johann Wolfgang von Goethe, Germany's greatest writer during the romantic era, was also a scientist and philosopher of science, as well as a friend and intellectual colleague of leading German scientists, including Alexander von Humboldt. As a natural philosopher, he occupied an intermediate position between Enlightenment Newtonianism, with its mechanistic approach, and German romantic *Naturphilosophie*. Goethe's scientific interests covered a broad range of disciplines. He is most remembered for his work in optics, in which he attempted unsuccessfully to dethrone Newton's theory of the composition of white light from rays of colored light. Newton's famous experiment with a prism, Goethe argued, was too complex, and required too many specific conditions, to truly establish the nature of light. Goethe's critique of Newton's optics was originally based on a misunderstanding of the implications of Newton's theories, but broadened into an alternative theory. Goethe believed that color phenomena needed to be examined in many contexts, including their psychological and aesthetic aspects. Despite its rejection by later scientists, who treated Goethe as the archetypal literary dabbler in science, Goethe's color theory did make some contributions, and prefigured the rise of wave theories of light in the nineteenth century.

Unlike some of the romantics, Goethe was never opposed to science per se, in which he took great intellectual delight. Goethe was a believer in the unity of knowledge and held that nature was a unity and that science and art could be united at the highest levels. His novel *Elective Affinities* employs a metaphor from chemistry—the "elective affinities" were the different ways that different substances had to unite with each other—to describe human relationships, although it also shows how inadequate such metaphors are. The characters actually engage in a lengthy discussion of current chemical theory.

Like other life scientists of the period, Goethe rejected the reduction of organic to mechanical processes. Goethe also distrusted

Naturphilosophie, which he believed resembled its seeming opposite, Newtonian mechanism, in its abstraction and remoteness from the empirical world of the senses. However, Goethe's biological thinking had some affinities with that of the *Naturphilosophs.* Goethe was a founder of "transcendental anatomy," the effort to understand living beings as variations on a single plan. In the late 1780s, after his studies of botany and anatomy at the University of Jena, Goethe engaged in the quest to establish the archetypal forms for plants and animals, a quest he was to christen with the name *morphology* in 1807. He went beyond the idea of plants as variations on archetypal forms to argue that the different parts of a plant—roots, flowers, and others—were themselves all variations on the archetypal form of the leaf. Goethe's belief in the archetypal form led him to the discovery of the intermaxillary bone of the human skull by analogy with its position in animal skulls. He never lost his interest in this subject, and as a very old man he followed avidly the struggles between the champion of transcendental anatomy, Étienne Geoffroy Saint-Hilaire, a scientist he admired, and his rival Georges Cuvier.

See also Literature; *Naturphilosophie;* Newtonianism; Optics; Romanticism.

References

Cunningham, Andrew, and Nicholas Jardine, eds. *Romanticism and the Sciences.* Cambridge: Cambridge University Press, 1990.

Stephenson, R. H. *Goethe's Conception of Knowledge and Science.* Edinburgh: Edinburgh University Press, 1995.

Hales, Stephen (1677–1761)

The Church of England clergyman Stephen Hales was the leader in applying Newtonian mechanics to plant and animal physiology. His earliest experiments, beginning around 1707, concerned the motion of the blood in animals, and his interest in science was such that he was elected a fellow of the Royal Society in 1718. His first public report of his research was a paper on the effects of the warmth of the Sun on the motion of sap read before the society in 1719. Hales's work on plants was published as a book, *Vegetable Staticks,* in 1727. He was an experimenter of genius, and most of *Vegetable Staticks* describes a series of classic experiments Hales performed on such subjects as the rising of sap, transpiration of fluids, and the expansion of vegetable matter through the absorption of water. It is the foundational work of eighteenth-century plant physiology.

In addition to his work on the transport of fluids through living bodies, Hales also investigated what was the new subject of the movement of gases. The nearly 100 pages of *Vegetable Staticks* regarding air were the starting point for one of the most important developments in eighteenth-century science, pneumatic chemistry. Hales's analysis of gases was Newtonian in that he considered them as composed of particles operating by short-range forces of attraction and repulsion. His innovation was to claim that under certain circumstances particles of air could become "fixed" or solid, through their attraction to particles of solids. He also systematically measured the quantity of air released from different substances by different processes, although he was unaware that atmospheric air is a combination of different gases. *Vegetable Staticks,* a work of wide influence on the Continent as well as in Britain, had perhaps the most illustrious pair of translators of any eighteenth-century scientific book, being translated into French by Georges-Louis Leclerc de Buffon and into German by Christian Wolff.

Hales's second scientific book, *Statickal Essays: Containing Haemastaticks* (1733), is devoted to animal subjects. It is also mostly composed of reports of experiments. *Haemastaticks* contains the first reports of the direct measurement of blood pressure. Hales draws on some of the "medical mechanism" associated with Giovanni Alfonso Borelli (1608–1679), Hermann Boerhaave, and some of Boerhaave's English medical disciples like George Cheyne, but he also criticizes some of its claims as reductionistic. He emphasizes the range of different blood pressures compatible with health in an animal.

After *Haemastaticks* Hales increasingly turned his attention away from pure science toward applying it to projects of social betterment. Hales was a zealous opponent of gin drinking, and his writings against it employ physiological arguments explaining the damage it causes to the body. Much of his energy went to devising ventilation systems for ships, mines, hospitals, and jails, some of which were actually effectively installed. Hales shared the common belief that disease was caused by bad air, and hoped that providing fresh air to these places would keep people healthy—his own brother had died of "gaol fever" while imprisoned. Hales's experimental work on the dissolution of kidney stones—a classic medical problem of the eighteenth century—won him the Copley Medal from the Royal Society in 1739 (although it did not result in a cure for the condition). His interest in practical applications of experimental science led him to be one of the founding members of the Society of Arts in 1754. Hales was also typical of many early English Newtonians in his natural theology, which strongly emphasized the ingenuity of God's providential design of living things.

See also Chemistry; Natural Theology; Newtonianism; Plant Physiology.

Reference

Allan, D. G. C., and R. E. Schofield. *Stephen Hales: Scientist and Philanthropist.* London: Scolar, 1980.

Haller, Albrecht von (1708–1777)

The Swiss physician Albrecht von Haller was the foremost life scientist of the mid-eighteenth century. A student of Hermann Boerhaave at Leiden, Haller was appointed professor of medicine at the University of Göttingen in 1737, helping to shape it into a leading medical institution. He had a wide range of interests, extending to anatomy, physiology, embryology, medical practice, physics, politics, and theology. A convinced methodological empiricist, Haller was one of the greatest experimentalists in the life sciences, investigating questions such as the formation of bone, the nature of embryos, and the forces inherent in living tissues. The devout Haller also viewed his work in the Newtonian tradition of demonstrating the glory of God.

Haller's voluminous publications include his eight-volume *Physiological Elements of the Human Body* (1757–1766), an unrivaled collection of physiological knowledge, and *Anatomical Images* (1743–1754), which employed the technique of injection of colored fluids into veins to masterful effect, as well as an enormous number of papers and book reviews.

Haller's most significant original scientific contributions were aimed not at settling strictly medical questions but at defining the nature of life. Originally a supporter of Boerhaave's mechanical approach, Haller found strict mechanism inadequate for explaining living processes. His eventual solution was to view animal organisms as aggregations of tiny parts called "fibers." The fibers possessed nonmechanical, self-acting properties, of which the most important was irritability, or contractility. Irritability was the ability of a muscle fiber to react to a stimulus. It was a property of animal material, but independent of life, as could be seen in the contractions of a dead animal's muscles. The heart, the most irritable of organs, beat due to the stimulus provided by the inflow of blood. Another property, restricted to the nervous fibers, was sensibility. Haller's developed theory of these forces was first presented in a paper before the Royal Society of Sciences of Göttingen in 1752, and later elaborated into a system. He was concerned that the ascription of these forces to matter would encourage materialists and emphasized that these and all other natural forces were derived from God. Despite his best intentions, materialists such as Julien Offroy de La Mettrie and Denis Diderot did use the idea of self-acting fibers. (Haller's relationship with La Mettrie was particularly bitter, as he accused La Mettrie, a fellow student of Boerhaave, of plagiarizing Haller's notes on Boerhaave's lectures.)

Haller's views on embryology are particularly interesting, as he successively held all three of the major interpretations of conception and fetal development. He began as a spermaticist, following Boerhaave in believing that the animal existed as a tiny, preformed being in the male sperm prior to conception. Influenced by the discovery of the regenerating polyp by Abraham Trembley (1700–1784) in the 1740s, Haller became an epigenesist, believing that the animal was formed or assembled out of preexisting materials in the process of conception, possibly by some form of natural attraction analogous to chemical or gravitational attraction. His final conversion, to ovist preformationism, which located the preformed animal in the female egg, was precipitated by the difficulty of finding a coherent mechanical model for epigenesis, as well as a series of experiments and observations performed on chicken eggs in 1755, 1756, and 1757. Haller believed that he discerned preformed embryos in the egg before fertilization. He claimed that the membrane covering the yolk is an extension of the intestine of the preformed embryo, and thus demonstrated its existence even when it was too small to be seen. Haller limited the role of the male seed in conception to stimulating the embryo's heart to activity by irritating it. Haller's conversion to preformationism, announced in another paper before the Royal Society of Göttingen in 1757, led to a protracted, although courteous, controversy with the epigenesist Berlin physician Caspar Friedrich Wolff (1733–1794). In the course of the nearly twenty-year controversy, carried out by correspondence as well as Haller's reviews of Wolff's books, Haller's ovist position grew more rigid.

See also Boerhaave, Hermann; Embryology; La Mettrie, Julien Offroy de; Vitalism.

References

Porter, Roy. *The Greatest Benefit to Mankind: A Medical History of Humanity.* New York: W. W. Norton, 1998.

Roe, Shirley. *Matter, Life, and Generation: Eighteenth-Century Embryology and the Haller-Wolff Debate.* Cambridge: Cambridge University Press, 1981.

Hauksbee, Francis (c. 1666–1713)

Francis Hauksbee became England's leading demonstrator and public experimenter after the death of Robert Hooke (1635–1703). Hauksbee was a Newtonian, and the ascendancy of Isaac Newton (1642–1727) to the presidency of the Royal Society in 1703 made Hauksbee the successor to Hooke as the person responsible for carrying out demonstrations and experiments at society meetings. Hauksbee, a small tradesman from a middle-class background who had never attended a university, was a gifted designer of experimental equipment who refined the use of the air pump. His presentations before the society (he became a fellow in 1705) were part of Newton's strategy of revitalizing the Royal Society, which had fallen into a period of quiescence before Newton's presidency.

Hauksbee's most significant innovations were connected with electricity, in which he carried out the first sustained series of experiments. Beginning with a series of demonstrations based on the fact that mercury sometimes glowed in the vacuum of a shaken mercury barometer, Hauksbee demonstrated that the effect could be produced without mercury, simply by rubbing and chafing. He produced an electrical machine that rapidly rotated an evacuated globe, and demonstrated the globe's electrification by showing it attracting fine threads, those positioned on both the inside and the outside of the globe. These threads became known as "Hauksbee's threads." This was the most powerful electrical generator known at the time, but elaborate and not very portable. Another Hauksbee innovation was a portable electrical generator consisting of a hollow glass tube about thirty inches long. Hauksbee's electrical devices and other experimental innovations were set forth in his book, *Physico-Mechanical Experiments on Various Subjects, Containing an Account of Several Surprising Phenomena Touching on Light and Electricity* (1709). *Physico-Mechanical Experiments* was based on the papers Hauksbee had given before the Royal Society, with an appendix in which he abandoned the idea that the glow produced by the mercury in the barometer was electric. It contains one of the first discussions of electrical repulsion, although Hauksbee, not theoretically inclined, did not seem to have a very clear idea of it.

Hauksbee also experimented on magnetism, unsuccessfully trying to ascertain a mathematical formula for the variation of magnetic force over distance analogous to Newton's inverse-square law for gravity. His other areas of experimental interest included freezing, capillary action, and the propagation of sound. His son, Francis Hauksbee the younger (1688–1763), followed in his footsteps as an experimenter, but John Theophilus Desaguliers succeeded him as experimenter to the Royal Society.

See also Electricity; Newtonianism; Popularization; Royal Society.

Reference

Heilbron, J. L. *Electricity in the 17th and 18th Centuries: A Study in Early Modern Physics.* 2d ed. Mineola, NY: Dover, 1999.

Heat

Although many facts about heat and its effects were discovered in the eighteenth century, its ultimate nature remained elusive. The principal theoretical division was between those who saw heat as a substance, or "subtle fluid," and those who saw it as the motion of the smallest pieces or corpuscles of the heated substance—the "mechanical theory." Although the mechanical theory had the support of many leading scientists in the seventeenth century, fluid theory dominated the eighteenth century. The expansion undergone by many substances when heated tended to support it, as the expansion was seen as a result of the absorption of the fluid of heat by the heated substance. Experiments performed to identify an increase in weight on the part of heated objects were more ambiguous (particularly since not all agreed that the fluid of heat had weight). The situation was further confused by the difficulty of measuring heat and the frequent identification of temperature with quantity of heat. When temperature and quantity of heat were distinguished, quantity of heat was usually seen as a simple function of temperature and volume, a theory endorsed by Hermann Boerhaave in his *Elements of Chemistry* (1732), or of temperature, volume, and density. These theories did not always fit the experimental evidence, though.

Many workers on heat ignored the theoretical debate as irrelevant to such experimental concerns as measuring different substances' rates of thermal expansion, often important for engineers. In England John Ellicott and the civil engineer John Smeaton (1724–1792) invented "pyrometers," very sensitive devices for the measurement of the thermal expansion of a bar of metal to, Smeaton claimed, one-twenty-thousandth of an inch. Another, mostly German and Scandinavian, experimental tradition studied the temperature of mixtures of substances originally at different temperatures. Workers in that field included Georg Wolfgang Krafft (1701–1754), Georg Wilhelm Richmann (1711–1753), Johann Carl Wilcke (1732–1796), and Johan Gadolin (1760–1852).

As heat was clearly related to chemical changes, students of heat tended to be chemists more than physicists. The Edinburgh chemist Joseph Black made fundamental discoveries about the relation of heat and substance, as well as clarifying the distinction between temperature and quantity of heat. He identified the fact that different substances had different capacities for absorbing heat (specific heat) and that the change of state from a solid to a liquid and from a liquid to a gas absorbed heat without raising the temperature of the substance heated (latent heat). Black explained heat phenomena in terms of a weightless substance he called "matter of heat." For example, Black thought water was a compound of ice and matter of heat, and that the addition of further matter of heat produced gaseous water. Many of Black's conclusions were arrived at independently by Wilcke.

Antoine-Laurent Lavoisier also believed in a weightless heat substance that entered into combinations. He called this substance caloric and integrated it into his new system of chemistry. Lavoisier moved beyond Black in inventing, along with Pierre-Simon de Laplace, a device for measuring the quantity of heat, the calorimeter. The calorimeter, along with the standardization of the thermometer, enabled the study of heat to be put on a much more precise quantitative basis. The prestige of Lavoisier and Laplace helped the caloric theory to dominate French science well into the nineteenth century. The weightlessness of Black's and Lavoisier's caloric was particularly useful, as it enabled Lavoisier to sidestep the experimental debate about whether objects gained or lost weight on being heated or cooled, a debate that was producing confusing and contradictory evidence—Georges-Louis Leclerc de Buffon found that iron became heavier on heating, while John Roebuck (1718–1794) found that it lost weight.

The mechanical theory of heat underwent a late-eighteenth-century revival in the English-speaking world, chiefly due to the experimental demonstrations of Benjamin Thompson, Count Rumford, whose understanding of heat's practical applications in chimneys, stoves, and other common household implements was unparalleled. Rumford's experiments, beginning in 1798 with his observations on the boring of guns, showed the prodigious quantities of heat that could be generated by friction. He found that dull implements actually produced more heat in boring than sharp ones. This was particularly interesting, as the caloric theory would have predicted that sharp instruments would more effectively release the caloric trapped in the metal being bored, and thus produce more heat. Rumford also conducted the most precise and rigorous experiments demonstrating that weight was neither gained nor lost by heat. He theorized that heat was caused by vibrations in an ether, which set particles in motion, and was followed by Sir Humphry Davy, whose endorsement further raised the prestige of the mechanical theory. Thomas Young (1773–1829) also set forth a mechanical theory of heat, which he believed was propagated in waves. However, the caloric theory, which was much more quantitative than the mechanical theory, remained dominant.

The book *Analytical Theory of Heat* (1822) by the French mathematician Jean-Baptiste-Joseph Fourier (1768–1830) raised the study of heat to its highest degree of mathematical sophistication. Fourier concentrated on problems of conduction, and the mathematical formalism of his work prevented it from contributing to the debate on the nature of heat. It did, however, help push the subject of heat from the realm of chemistry to that of mathematical physics. The mechanical theory vanquished caloric only with the rise of thermodynamics in the mid-nineteenth century.

See also Black, Sir Joseph; Calorimeters; Chemistry; Lavoisier, Antoine-Laurent; Thermometers; Thompson, Benjamin (Count Rumford).

References

Hankins, Thomas L. *Science and the Enlightenment.* Cambridge: Cambridge University Press, 1985.

Wolf, Abraham. *A History of Science, Technology, and Philosophy in the Eighteenth Century.* 2d ed. Revised by D. McKie. Gloucester, MA: Peter Smith, 1968.

Herschel Family

Friedrich Wilhelm Herschel (1738–1822) and his sister and assistant Caroline Herschel (1750–1848) revolutionized observational astronomy in the late eighteenth century. The Herschel family originated in the German electorate of Hanover, then a possession of the king of Great Britain. The young Friedrich Wilhelm entered the Hanoverian Guards, a regiment in British service, as an

William Herschel was a renowned astronomer in England who conducted the first in-depth studies of our solar system and the universe beyond. In his later years, he was the royal astronomer for King George III. (Perry-Castaneda Library)

oboe player, and fled to England after his regiment suffered defeat at the hands of the French. He lived for several years in England as a musician and composer, eventually becoming organist to the cathedral at Bath. (The English knew him as William Herschel, the name by which he is usually referred.) After the death of their father in 1763, William brought his younger brother Alexander, a talented instrument maker, over to England. Caroline came over in 1772. By this time, William had become fascinated by telescopes and astronomy, a subject in which he was largely self-taught. Caroline, whose education had been entirely in housewifely duties, also taught herself astronomy with William's assistance.

William determined to build a reflecting telescope larger than any available. With the assistance of Alexander and Caroline, who later described herself as stuffing bits of food in William's mouth because he could not interrupt his lens grinding to eat, he built first a five-and-one-half-foot telescope, then a seven-foot, a ten-foot, and finally a twenty-foot telescope. All this time, he continued full-time work as a musician and composer, in which Caroline also assisted him. William determined on using his excellent telescopes for a systematic study of the entire heaven as viewed from England—a "sweep" of the heavens. He would conduct four sweeps of the heavens in his career.

Herschel made many discoveries and began to communicate them to the Royal Society. The most dramatic discovery occurred on 13 March 1781, when Herschel detected a previously unrecorded object in the constellation Gemini. At first he suspected it to be a comet. So high was the quality of Herschel's telescopes that other astronomers, including Nevil Maskelyne at Greenwich, were unable to verify the position of the new object. However, that summer Anders Johann Lexell (1740–1784), the imperial astronomer of Russia, was able to compute the orbit of the new object from Herschel's observations and determine that it was a planetary orbit rather than a comet's. There was no precedent for the discovery of a new planet. It was not clear even how it should be named. Herschel wanted "Georgium Sidus," George's star, after George III of Great Britain and Hanover (r. 1760–1820). Some in France, not wanting a British king among the celestial deities, called it Herschelium, but the ultimate solution, naming the planet after Uranus, father of the gods in classical Greek mythology, was put forth by the German astronomer Johann Elert Bode (1747–1826).

Uranus made Herschel the most famous astronomer in Europe. He was quickly admitted as a fellow of the Royal Society. In return for some entertaining astronomy demonstrations before the royal family, Herschel received a pension from George III, enabling him to quit his music job and move to the area of London. The king also financed an enormous forty-foot reflector for Herschel that proved extraordinarily difficult to build. When finally put into operation in 1789, it enabled Herschel to almost immediately make another significant discovery, the identification of a sixth moon of Saturn, Enceladus. He also discovered two moons to Uranus, Oberon and Titania, and another Saturnian moon, Mimas.

In 1783 William Herschel married Mary Pitt (d. 1832). This led to Caroline's moving to a separate residence, although the two continued to work together, Caroline taking notes of William's observations and discovering eight comets on her own. She discovered her first comet in 1786. It received a great deal of publicity, being erroneously identified as the first comet discovered by a woman. (Maria Winkelmann [1670–1720] had discovered a comet in 1702.) The following year Caroline received a pension (much smaller than her brother's) from the crown. She was one of the very few women of the period who can be described as a professional research scientist. The Herschel astronomical tradition was carried into the next generation by William's son, John Frederick William Herschel (1792–1871).

William Herschel's interest in the late eighteenth century was turning from the solar system to the stars. In 1783, inspired by the catalog of nebulae created by Charles Messier (1730–1817), Herschel began his own search. In his career he charted more than 2,000 new objects. Another area in which Herschel was extraordinarily productive was the identification of double stars, close pairs of stars. The study of double stars was pioneered by the Herschels' friend John Michell (1724–1793). Herschel identified many new double stars, and published catalogs of them in 1782, 1785, and 1821. He established that the stars were indeed close in astronomical terms, and not just pairs of stars on the same line of sight from Earth. Herschel found additional evidence for the "proper motion" of the stars—the assertion that the stars of the Milky Way were not static but moving relative to each other and the Sun. This idea did not originate with Herschel, but he provided much stronger evidence for it. He also showed that the Sun itself was moving toward a point in the constellation Hercules. The full implications of Herschel's discoveries, however, were not followed up for decades. He also surveyed the Milky Way galaxy and gave an approximate description of its shape. The work of Herschel is the foundation of stellar astronomy.

Herschel was one of the few astronomers doing solar observations. In 1800 he noticed that the heat of the image projected by his lens did not correspond with the light of the image. Further experimental investigation of this phenomenon led him to the discovery of infrared light that same year.

Caroline Herschel published a revision and update of the star catalog by John Flamsteed (1646–1719) in 1798. The following period was spent assisting William rather than in her own research. After her brother's death in 1822 Herschel moved back to Hanover. She resumed her independent research, producing a catalog of nebulae in 1828 for which she received a gold medal from the Royal Astronomical Society. In the many years of life left to her, Herschel was known for her fierce devotion to her brother's memory and as a pioneer for scientific women. Along with the mathematician Mary Somerville (1780–1872), she was admitted as an honorary, nonvoting member of the Royal Astronomical Society in 1835.

See also Astronomy; Comets; Telescopes; Women.

References

Lodge, Sir Oliver. *Pioneers of Science and the Development of Their Scientific Theories.* 1926. Reprint, Mineola, NY: Dover, 1960.

North, John. *The Norton History of Astronomy and Cosmology.* New York: W. W. Norton, 1995.

Hospitals

The hospital came into its own as a center of scientific and medical research and teaching by the late eighteenth century. Although clinical teaching had a long history in medical education, the eighteenth-century emphasis on bedside experience in the training of physicians meant that hospitals drew steadily closer to university medical professions. The University of Edinburgh was one of many universities to use the town hospital to provide its medical students with practical training. Edinburgh professors took groups of students through the hospital, discussing the cases.

Hospitals themselves were rapidly changing and expanding, becoming institutions for sick people rather than catchall institutions serving the poor and destitute in general. Medical hospitals themselves specialized more, with the rise of institutions such as the hospital for the insane, the hospital for sexually transmitted diseases, and the lying-in hospital for pregnant women to deliver. Much hospital change was due to the drive of absolutist states to get hospitals, usually run by religious orders in Catholic countries, under government control. Although the most dramatic example of this phenomenon is the government takeover and expansion of the hospital sector in revolutionary Paris, the revolutionaries were continuing a trend begun in the last two decades of the monarchy. A commission of the Royal Academy of Sciences was appointed to examine Paris's largest hospital, the notoriously overcrowded and unsanitary Hôtel-Dieu. Pierre-Simon de Laplace, one member of the star-studded commission, rated hospitals by the probability of a patient dying in one, and rated the Hôtel-Dieu poorly, pointing out that two of nine patients admitted died there. In addition to the report, another document associated with the commission is *Memoirs on the Hospitals of Paris* (1788) by another member of the commission, the surgeon Jacques Tenon (1724–1816), a classic discussion of the functions of hospitals in the context of eighteenth-century medicine.

The founding or takeover of hospitals was considered an appropriate activity for reforming absolutist monarchs throughout Europe. (In England, private charity played the leading role in hospital expansion.) Among the most important of these absolutist foundations was the General Hospital of Vienna, reorganized and rebuilt in 1784 by the Hapsburg emperor Joseph II (r. 1765–1790), one of the keenest secularizers among the enlightened absolutists. Rationally organized, its staff included a physician in charge of teaching, and it had eighty-six clinical beds. Vienna was the first city where the hospital became the dominant institution in medical education and research.

It was quickly followed by revolutionary Paris. The revolutionary insistence on secularization and centralization—hospitals became state property in 1790—enabled the construction of larger and better-run institutions. (The state of the prerevolutionary hospitals can be seen in the decree of the Constituent Assembly in 1793 that every patient should have his or her own bed.) The leading physicians of revolutionary and Napoleonic Paris, such as Marie-François-Xavier Bichat, Philippe Pinel (1745–1826), and René-Théophile-Hyacinthe Laënnec (1781–1826), were all on staff at hospitals, and much of their medical knowledge was based on clinical experience. The concentration of sick and injured people in the Paris hospitals, the largest in the world, meant that physicians had enough living and dead bodies (autopsies were fundamental to medical research in the hospitals) to study diseases systematically. Paris physicians flaunted their contempt for "bookish" medicine acquired from texts rather than from the bodies of the sick. Paris became the center of medical research and training, replacing Montpellier in France and Edinburgh in Europe as a whole. In addition to the prestige of its leading physicians, its key advantage for students was freedom of access to hospital patients, whether living or dead. French medicine for most of this period was known for its clinical studies of specific diseases or other phenomena, such as Laënnec's 1802 articles on peritonitis in the *Journal of Medicine, Surgery, and Pharmacy,* rather than works of theory or the application of knowledge from other sciences to medicine. It also became known for its passion for diagnosis and relative lack of interest in curing the sick.

See also Bichat, Marie-François-Xavier; Madness; Medicine.

References

Ackerknecht, Erwin H. *Medicine at the Paris Hospital, 1794–1848.* Baltimore: Johns Hopkins University Press, 1967.

Gillispie, Charles Coulston. *Science and Polity in France at the End of the Old Regime.* Princeton: Princeton University Press, 1980.

Porter, Roy. *The Greatest Benefit to Mankind: A Medical History of Humanity.* New York: W. W. Norton, 1998.

Humboldt, Alexander von (1769–1859)

The German Alexander von Humboldt was among the most versatile scientists of his time, and certainly the most well traveled. From a noble Prussian family, Humboldt was educated at home, acquiring interests in botany and physics from a young age. In 1789 he entered the University of Göttingen, Germany's leading institution, where he studied under Johann Friedrich Blumenbach for a brief period. (Humboldt's lifelong disdain for hierarchical arrangements of human races may have originated in Blumenbach's teaching.) In 1791, hoping to fulfill his family's wish for an administrative career, he entered the Freiberg School of Mining, which would open the way for a position as a Prussian mining official. At Freiberg Humboldt learned geology and mineralogy according to the doctrine of the school's head, Abraham Gottlob Werner, which he later repudiated in favor of a more dynamic view. In 1792 he entered the Prussian state service, and the following year he established an innovative training institution for miners themselves. He maintained his interest in botany and in 1793 published his first major scientific book, *Freiberg Flora.*

Freiberg Flora, a study of the hitherto largely unexplored subject of those plants, lichens, and mosses that grow underground, away from the light, greatly increased Humboldt's fame in the scientific world. It was clear that his destiny was not a steady rise in the Prussian bureaucracy. Instead, he left the state service and took a geological and botanical trip through the Swiss Alps in 1795. He also carried on an extensive series of experiments in the relation of electricity to physiology, subjecting his own body to many painful shocks. This work was published in two volumes in 1797.

Accompanied by a French botanist, Aimé-Jacques-Alexandre Bonpland (1773–1858), Humboldt planned to join the group of savants in Egypt during the Napoleonic invasion, but unsettled political conditions in North Africa forced the two to abandon their plan. Instead, they set out for Spanish America, with a commission from the king of Spain to collect for the king's museum, as well as make geographical and astronomical observations. Although the Spanish had sent the Royal Botanical Expedition to Mexico in 1787, much of Spanish America had never been systematically examined by European scientists. After dodging the British blockade of Spanish ports and surviving a shipboard typhoid epidemic, the expedition landed in Venezuela. From there Humboldt and Bonpland explored up the course of the Orinoco River, finding many new specimens and observing the life of the Native Americans in the area, as well as suffering intensely from mosquitoes. After spending the winter of 1800–1801 in Cuba—where close observation reinforced Humboldt's hatred of slavery—the two set out again, this time along the Andes.

There was much to see in the mountains and adjoining regions. Humboldt's observations of the different plants found at different elevations contributed to his later development of plant geography. Along with Bonpland and two others, he climbed the previously unknown peak of Chimborazo, then thought the highest mountain in the Americas. He observed the impressive remains of the Inca civilization of Peru. The expedition then went to Mexico, where Humboldt enjoyed the company of more educated people and indulged his growing interest in pre-Columbian American history and archaeology. From there he and Bonpland went to the

United States by way of Cuba. In the United States Humboldt was treated as a scientific celebrity and admitted to the American Philosophical Society. He visited Thomas Jefferson (1743–1826) at Monticello and discussed the future of the Americas, a subject in which he was always keenly interested, with American statesmen.

The scientific fruits of Humboldt and Bonpland's expeditions in South America were enormous. They discovered and collected thousands of species, and Humboldt produced the first accurate maps of the continent, with the aid of the state-of-the-art instruments he had brought from Europe. Humboldt also pioneered what became one of his lifelong interests, the mapping of magnetic variation. His publications also laid the foundations for the study of pre-Columbian America, the greatness of which he brought to European attention.

The best place from which to publish and disseminate these findings was the center of European science, Paris, where Humboldt lived after his return to Europe in 1803. There he became a member of the Society of Arcueil, the elite group of French physical scientists, and was admitted to the First Class of the Institute of France, France's leading "official" scientific body, as a foreign associate. Humboldt became particularly close friends with Joseph-Louis Gay-Lussac, with whom he collaborated on *Analysis of Atmospheric Air* (1805). (With the exception of that with his sister-in-law, Caroline, all of Humboldt's close emotional relationships were with men.) But the restless Humboldt could not stay long in one place, and the authoritarianism of Napoleonic Paris was not congenial. He was soon off to Italy and Berlin, from which he returned to Paris in 1808. These years were marked by a steady stream of publications, caused by Humboldt's desire to disseminate what he had learned of the Americas, and also by a financial pinch, as the Humboldt family lands had suffered greatly during the Napoleonic Wars. *Views of Nature* (1807) was a collection of popular essays on natural phenomena, originally published in German and translated into French and English. More specialized was *On Isothermic Lines and the Distribution of Heat in the World* (1817), which introduced the concept of isothermic mapping.

The bulk of Humboldt's labors went into the vast thirty-volume report on his travels in Spanish America. On specific parts of this work, he had the collaboration of other scientists, including Bonpland and the German astronomer Jabbo Oltmanns (1783–1833). But most of it was Humboldt's, including three of the most celebrated sections: the single-volume *Essay on the Geography of Plants* (1807), which set forth his theories on the influence of height, latitude, temperature, and rainfall on the distribution of plants; the four-volume *Political Essay on the Kingdom of New Spain* (1811), the first elaborate and accurate statistical and geographical description of Mexico; and the seven-volume *Personal Narrative* (1815–1826), which recounted Humboldt's experiences from 1799 to 1803. Charles Darwin (1809–1882) would later find the *Personal Narrative* an inspiration in his own scientific journey to South America.

In 1827 the Russian government invited Humboldt, then living in Berlin, to carry out a long-held ambition to travel in Central Asia. The journey took place in 1829, and took Humboldt as far as the borders of China. It was primarily mineralogical, as the Russian government was interested in more efficiently exploiting the country's mineral wealth, although Humboldt also made many geographical, geological, climatological, and magnetic observations. He did not make the social observations he made in South America, for fear of offending the repressive Russian government. The results of the journey were published in three volumes as *Central Asia* (1843).

Most of Humboldt's scientific energies in the closing decades of his life were taken up by serving as a patron and correspondent to younger scientists and by writing a huge work of natural philosophy, the five-volume *Cosmos* (1845–1861). *Cosmos* was an attempt to both popularize science and integrate the science of the day—and Humboldt had kept up with new scientific developments—into a universal philosophy of nature. This philosophy was romantic, based on the historical development of the universe, and particularly Earth, and its dynamic changes. Humboldt died before he could finish *Cosmos,* but it was very popular throughout the nineteenth century.

See also Arcueil, Society of; Botany; Exploration, Discovery, and Colonization; Gay-Lussac, Joseph-Louis; Geology; Meteorology; Romanticism.

References

Crosland, Maurice. *The Society of Arcueil: A View of French Science at the Time of Napoleon I.* Cambridge: Harvard University Press, 1967.

De Terra, Helmut. *Humboldt: The Life and Times of Alexander von Humboldt, 1769–1859.* New York: Alfred A. Knopf, 1955.

Hunter Family

Two brothers from Scotland, William Hunter (1718–1783) and John Hunter (1728–1793), were the intellectual and professional leaders of the surgical community in London in the mid-eighteenth century. Originally intended for the church, William Hunter decided he would not conform to the orthodoxy of the Church of Scotland. He became an assistant to William Cullen (1710–1790), then practicing medicine at Hamilton. William Hunter showed a facility for surgery and obstetrics, and in 1740 went to London as an assistant to the man-midwife William Smellie (1697–1763), a friend of Cullen. The community of Scottish medical men in London was an extensive one, and Hunter soon transferred his loyalties from Smellie to James Douglas, an anatomist, fellow of the Royal Society, and fellow of the Royal College of Physicians. Hunter moved in with the Douglas household, staying there after Douglas's death. He made his career in London as a surgeon, man-midwife, and anatomical lecturer, attaining a leading position in all these areas. In 1748 John, without a profession, joined William in London, where he spent ten years as William's assistant.

William Hunter founded a famous school of anatomy in 1768. This institution was also the site of collaborative research, the most important of which concerned the lymphatic system and the circulation of blood between a pregnant woman and her fetus. Since his individual knowledge and reputation are what attracted pupils, it was to Hunter's advantage not to make much of the knowledge public. Much of his anatomical teaching was given in the form of lectures rather than published treatises. This practice led to endless priority disputes when another researcher published discoveries, often made collaboratively, that Hunter as the head of the school claimed as his own. The most important dispute was between William and John over the utero-placental circulation in 1780. The brothers, who had been drifting apart for years, seldom spoke after that. William Hunter's most important medical publication was a lavishly illustrated elephant folio, *Anatomy of the Human Gravid Uterus* (1774). His writings on obstetrics recommended a less interventionist practice than Smellie's, although he allowed for the use of forceps in specific situations.

John Hunter's experimental investigations covered such diverse subjects as the teeth, gunshot wounds (extremely important to surgeons, as military practice employed hundreds of them), and venereal disease, in which he concluded incorrectly that syphilis and gonorrhea were really different forms of the same disease. He supervised the first recorded case of artificial insemination in 1776, although it was kept secret for many years. Hunter performed a huge number of dissections of both humans and animals, and published some of the most important work

This portrait of John Hunter shows him as both author and collector. (National Library of Medicine)

on comparative anatomy before Georges Cuvier. He established the slow rate of digestion of hibernating animals by forcing meat in small sacks tied to strings into their stomachs, and then pulling it out later. He also disputed Lazzaro Spallanzani over digestion. Hunter was an extreme vitalist, who saw blood as alive and almost as conscious.

Both brothers were financially and socially successful (they were admitted, separately, to the Royal Society in 1767) and avid collectors. John Hunter's massive collection of specimens became the basis of the Hunterian Collection of the Royal College of Surgeons, whereas William Hunter's anatomical and natural history collections were bequeathed to the University of Glasgow.

The brothers' nephew, Matthew Baillie (1761–1823), also a surgeon, inherited William's anatomy school and became a physician extraordinary to George III (r. 1760–1820). Baillie was the author of an influential textbook of pathology, *Morbid Anatomy of Some of the Most Important Parts of the Human Body* (1793). Including detailed copperplate engravings by William Clift (1775–1849), it contains classic descriptions of cirrhosis of the liver, ovarian cysts, and gastric ulcers. It went through several editions in England and America and was translated into the major European languages. The anatomy school closed in 1831.

See also Anatomy; Medicine; Midwives; Surgeons and Surgery.

References

Porter, Roy. *The Greatest Benefit to Mankind: A Medical History of Humanity.* New York: W. W. Norton, 1998.

Porter, Roy, and W. F. Bynum, eds. *William Hunter and the Eighteenth-Century Medical World.* Cambridge: Cambridge University Press, 1985.

Hutchinsonianism

The natural philosophy of John Hutchinson (1674–1737) was the principal rival to Newtonianism in eighteenth-century Britain. Hutchinson, who had served as an assistant to the fossil hunter John Woodward (c. 1665–1728), published *Moses's Principia,* a history and theory of the Earth, in 1724 and followed it the next year with *An Essay toward a Natural History of the Bible.* Hutchinson's natural philosophy was based on his unique approach to biblical interpretation. He claimed that Hebrew, the divine language, properly interpreted, revealed natural truths. The Bible, every letter of which was inspired and revealing of truth, was the only legitimate textual source for natural philosophy, and Hutchinson attacked Isaac Newton (1642–1727) for deriving his ideas from ancient pagan authors while ignoring Hebrew. Hutchinsonians achieved their biblical interpretations in a remarkably flexible manner, by insisting that only the consonants of the Hebrew text were divinely inspired, the vowels being Jewish corruptions. Hutchinson identified the Trinity with the three elements that he believed governed the Earth:

the Father with fire, the Son with light, and the Holy Spirit with air. He argued that the workings of the world were purely mechanical, and that one thing influenced another only by direct contact—thus, he shunned the Newtonian theory of gravity, which involved action at a distance. By invoking these "occult" forces, Hutchinson claimed, Newton's theory tended toward paganism and, if God were claimed to be the immediate cause of gravity, toward pantheism. Hutchinsonianism identified the universe as a self-maintaining creation and disapproved of some Newtonians' willingness to invoke occasional divine intervention. In many ways Hutchinsonianism was an English version of Cartesianism, although its intellectual origins are difficult to reconstruct, as Hutchinson acknowledged no sources but the Bible.

Interest in Hutchinson's work was raised by the publication of his collected works in 1748–1749. Hutchinsonians tended to be those who attacked the established Church of England, dominated by Whig Newtonians, from a High Church, Tory position. Oxford University, the more High Church of the two English universities, was a Hutchinsonian stronghold in the mid-eighteenth century, as was the disestablished Scottish Episcopal Church. The Anglican minister William Jones of Nayland (1726–1800) wrote the most important work of Hutchinsonian theory after Hutchinson, *Essay on the First Principles of Natural Philosophy* (1762). (Jones and some other later Hutchinsonians did not share Hutchinson's disdain for Newton, speaking of him with respect and incorporating some aspects of Newtonianism.) Hutchinsonianism was introduced to America by the Reverend Samuel Johnson (1696–1772), the founder of King's College, New York, the future Columbia University.

Hutchinson's identification of the Trinity as manifested in nature was particularly reassuring to those churchmen concerned over the rise of Unitarianism, but the Hutchinsonians were always the weaker party in the church, receiving no promotions under George II (r. 1727–1760). Hutchinsonianism did inspire some scientific work, as in the case of the clergyman and fossil hunter Alexander Catcott (1725–1779), whose geological fieldwork was inspired by the Hutchinsonian doctrine of the stability of the Earth, which Catcott applied to the geological strata. Like many Hutchinsonians, he ascribed any interruptions or distortions in the strata to the Flood of Noah, an assertion he put forth in *Treatise on the Deluge* (1761). Despite increasing evidence of textual variations in the Hebrew Old Testament text, Hutchinsonianism persisted in Britain, particularly in Scotland, to the early nineteenth century, and is one root of the pronounced interest in biblical issues by many nineteenth-century British geologists, in marked contrast to their continental peers.

See also Cartesianism; Geology; Religion.

References

Katz, David. "The Occult Bible: Hebraic Millenarianism in Eighteenth-Century England." In *The Millenarian Turn: Millenarian Contexts of Science, Politics, and Everyday Anglo-American Life in the Seventeenth and Eighteenth Centuries,* edited by James E. Force and Richard H. Popkin, 119–132. Dordrecht, Netherlands: Kluwer, 2001.

Porter, Roy. *The Making of Geology: Earth Science in Britain, 1660–1815.* Cambridge: Cambridge University Press, 1977.

Schofield, Robert E. *Mechanism and Materialism: British Natural Philosophy in an Age of Reason.* Princeton: Princeton University Press, 1970.

Wilde, C. B. "Hutchinsonianism, Natural Philosophy, and Religious Controversy in Eighteenth-Century Britain." *History of Science* 18 (1980): 1–24.

Hutton, James (1726–1797)

The Scottish physician and natural philosopher James Hutton promoted an original theory of the Earth, although his influence on the geologists of his own time was limited. Hutton earned a Leiden M.D. in 1749, and moved from his rural Scottish home to

Edinburgh in 1768. There he participated in the Scottish Enlightenment as a member of the Royal Society of Edinburgh. He became a particularly close friend of the economist and moral philosopher Adam Smith (1723–1790) and Joseph Black, who greatly influenced Hutton's ideas about heat. It was before the Royal Society of Edinburgh that he read the paper that propounded his geological theory in 1785, publishing it three years later.

Hutton's theory of the Earth was a striking departure from both the dominant Wernerian tradition being established on the Continent and the British Isles and the Bible-influenced tradition of much older British theorizing. Hutton, a Deist in religion, was much less interested in the beginning of the Earth than were Wernerians or biblical geologists, and was willing to use, at least as a working idea, the old Aristotelian concept of the eternity of the Earth. Unlike his Scottish geological contemporaries, he was not a university professor, nor did he view geology as a means for economic development. Nor, despite his later reputation as a geological empiricist, did he derive his theories from observations. Hutton put his excellent skills as a geological observer to work not to gather data from which he then theorized, but to test the theories he had arrived at through rational deductions. Indeed, many of his early opponents criticized his work for its lack of induction.

Hutton's claim was that the Earth had been formed in its present habitable state by the working of gradual processes, which can still be observed. Heat, light, water, and living things were bound together in an endless cycle whereby inorganic substances maintained plants, plants produced vital air and phlogiston (Hutton always remained loyal to phlogistic chemistry) for animals, and animals by dying replenished the stock of inorganic substances. Erosion, which Hutton, unlike his contemporaries, treated as a central issue, was a necessary part of this unending process. One of the most famous statements in the history of geology is that which concludes Hutton's 1788 paper, "The result, therefore, of our present enquiry is that we find no vestige of a beginning—no prospect of an end." Hutton's Deist God made the Earth for humanity not by an arbitrary act of creation, but by setting up a system of natural laws that maintained habitability. Hutton's conception of his intellectual task was fundamentally Newtonian, in that he sought fundamental laws that governed the system, rather than piling up observations of local conditions.

The principal engine of specifically geological processes for Hutton was a central fire in the Earth. For this reason his theory is often known as plutonism, after the Roman god of the underworld. The heat of the central fire, working on rock eroded from the continents and falling to the seafloor, created new rock, and was responsible for elevations, veins, earthquakes, and volcanoes. The igneous origin of granite and whinstone, which he had observed at Salisbury Crags, was central to Hutton's theory. He considered his observations of veins of red granite in the southern Scottish Highlands, veins that had intruded from elsewhere, as powerful evidence for his theory.

The principal debate inspired by Hutton's theory was not with Wernerian neptunists, but with British and Irish geologists suspicious of its implications for the biblical account of the Earth's history. It did him no good that the decade after the publication of the first version of *Theory of the Earth* was a time of violent cultural and religious reaction in Britain, sparked by the French Revolution. Skepticism about Christianity and the Bible, which was acceptable in Hutton's Scottish Enlightenment circles, was no longer so in a Britain combating the forces of revolutionary infidelity emanating from France. Hutton's leading opponent was the Irish chemist and mineralogist Richard Kirwan (1733–1812). In 1793 Kirwan savagely attacked Hutton on the irreconcilability of Hutton's Earth theory with Genesis and the igneous origins of granite and other rocks. This prompted Hutton's publication of an expanded version of his theory in 1795, the two-volume *Theory of the Earth, with Proofs and Illustrations.*

More important than the Genesis issue in limiting the impact of Hutton's theory on continental geology were its difficulties in explaining how heat led to the formation of different rocks, such as limestone, whose formation seemed explainable by the action of water. The dominant Wernerian school of geologists rejected Hutton's theory as not in accordance with their observations. Parts of Hutton's theory, such as the igneous origin of granite, did win acceptance, but the theory itself won few disciples even in Scotland. The most important was John Playfair (1748–1819), whose *Illustrations of the Huttonian Theory of the Earth* (1802) became the classic statement of Huttonian doctrine. Another Huttonian was the experimentalist Sir James Hall (1761–1832), whose efforts to duplicate Hutton's ideas about the formation of rocks by heat under laboratory conditions seemed to support Hutton's claims. But the Huttonians remained a minority in the early nineteenth century.

See also Geology; Phlogiston; Religion; Royal Society of Edinburgh; Werner, Abraham Gottlob.

References

Laudan, Rachel. *From Mineralogy to Geology: The Foundations of a Science, 1650–1830.* Chicago: University of Chicago Press, 1987.

Porter, Roy. *The Making of Geology: Earth Science in Britain, 1660–1815.* Cambridge: Cambridge University Press, 1977.

Hygrometers

Instruments to measure the moisture of the air took several different forms in the Enlightenment. Hygroscopic hygrometers and condensation hygrometers were both refined, while the psychrometer was invented during the period.

Hygroscopic hygrometers used the changes and deformations atmospheric moisture induced in different substances, mostly organic. The simplest, originating in the fifteenth century, measured the weight increase of a sponge or ball of wool exposed to the air. Many in the eighteenth century used volume increases or twistings. A popular model invented by Johann Georg Leutmann (1667–1736) in the early eighteenth century used a short length of catgut. Two were announced in 1773, the work of the great rival meteorologists Jean André Deluc (1727–1817) and Horace-Bénédict de Saussure (1740–1799). De Saussure used a specially prepared human hair whose length varied with moisture, whereas Deluc used an ivory vessel filled with an indicator fluid that mounted a scale as the ivory expanded. Deluc later switched to whalebone, and Noel Retz came up with a much cheaper model employing the same principle but using goose quills. (Goose-quill hygrometers were the ones used in the Mannheim Meteorological Society project to record standardized readings.) The controversy between Deluc and de Saussure was bitter, but the superior sensitivity and cheapness of de Saussure's hygrometer won the day, and by 1820 the human-hair hygrometer was the only hygroscopic hygrometer in common use.

Condensation hygrometers, invented in the mid-seventeenth century, were vessels filled with cold water or ice. The experimenter measured the condensation of atmospheric moisture on the outside of the vessel. The Montpellier medical professor Charles Le Roy (1726–1779) used a condensation hygrometer to establish the dew point, the temperature at which atmospheric water condenses, around 1751. Le Roy's method was based on trial and error, with repeated fillings of the vessel with water at different temperatures. Later condensation hygrometers incorporated thermometers and worked more efficiently.

The psychrometric hygrometer was based on an observation from the early days of thermometry: that thermometers with wet bulbs record slightly lower temperatures than thermometers with dry bulbs. William Cullen (1710–1790) correctly identified the cause of the cooling as evaporation. Wet-bulb thermometers could then be used to measure the rate at which water evaporated, a process related to the amount of moisture already in the air. Sir John Leslie (1766–1832) combined wet and dry bulbs into a single hygrometer in 1799.

See also Instrument Making; Meteorology.

Reference

Middleton, W. E. Knowles. *Invention of the Meteorological Instruments.* Baltimore: Johns Hopkins University Press, 1969.

I

Illustration

The eighteenth and early nineteenth centuries were the golden age of scientific illustration. Many techniques became more widely diffused, such as copperplate engraving and color printing. Lithography was invented at the end of the eighteenth century, and the ancient technique of woodcutting was made more capable of detailed representation. Technical advances fed the expanding market for scientific books, prints, and catalogs. Illustrators accompanied voyages of exploration, such as those of Captain James Cook, to capture newly found plants and animals in a way resistant to decay. Lavishly illustrated works were produced to record the gardens and collections of the famous and powerful, or for political reasons as part of a strategy of claiming dominance over an area. The *Encylopédie* alone contained many volumes of plates, sold separately from the text, as was common for large works. The demand for illustration was such that alongside new illustrations, many old ones continued to be reused, even as their scientific accuracy was out of date. The famous illustration of a flea by Robert Hooke (1635–1703), from *Micrographia* (1665), appeared in a work of popular science as late as 1825. Illustration itself became a more expensive and complex process, with workers increasingly specializing in one stage, such as producing the initial image, engraving, or coloring.

Illustrators occupied a particularly marginal position among science, craft, and art. The training required was not part of the normal curriculum of European educational institutions or scientific academies. One result was that illustration was among the aspects of science most open to women. The only woman on the staff of the eighteenth-century Royal Botanical Garden of Paris, a major center of botanical art, was an illustrator, Magdeleine Basseporte.

Artistically, the most important illustrators worked in botany and natural history. Preeminent among them was the German flower painter Georg Dionysius Ehret (1708–1770), a friend and coworker of Carolus Linnaeus who later settled in Britain, where he became a fellow of the Royal Society. The German brothers Ferdinand Bauer (1758–1840) and Franz Bauer (1756–1826) also made careers in England, where Franz was appointed flower painter to Kew Gardens in 1790. Another great flower painter was Pierre-Joseph Redouté (1759–1840), whose

This anatomical textbook title page includes angels at the top, and at the bottom, dogs eating body parts that have fallen off the dissector's table. (National Library of Medicine)

artistically most important work was *The Lilies* (1802–1816). A host of lesser illustrators were on the staff of the botanical gardens of Europe, notably the Royal Botanical Garden, which had had a staff artist since the seventeenth century and developed into a center of botanical art.

Botanical illustration and much else would be transformed in the nineteenth century by the 1797 invention of lithography by the German Aloys Senefelder (1771–1834). The technique, whereby different sections of stone were treated to attract or repel the colored inks, was used in botanical works by 1812, when Rudolf Ackermann (1764–1834) created *Series of Thirty Studies from Nature*. Around the same time woodcutting was transformed by the Newcastle engraver Thomas Bewick (1753–1828) in *A General History of Quadrupeds* (1790) and *History of British Birds* (1797–1804). Bewick led the revival of the ancient technique of cutting across the grain in harder wood rather than with the grain in softer wood. This made woodcuts capable of much finer detail. Zoological illustration in general was less developed than botanical, but there was an extensive market for it in the great encyclopedic works of natural history produced in the period, from Georges-Louis Leclerc de Buffon's *Natural History* to the works of Georges Cuvier. The greatest bird illustrator, John James Audubon (1785–1851), worked after this period, but his work was preceded by the lavishly illustrated nine volumes by Alexander Wilson (1766–1813), *American Ornithology* (1808–1814). Another area where professional illustrators were increasingly important was anatomy, where copperplate engraving dominated. Anatomists like London's Hunter brothers worked closely with professional illustrators.

Although not as aesthetically pleasing or as technically developed as botanical or anatomical illustration, illustrations also played an important role in experimental physics. Many textbooks contained illustrations of experiments as an aid for those wishing to repeat them. Illustrations often brought experiments into the domestic sphere by placing them in the surroundings of a home rather than a laboratory. The fact that many times illustrations were carried from book to book sometimes posed problems for home experimenters when text and illustrations related to different experiments. The role of the experimenter in the illustrated experiments was often indicated by disembodied hands carrying out the procedures. The hands were successors to the putti, the infantlike angels who frequently carried out experiments in seventeenth-century illustrations. The use of putti continued during the Enlightenment in Catholic regions, but disappeared after the suppression of the Jesuit order in 1773. Other illustrations, such as those in *Opticks* (1704) by Isaac Newton (1642–1727), were more schematic, emphasizing the physical principles demonstrated by the experiment rather than the experiment itself. In both experimental physics and astronomy, the illustration of instruments was a way of legitimating their use and results.

Not all disciplines developed in the direction of more elaborate illustration. Some went the opposite way. Mathematical mechanics, as it became more purely mathematical and analytical rather than geometrical, also became more textual. Joseph-Louis Lagrange boasted that his *Analytical Mechanics* (1787) contained not a single diagram, only equations. Chemistry was another discipline that moved away from illustration, as the complex symbolic language of alchemy, which lent itself so well to visual presentation, was abandoned. The most popular chemical reference of the eighteenth century, Pierre Joseph Macquer's *Dictionary of Chemistry* (1766), was unillustrated. The visual presentation of information did continue with the display of tables of affinity in chemical texts.

See also Anatomy; Botany; Popularization; Women.

References

Ford, Brian J. *Images of Science: A History of Scientific Illustration.* New York: Oxford University Press, 1993.

Mazzolini, Renato G., ed. *Non-verbal Communication in Science prior to 1900.* Florence: Leo S. Olschki, 1993.

Shea, William R., ed. *Science and the Visual Image in the Enlightenment.* Canton, MA: Science History Publications, 2000.

Imperial Academy of Sciences of St. Petersburg

Russia's first academy of sciences was founded in 1724, the culmination of the efforts of Czar Peter the Great (r. 1682–1725) to bring Western knowledge and techniques to Russia. Unlike Western institutions such as the Berlin Academy, the Imperial Academy of Sciences was a vast complex, including a university, high school, observatory, library, botanical garden, and printing press and translation bureau for Western scientific books. The academy had an almanac monopoly and published Russia's only newspaper, the *St. Petersburg Gazette.* The scientific academy proper, known as the "Archives Conference," was only part of this massive institution, designed to build a Russian scientific culture virtually from scratch. The Archives Conference was built on the model of the Paris Royal Academy of Sciences, with paid academicians. The members were divided into twenty academicians and twenty associates, and met weekly. The whole operation was overseen by a government-appointed director and vice director.

The first task to set up the academy was to recruit European academicians. With the aid of Christian Wolff, who politely declined an offer to come to St. Petersburg himself, sixteen European scholars were recruited to provide the nucleus of the academy. The leaders among the first group were the astronomer Joseph-Nicolas Delisle (1688–1768) from France and the mathematicians Nikolaus II Bernoulli, Daniel Bernoulli, and Jakob Hermann (1678–1733) from Switzerland. Factionalism along national lines, usually pitting the Germans against the Russians and French, would plague the Imperial Academy for decades. Few of the foreign academicians bothered to learn Russian. One exception was the most important foreign academician in St. Petersburg during the eighteenth century, Leonhard Euler, who arrived in St. Petersburg in 1727 to fill the chair vacated by the death of Nikolaus II Bernoulli and whose association with the Imperial Academy continued throughout his long career.

In its early years the academy suffered from intermittent government support, and its educational institutions withered as the Russian government preferred to send young Russian men to universities in Europe. Nonetheless, the Imperial Academy managed to establish a record of scholarly production. The first volume of *Commentaries* appeared in 1726, and its successors appeared on a regular basis. The Imperial Academy sought to overcome its isolation by building relationships with other scientific societies. Its strongest early relationship was with the Royal Society, which sent a number of meteorological queries for the Imperial Academy and with which the Imperial Academy set up an official exchange of publications in 1729. This was the first official exchange of publications between scientific societies, and a significant step in building international cooperation. Communications between St. Petersburg and London were good enough that the Imperial Academy tried to acquire the famous natural history collection of Sir Hans Sloane. The Imperial Academy was also in regular communication with a Jesuit scientific circle at the Chinese court in Beijing.

Given Russia's vast size, and the complete

lack of geographical knowledge about much of it, geography and cartography were high on the Imperial Academy's agenda. It sponsored the second expedition of Vitus Johassen Bering (1681–1741), from 1732 to 1743. The academy produced the first map of Russia in 1734, and the first Russian atlas in 1745. It published maps of the Pacific in 1758 and 1773.

In 1747 the academy finally received a formal charter, and in 1749 it launched its first prize competition. The subject was the relation of variations in the lunar motion to Newtonian theory, and the winner was the French mathematical physicist Alexis-Claude Clairaut (1713–1765). The Imperial Academy cooperated with the Paris Academy of Sciences in the observation of the transit of Mercury in 1753 and cosponsored an observing trip to Tobol'sk by the French astronomer Jean-Baptiste Chappe d'Auteroche (1722–1769) in connection with the transit of Venus in 1761. For the 1769 transit the academy sent out eight expeditions, including one to the farthest reaches of Russian Asia in Kamchatka. Conflicts between foreign and Russian-born academicians continued, particularly when the Russians acquired a distinguished scientist as leader, Mikhail Vasilyevich Lomonosov. The academy's high salaries and growing reputation attracted first-class savants (despite the handicap of the Russian weather), such as the German electrician Franz Maria Ulrich Theodor Hoch Aepinus (1724–1802), who arrived in 1757. It benefited from the patronage of Czarina Catherine the Great (r. 1762–1796), who in 1783 appointed Princess Yekaterina Dashkova (1744–1810) as the first woman head of a European scientific academy. The Imperial Academy suffered in the late eighteenth century from the accession of the anti-intellectual Czar Paul I (r. 1796–1801). In the early nineteenth century the Russian reaction against all things Western, inspired by revulsion at the French Revolution, reduced the Imperial Academy to marginality.

See also Academies and Scientific Societies; Bernoulli Family; Euler, Leonhard; Lomonosov, Mikhail Vasilyevich.

Reference

McClellan, James E., III. *Science Reorganized: Scientific Societies in the Eighteenth Century.* New York: Columbia University Press, 1985.

Industrialization

One of the most vexed questions in the entire historiography of eighteenth-century science is the relation of science to the burst of industrialization, later named the Industrial Revolution, which began in mid-eighteenth-century Britain and spread to the Continent in the early nineteenth century. Although no one regards technological innovation as a simple consequence of scientific progress, there is a range of positions, from those who hold that technology and science were largely separate in their development until the mid-nineteenth century to those holding that their development was already deeply interconnected in the eighteenth century.

The idea found in some of the older histories that the founders of British industry were uneducated "tinkerers" ignorant of the scientific developments of their time has been discredited. The single most important invention of the British Industrial Revolution was the steam engine with a separate condenser invented by James Watt (1736–1819). Watt came from a Scottish family with a tradition of knowledge of applied mathematics, mechanics, and basic science, and early in his career he was employed as a scientific instrument maker for the University of Glasgow where he was acquainted with Joseph Black's new discoveries about heat. He became a fellow of the Royal Society and maintained an extensive scientific correspondence in both Britain and France. Like his business partner, Matthew Boulton (1728–1809), Watt perceived his work as science-based, a perception shared by his scientific contemporaries. Like other technological innovators, Watt and Boulton frequently appealed to the improvement of science as a precedent for successful innovation. Both were members of the Lunar Society of Birmingham, and both educated their sons in modern science as preparation for taking over their fathers' businesses.

Newtonian mechanics as set forth in textbooks provided a common language and set of concepts for British engineers, and for many of their clients as well. Paradoxically, Britain's increasing backwardness in theoretical mechanics as the century progressed helped its engineers think in practical terms, as British mechanics textbooks tended to exemplify concepts in models and illustrations rather than equations. (The French moved toward greater emphasis on models in the late eighteenth century, in part in hopes of emulating British economic success.) British natural philosophers and demonstrators also gave lectures and wrote books specifically on the application of mechanics to industry, such as the northern English lecturer John Banks's *Treatise on Mills* (1795). British employers could also draw upon a pool of literate artisans and workers with some knowledge of mathematics and mechanics—a much smaller group in most continental countries, where literacy rates were lower, class divisions wider, and opportunities for self-education fewer.

The methods as well as the content of science influenced industrial development. Many engineers and inventors in the eighteenth century took a more "scientific" approach to innovation, performing repeated experiments and emphasizing precise measurement, such as John Smeaton (1724–1792), whose experiments on the waterwheel won him the Royal Society's Copley Medal.

The impact of science on technological innovation varied by sector. Although science had great impact on power industries, whether of water or steam, it had far less on mechanized cotton spinning and weaving, a key element in British industrial expansion. Innovators like Sir Richard Arkwright (1732–1792), inventor and improver of the water-frame spinning machine, were considerably less "scientific" than Watt, and their innovations made little or no use of new scientific principles.

The booming fabric industry also opened new possibilities for chemical technology. A university chemist, Francis Home (1719–1813) of Edinburgh, wrote the first scientific study of bleaching, *The Art of Bleaching* (1756). Home recommended sulfuric acid over sour milk as a bleaching agent. Although the manufacture of sulfuric acid expanded greatly in the eighteenth century, the ultimate winner was chlorine bleaching, invented by the French but used most widely and successfully by the new British textile industries. Another French innovation was the "Leblanc process" for making soda, a chemical used to clean textiles as well as in making soap and glass. Nicolas Leblanc (1742–1806), a French surgeon, patented his process in 1791. Dyeing also grew more "scientific" in this period, although the direct application of chemistry to dyeing had to wait until the mid-nineteenth century. Outside textiles, the Manchester chemist Thomas Henry (1734–1816) discovered a new process for making magnesium oxide and put the discoveries of different "airs" by chemists to work by starting a thriving business in artificial mineral waters.

Industry set new problems for science as well as drawing from it. Steam-engine theory, which eventually led to nineteenth-century thermodynamics, was carried on at the highest level by the French. Gaspard-Clair-François-Marie Riche de Prony (1755–1839) wrote *New Hydraulic Architecture* (1790) and Sadi Carnot (1796–1832) *Reflections on the Motive Power of Fire* (1824). The French also pioneered the industrial exhibition, first held in Paris in 1798. The leader of the jury that judged the competition was a scientist, the chemist Jean-Antoine Chaptal (1756–1832). It was Chaptal and fellow scientists including Claude-Louis Berthollet and Gaspard Monge (1746–1818) whom Napoléon turned to in his effort to create a French industrial system to rival the British.

See also Berthollet, Claude-Louis; Böttger, Johann; Desaguliers, John Theophilus; Instrument Making; Lunar Society of Birmingham; Technology and Engineering.

References

Cardwell, Donald. *The Norton History of Technology.* New York: W. W. Norton, 1994.

Jacob, Margaret. *Scientific Culture and the Making of the Industrial West.* New York: Oxford University Press, 1997.

Musson, A. E., and Eric Robinson. *Science and Technology in the Industrial Revolution.* Toronto: University of Toronto Press, 1969.

Insanity

See Madness.

Instrument Making

The eighteenth century saw a tremendous expansion of the business of making scientific instruments. This business was centered in London, although Germany was displacing England in the early nineteenth century. Scientific instruments could be roughly divided into three basic types: those made for functionality, such as navigational and surveying aids; those made for carrying on scientific research, such as the most advanced telescopes and microscopes; and what were called in the eighteenth century "philosophical instruments," those made to demonstrate scientific principles or for educational or entertainment purposes. (In London the practical instrument makers congregated in the east, and the makers of philosophical instruments in the west.) Instruments in all three categories saw steady improvement in consistency and precision. Wood and leather were replaced by brass. Other instruments, such as the calorimeter and the Leiden jar, were invented during this period.

The foundations of London's supremacy were laid in the early eighteenth century by, among others, the clock maker and fellow of the Royal Society George Graham (1673–1751) and the microscope maker John Marshall. These men mainly worked in small shops in a craft tradition. Businesses were often passed down in families. George Adams (c. 1714–1773) and his two sons, George the younger (1750–1795) and Dudley (c. 1760–1826), were in the instrument business for nearly a century. The London makers worked for a Europe-wide market, and many issued their promotional materials in French as well as English. Their workshops were attractions for scientifically minded visitors to London from the Continent and America.

Many wealthy Europeans, and some of more modest means, built collections, or "cabinets," of philosophical instruments. The Adamses built a close connection with the British royal family, which included one great collector, George III (r. 1760–1820). Special instruments meant more for display than for use, such as George Adams the younger's silver microscope, were created for rich collectors. Another aristocratic collector, Charles Boyle, fourth earl of Orrery (1676–1731), actually had an instrument named after him—the orrery, a clockwork device that represents the movements of the solar system's planets and moons. One group for whom good philosophical equipment was not a luxury but a necessity was that of public scientific lecturers and demonstrators. Lecturers and instrument makers often formed close alliances. Benjamin Martin (1705–1782) combined lecturing, textbook writing, and instrument retailing. The foremost Dutch instrument maker of the early eighteenth century, Jan van Musschenbroek (1687–1748), was the brother of the famous demonstrator Pieter van Musschenbroek (1692–1761). University professors of physics also needed good equipment, and at Protestant universities frequently had to pay for it out of pocket.

Around midcentury the London instrument business changed from one dominated by small shops to one where large firms, some employing more than fifty workers, controlled much of the market. The number of manufacturing concerns also increased.

A minister and schoolteacher, Joseph Priestley, whose chemical balance and weights are shown here, could not afford the precision balances used by his wealthy aristocratic contemporaries Henry Cavendish and Antoine-Laurent Lavoisier. (Edgar Fahs Smith Collection, University of Pennsylvania Library)

Whereas for most of the century around twenty-five to thirty new firms were founded per decade, this figure nearly doubled in the last two decades. Most large firms sold a broad range of different kinds of equipment, although one maker, James Short (1710–1786), took the opposite tack and specialized completely in reflecting telescopes. Dominant firms in the late eighteenth century included those of the Dollond family, originally French Protestant immigrants, and their relative by marriage Jesse Ramsden (1735–1800), generally considered the greatest maker, if not businessman, in the industry. Ramsden made one of the first precision balances for Henry Cavendish, and also the first plate generator of electricity. His perfectionism, however, meant that he often delayed the fulfillment of contracts, or even failed to fulfill them.

Many English makers, including Short, Peter Dollond (1730–1820), and Ramsden, were fellows of the Royal Society. This was a central difference between the London (and the Dutch, which was also expanding) and Paris communities of makers. French makers were treated as hired artisans by French scientists, and many were not even literate. Some were skilled nonetheless, notably Jean-Nicolas Fortin (1750–1831), who made another early precision balance, this one for Antoine-Laurent Lavoisier. The advent of precise weighing was a central aspect of Lavoisier's chemical revolution. (Far less is known about chemical equipment than about that of other branches of science, because so much was made of fragile glass. Balances are one exception.) French scientists and the French government looked with envy on London's instrument industry. In the 1780s there was a concerted effort to create a French industry rivaling England's led by the head of the Paris Observatory, Jacques-Dominique Cassini. Cassini's reorganization of the observatory was based on using instruments of French manufacture. His effort to find qualified French manufacturers was handicapped by Paris's archaic guild structure, which had no defined place for instrument makers. Cassini and other French scientists attempted to get around this by creating a licensed body of instrument makers under royal letters patent. The body was created in 1787. Cassini, however, gave up and ordered the observatory's great quadrant from Ramsden.

The French effort was associated with one great achievement: the repeating circle invented by the academician and naval officer Jean-Charles de Borda (1733–1799). Borda's surveying instrument allowed for triangulations of unprecedented accuracy, and held its own against a far more cumbersome Ramsden theodolite in the joint Anglo-French effort to establish the exact geographical relation of the Paris and Greenwich observatories in 1787. It was later used on the expeditions to define the size of Earth that established the metric system. French efforts to build an instrument industry rivaling England's continued throughout the revolutionary and Napoleonic periods with some further success. However, the effective challenge to the English came from a Munich-based group, led by the brilliant optician Joseph von Fraunhofer (1787–1826) and the mechanical instrument maker Georg von Reichenbach (1772–1826). America, which had long imported most of its philosophical instruments from England, also began to build an instrument industry centered in Philadelphia and aided by the immigration of French and English craftsmen.

See also Calorimeters; Hygrometers; Leiden Jars; Metric System; Microscopes; Rain Gauges; Telescopes; Thermometers.

References

Bedini, Silvio. *Early American Scientific Instruments and Their Makers.* Rancho Cordova, CA: Landmark Enterprises, 1986.

Daumas, Maurice. *Scientific Instruments of the Seventeenth and Eighteenth Centuries.* New York: Praeger, 1972.

Gillispie, Charles Coulston. *Science and Polity in France at the End of the Old Regime.* Princeton: Princeton University Press, 1980.

Heilbron, J. L. *Electricity in the 17th and 18th Centuries: A Study in Early Modern Physics.* 2d ed. Mineola, NY: Dover, 1999.

Turner, Gerard L'Estrange. *Scientific Instruments and Experimental Philosophy, 1550–1850.* Aldershot, UK: Variorum, 1990.

J

Japan, Western Science in

During the eighteenth and early nineteenth centuries, Japan was the non-Western society that took the most interest in Western science. This was paradoxical, as the Japanese government had shut itself off from the outside world in the seventeenth century, restricting Western contact to a Dutch trading post on a small artificial island in the harbor of Nagasaki, Deshima. By contrast with China, which when it wanted Western technical expertise could hire Westerners (the head of the imperial Chinese astronomical bureau was a Western Jesuit until word of the suppression of the order reached China in 1774, and the Jesuits were succeeded by other Westerners), if the Japanese wanted to employ Western knowledge, they would have to learn it themselves. Eighteenth-century Japanese interest in *rangaku,* or "Dutch learning," was the greatest in the areas of medicine and astronomy. Japanese *rangaku* study was a slowly building process, beginning with activities of Shogun Yoshimune (r. 1716–1745), who was interested in the possibility of using Western learning for devising a more accurate calendar. Advised by the samurai astronomer Nakane Genkei (c. 1661–1733), Yoshimune relaxed the government ban on imported Western books in 1720. This did not result in an improved calendar, as the next Japanese calendar, prepared in 1754, continued to rely on traditional Chinese methods. However, the possibility of studying Western texts now existed.

Emphasis gradually shifted away from the Chinese texts produced by Western Jesuits to the more modern texts available from the Dutch at Deshima. In 1740 Yoshimune commissioned two scholars to make a study of Dutch medical works. This project initially had little impact on Japanese medicine, but one scholar, the Confucian Aoki Konyo (1698–1769), produced the first Japanese dictionary of the Dutch language. Although the shogunal government after Yoshimune's retirement in 1745 was suspicious of Western learning, by the early nineteenth century the country boasted several *Rangakujuku,* private academies for the study of the Dutch language and Western learning. The usefulness of Western learning became apparent with the calendar reform of 1798, and in 1811 the shogunate opened an office of translation to promote Dutch studies, expanding beyond the fields of astronomy and medicine. Most astronomers focused on Western techniques of astronomical calculation, seen as of practical value, but some Japanese translated Dutch works relating to Copernican astronomy and Newtonian physics.

Interest in Dutch medicine grew among a circle of physicians, including Konyo's student Maeno Ryotaku (1723–1803), who was impressed by Western anatomy. Interest in dissection was growing among physicians in Japan at this time, and Western texts seemed to give a much more accurate picture of the human body than the traditional Chinese medical canon. Ryotaku and his associates labored greatly to produce an illustrated translation of a Dutch translation of *Anatomical Tables,* a textbook by Johann Adam Kulmus (1689–1745). Their work was published in 1774 as *New Anatomical Text.* This provoked tremendous excitement among Japanese physicians and led to the creation of a new medical tradition, "Dutch medicine," to compete with the existing schools based on Chinese medicine. Several other Dutch works in different branches of medicine were also translated. The translations showed a tendency to improve in quality and detail—in 1808 the first Japanese anatomical texts with illustrations in copperplate rather than woodblock appeared, a change that permitted the display of much more accurate detail. The arrival of the German physician Phillip Franz Balthasar von Siebold at Deshima in 1823 brought the personal influence of a Western-trained scientist with a missionary zeal. Siebold gathered a large group of Japanese students and lectured on clinical procedures as well as zoology and botany. He also set his pupils to writing essays in the Western scientific tradition on issues in Japanese medicine and natural history. However, Siebold's aggressiveness got him expelled from Japan as a spy in 1829, and government measures against his students set back Dutch medicine and Dutch learning generally for several years.

See also Medicine; Nationalism.

Reference

Sugimoto, Masayoshi, and David L. Swain. *Science and Culture in Traditional Japan,* A.D. *600–1854.* Cambridge: Massachusetts Institute of Technology Press, 1978.

Jewish Culture

Jewish awareness of and involvement in European science increased during the eighteenth century. New scientific knowledge posed problems for traditional Jewish laws and practices. One example is the dispute between the Ferrara physician Isaac Lampronti (1679–1756) and Rabbi Judah Briel on the relation of natural science and Jewish law, as interpreted by the rabbinical tradition. Rabbinical authorities allowed Jews to kill lice, but not fleas, on the Sabbath, justifying the distinction by claiming that fleas were true animals that reproduced sexually, whereas lice were spontaneously generated from decaying matter. Lampronti pointed out that modern scientists had disproved the idea that lice and other small insects were spontaneously generated and that rabbinical law needed to change to recognize that. Briel argued that the wisdom of Gentile scientists could never be cited against rabbinical authority, which was from God. In Germany the Talmudist Israel Zamosc (c. 1700–1772) also criticized some Talmudic assertions in the light of natural science, although the vast majority of observant Jews in the period continued to hold rabbinical above scientific authority.

Astronomy was particularly problematic. Rabbis liked to point out that the ancient Jewish sources referred to the Sun as immovable, thus preceding the same discovery by Gentile scholars by hundreds of years. However, some Jewish scientific authorities were reluctant to accept Copernicanism, which seemed to contradict the Torah. One possibility was to endorse the increasingly unsatisfactory Tychonic compromise, which kept Earth at the center and the Sun rotating around it, while the other planets rotated around the Sun.

Barriers to Jewish participation in European scientific institutions were coming down, albeit slowly and unevenly, during the eighteenth century. More Jews from the middle and upper classes were receiving education in secular subjects, including science. The number of universities accepting Jewish students rose, particularly in the Protestant world, although Catholic Padua, which had been admitting Jews into its medical school since the sixteenth century, remained dominant. Jews were admitted into a few scientific societies. The British Royal Society admitted its first Jewish fellow, Moses da Costa, in 1736. The philosopher Moses Mendelssohn (1729–1786) was proposed and elected to membership in the Berlin Academy, but his admission was vetoed by the king of Prussia, Frederick II (r. 1740–1786). Jewish and Christian enlightened intellectuals and persons interested in science also socially interacted in Masonic lodges, particularly in England.

As had been true for centuries, it was primarily as physicians that Jews interacted with scientific advances. Physicians had high status in Jewish communities. Many were rabbis, and brought the scientific knowledge they had acquired in their medical studies to bear upon religious topics. A major compendium of modern scientific information directed at a Jewish audience was the Hebrew textbook of medicine, first published in 1707 and republished five times to 1850, by the Padua-educated physician Tobias Cohn (1652–1729). Cohn's work became increasingly obsolete over this time—he had referred to Copernicus as a child of the devil.

Newtonianism was first absorbed into Jewish thought by Jewish intellectuals living in England. In addition to residing in the homeland of Newtonianism, Anglo-Jewish thinkers had the advantage of living in the society that posed the fewest barriers between Jews and Gentiles. Many Jewish writers were attracted to Newtonianism for similar reasons that attracted Christian writers—the adaptability of Newton's thought for natural theology. The first Jewish writer to incorporate Newtonianism into a Jewish natural theology in his writing was Rabbi David Nieto (1657–1728) of the Bevis Marks Synagogue in London. Probably the most important Jewish Newtonian was Mordecai Gumper Schnaber Levinson (1741–1797), a cosmopolitan who lived in England, Sweden, and Germany and published in Hebrew, German, and English. The enormously prolific Levinson wrote the first book in Hebrew fully expounding Newton's scientific theories as well as their relevance for Judaism, *A Dissertation upon the Law and Science* (1771).

Few Jews were accepted as original scientists by the European scientific community. One who was was Emanuel Mendes da Costa (1717–1791), a natural historian, conchologist, and mineralogist. Costa was admitted to the Royal Society in 1747 and maintained an enormous scientific correspondence in many languages throughout Europe and America. His association with the Royal Society came to a disastrous end in 1767 when it was discovered that he had embezzled money from it. Costa was briefly imprisoned in 1768, and his scientific career never recovered.

With the growth of the *haskalah,* or Jewish Enlightenment movement, among German-speaking Jews like Mendelssohn in the late eighteenth century came a new emphasis on modern knowledge, including science, in Jewish education. *Maskilim,* enlightened Jews, wrote textbooks in Hebrew on elementary science and mathematics for use in Jewish schools in Germany and eastern Europe. Entry of Jews into scientific professions on a large scale, however, did not occur until the mid-nineteenth century.

See also Natural Theology; Newtonianism; Religion; Royal Society.

References

Graupe, Heinz Moshe. *The Rise of Modern Judaism: An Intellectual History of German Jewry, 1650–1942.* Translated by John Robinson. Huntington, NY: Krieger, 1978.

Ruderman, David B. *Jewish Enlightenment in a New Key: Anglo-Jewry's Construction of Modern Jewish Thought.* Princeton: Princeton University Press, 2000.

———. *Jewish Thought and Scientific Discovery in Early Modern Europe.* New Haven: Yale University Press, 1995.

Journals

See Periodicals.

K

Kant, Immanuel (1724–1804)

Immanuel Kant, usually considered the Enlightenment's greatest philosopher, was deeply influenced by the science of his time. He was trained at the University of Königsberg in Newtonian physics as well as the philosophy of Gottfried Wilhelm Leibniz (1646–1716), as systematized by Christian Wolff. He spent his working career at Königsberg, being appointed to the chair of logic and metaphysics in 1770 and retiring in 1797. Much of his early intellectual effort dealt with the differences between the systems of Isaac Newton (1642–1727) and Leibniz, which Kant tried to reconcile by transcending both, rather than compromising between the two. Kant transcended the conflict between Newtonian absolute space and time and Leibnizian relational space and time by identifying both space and time as mental categories.

Kant's opinions on science changed and evolved both with current scientific developments, which he followed closely, and with his own philosophical development. His early writings addressed scientific questions directly. His *Universal Natural History and Theory of the Heavens* (1755) set forth the nebular hypothesis of the origin of the solar system, which was later known as the Kant-Laplace hypothesis. His Latin *Physical Monadology* (1756) set forth an atomistic matter theory he later renounced in favor of the idea that matter was infinitely divisible and filled space. Although his writings beginning with the *Critique of Pure Reason* (1781) focused on philosophical issues rather than science, he continued to teach many scientific subjects, including physics, at Königsberg.

Kant's epistemology rested on a distinction between those truths known through perception, a posteriori, and those known prior to perception, a priori. In the sciences Kant distinguished between those based on a priori and a posteriori knowledge, doubting if the latter truly deserved the name of science. To be a true science a body of knowledge must be known a priori, with certainty, and as an ordered system, which usually meant a mathematical system. The best example of a true science was mathematical Newtonian physics, which Kant labored to show was a priori rather than based on observations. Other sciences, such as chemistry and psychology, were not true sciences, which did not mean that they were not worth doing. Kant's work on the relations of science to philosophy was *Metaphysical Foundations of Natural Science* (1786). In the last decade of his life he was working on a further treatment of the issue, which would incorporate the chemistry of Antoine-Laurent Lavoisier, of which Kant was an early German supporter.

The writings of German philosopher Immanuel Kant profoundly shaped the philosophical climate of Europe during the eighteenth century. (Library of Congress)

Kant attacked natural theology and the idea that God's existence was demonstrable from science. In the struggles of the German university world, Kant, despite his Pietist upbringing, was a champion of the philosophical faculty against the theological. His philosophy is often seen as clearly distinguishing between science and religion without subordinating either to the other. Kantianism replaced Wolffianism as the dominant German academic philosophy in the late eighteenth century.

See also Laplace, Pierre-Simon de; Ørsted, Hans Christian; Wolff, Christian.

References

Friedman, Michael. *Kant and the Exact Sciences.* Cambridge: Harvard University Press, 1992.

Watkins, Eric, ed. *Kant and the Sciences.* Oxford: Oxford University Press, 2001.

Kew Gardens

During the late eighteenth century, the pleasure gardens belonging to the British royal family at their country residence at Richmond were transformed into Kew Gardens, the world's leading botanical gardens. Princess Augusta, mother of the future George III (r. 1760–1820), was a botanical enthusiast and established a small botanical garden there in 1759, with the collaboration of the Scottish nobleman John Stuart (1713–1792), third earl of Bute, also a keen botanist as well as a close friend and political ally of George. The Scottish gardener William Aiton (1731–1793) and other professionals were given the task of expanding the garden and creating the conditions for raising plants from different climates and parts of the world. A 114-foot-long greenhouse was constructed at Kew. The plants were arranged on the Linnaean system of botanical classification, as was Aiton's catalog, *Hortus Kewensis* (1789). On his death Aiton was succeeded by his son, William Townsend Aiton (1766– 1849), who had been born at Kew and brought out a new and expanded edition of his father's catalog.

What really gave the impetus for the expansion of Kew, however, was George III's appointment of Sir Joseph Banks as director of the gardens when the king inherited them from his mother in 1772. Kew became the botanical center of the British Empire and grew with that empire as Banks's men sent back plants from all over the world. Particular attention was paid to plants useful in such pursuits as medicine or dyeing. Banks's position as director of Kew, which he held for forty-eight years until his death, made him the center of an international network of the exchange of plants and botanical information. The first in what would be dozens of men he sent out to find plants for Kew was the Scotsman Frederick Masson (1741–1805). Banks sent Masson to the Cape of Good Hope, where he spent three years and sent about 400 different species of plants to Kew.

Kew continued to expand throughout the late eighteenth and early nineteenth centuries. From around 5,500 plants in 1789, its holdings doubled to 11,000 by 1814. With the growth of the collection, the plants were also organized more systematically, with the establishment of a registry of accessions in 1793. Kew under Banks was conceived as a research garden closed to the public, with experiments in the acclimation of different plants. This research was expected to contribute to the economic development of the British Empire. Banks hoped to supplement the garden with a botanical library and a herbarium, or collection of dried plants, also to be located at Kew, but these plans did not come to fruition. Kew Gardens went into a decline after Banks's death in 1820, but was revived later in the nineteenth century as a public institution, the Royal Botanic Garden.

See also Banks, Sir Joseph; Botanical Gardens; Exploration, Discovery, and Colonization.

Reference

Green, J. Reynolds. *A History of Botany in the United Kingdom from the Earliest Times to the End of the 19th Century.* London: J. Dent and Sons, 1904.

The Rose apothecary shop in Berlin, where Martin Heinrich Klaproth worked from 1771 to 1780. (Schoenberg Center for Electronic Text & Image at the University of Pennsylvania Library)

Klaproth, Martin Heinrich (1743–1817)

Martin Heinrich Klaproth was the leading German chemist of the late eighteenth and early nineteenth centuries. The son of a tailor, Klaproth was originally intended to be a Lutheran minister. Instead, he became an apothecary in Hanover and a self-educated chemist. Rather than the university hierarchy, Klaproth moved up through the Prussian state service. He moved to Berlin in 1768 and studied under the Prussian chemist Andreas Sigismund Marggraf, whose niece he married in 1780. He received his first appointment as a Prussian medical official in 1782. Klaproth taught at the Berlin Mining School and the Royal Artillery Academy. As a chemist he was known throughout Europe for his great skills in chemical analysis, isolating uranium from pitchblende and zirconium from zircon. He also studied titanium and tellurium, and in 1784 launched a hydrogen balloon in honor of the birthday of King Frederick II (r. 1740–1786). In 1788 he was admitted to the Berlin Academy.

Klaproth's insistence that precipitates be thoroughly dried before weighing contributed to the rise in standards of chemical practice in the late eighteenth century. His conversion to the antiphlogistic "French chemistry" of Antoine-Laurent Lavoisier in 1792 was decisive in the struggle of the German antiphlogistonists, led by his colleague at the mining school Sigismund Friedrich Hermbstaedt (1760–1833). Klaproth and Hermbstaedt were the leading champions of Lavoisier's chemistry in the German debate over the reduction of mercuric oxide in the early 1790s, which ended in the triumph of antiphlogistic chemistry in Germany. Klaproth himself received the chair of chemistry at the newly founded University of Berlin in 1810.

See also Ballooning; Berlin Academy; Chemistry.

Reference

Hufbauer, Karl. *The Formation of the German Chemical Community, 1720–1795.* Berkeley: University of California Press, 1982.

L

La Mettrie, Julien Offroy de (1709–1751)

Julien Offroy de La Mettrie was a leader in taking science in a materialist and mechanist direction in the early Enlightenment. A physician, La Mettrie trained at the University of Leiden under Hermann Boerhaave. Like many Enlightenment physicians, La Mettrie combined an exalted view of the potential of medicine as a discipline for understanding humanity and society with a poor view of many contemporary doctors. This was first manifested in the late 1730s, when he was the only physician to participate on the surgeons' side in the ongoing pamphlet war between physicians and surgeons over medical reform. La Mettrie defended the empiricism of the surgeons against the theoretical approach of the physicians. His attacks on conservative Parisian physicians, beginning in 1737 and ending in 1750, shortly before his death, often took the form of stinging satire, so savage that he was forced to leave France and move to the Dutch Republic. La Mettrie seems to have viewed himself as a missionary in the backward French medical scene for the more advanced medicine of Leiden and Boerhaave, whose works he translated into French and whose biography he wrote. He was not a slavish follower of Boerhaave, however, and was much more skeptical of the medical use of chemistry. He also introduced some of George Cheyne's ideas on maintaining health through diet and environment into France.

La Mettrie's most influential and scandalous work was *L'Homme-machine* (Man, a machine) (1747). As a philosophical materialist La Mettrie rejected both the vitalism of Georg Ernst Stahl and his followers and the Christian notion, followed by René Descartes (1596–1650) and the Cartesians, of the union of a material body and an immaterial soul. Against the vitalists and the Cartesians La Mettrie employed the doctrine of the "irritability" of bodily fibers as developed in the mid-eighteenth century by Albrecht von Haller to demonstrate that matter was capable of self-activation, requiring no assistance from a soul or a vital spirit. He also rejected the orthodox Cartesians' absolute distinction between humans, possessed of souls, and animals, mere machines. La Mettrie pointed to examples of animals doing things that were traditionally ascribed only to human beings, such as communicating. Analogously, many specifically human achievements were likened to the activities of trained animals. *L'Homme-machine* was driven by a missionary zeal to dethrone the metaphysicians and theologians from their position of arbiters of human nature and to replace them with the physician, whose knowledge of the body was the only sure foundation for knowledge of the human. La Mettrie was probably an atheist and mocked many of the traditional arguments for God's existence. This made *L'Homme-machine* a tremendous intellectual scandal, and even the tolerant Dutch Republic grew too hot for its author, who fled to the court of Frederick II of Prussia (r. 1740–1786) at Berlin. He died there from eating tainted pâté.

La Mettrie's writings after *L'Homme-machine* emphasize the physiological basis of human life, including human morality, which he steadfastly refused to ground on any transcendent good. This not only subjected him to vehement attacks from defenders of Christianity, but also made him suspect to Enlightenment thinkers in the second half of the century who viewed him as an advocate of a fixed and selfish human nature, which ruled out any possibility of effective reform. Despite his influence on subsequent materialists like Denis Diderot, Claude-Adrien Helvétius (1715–1771), and Baron Paul-Henri-Dietrich d'Holbach (1723–1789), they seldom praised him.

See also Boerhaave, Hermann; The Enlightenment; Haller, Albrecht von; Materialism; Medicine; Surgeons and Surgery.

Reference

Wellman, Kathleen. *La Mettrie: Medicine, Philosophy, and Enlightenment.* Durham: Duke University Press, 1992.

Lagrange, Joseph-Louis (1736–1813)

Joseph-Louis Lagrange was a leader in mathematics and terrestrial and celestial mechanics. Although he became one of France's leading mathematicians, he was born in the Italian kingdom of Piedmont. Inspired by the teaching of Giambatista Beccaria (1716–1781) at the University of Turin, he decided to become a mathematician. Although Lagrange's very early work did not announce his later genius, by 1754 he had done some work on the cycloid curve, which won the approval of Leonhard Euler. Lagrange also won a position as an instructor at the Royal Artillery School in Turin in 1755, turning down a paid position Euler had arranged for him at the Berlin Academy. Lagrange was a founding member of the Royal Academy of Sciences of Turin, and some of his work on the calculus of variations, dynamics, and the mathematics of vibrating strings was published in its journal, *Melanges de Turin.*

In 1764 Lagrange entered a competition sponsored by the Royal Academy of Sciences over a problem in celestial mechanics, the libration of the Moon. He also visited Paris for the first time, making the acquaintance of Jean Le Rond d'Alembert, who took the young mathematician into his considerable clientage. In 1766 he won another contest held by the Royal Academy, on the perturbations of the orbits of Jupiter and Saturn. In 1772 he shared the prize with Euler for work on the mutual relations of the Moon and Earth, a case of the "three-body problem." In 1774 he won another academy prize for a study of how the shapes of the Moon and Earth affect their motions. The last in this brilliant sequence of victories occurred in 1780, with a study of the effects of planetary gravitational pulls on the orbits of comets. After that Lagrange preferred to work on problems he selected himself rather than those set by the academy. Along with his friend and sometime collaborator Pierre-Simon de Laplace, Lagrange created classical celestial mechanics.

D'Alembert was pushing for his protégé to be recruited to the Berlin Academy. In 1766 he succeeded Euler, who had left for St. Petersburg, as head of the mathematical section of the Berlin Academy. In addition to his work in celestial mechanics, Lagrange at Berlin made important innovations in mechanics, probability, calculus, and number theory. Among other discoveries, Lagrange proved that every positive integer can be written as the sum of four square numbers. Despite the significance of his own work,

Lagrange was pessimistic about the future of mathematics. In a letter to d'Alembert in 1781, he expressed the fear that mathematics, like a mine, had yielded all its available resources, unless a new vein of ore was discovered. Lagrange's *Analytical Mechanics* (1787) fully mathematized mechanics and crowned the tradition of eighteenth-century rational mechanics. Lagrange boasted that he had no need for figures or geometric constructions, only equations. *Analytical Mechanics* is particularly noted for the skill with which Lagrange applied differential equations to mechanics.

Lagrange left Berlin after the death of Frederick II (r. 1740–1786), when the city was less welcoming to foreign savants. Although several Italian academies tried to recruit him, he took an offer from the Royal Academy of Sciences in Paris. Lagrange was put on the committee of the Royal Academy of Sciences for the standardization of weights and measures in 1790. He managed to survive the entire period of the French Revolution, as his colleague on the committee Antoine-Laurent Lavoisier did not. Lagrange is credited with saying that the executioner took only a moment to cut off the head of Lavoisier, but it would take a century to replace it. (He is also credited with the statement that there could be only one Newton because there is only one universe, and only one man could discover its laws.) Lagrange was appointed professor of analysis at the newly founded Polytechnic School in 1794, and although he was not an inspiring teacher he helped found the Polytechnic's tradition of excellence in mathematics. The Polytechnic seems to have rekindled Lagrange's interest in pure mathematics; he published two volumes on the foundations of the calculus. Lagrange was less active in the early nineteenth century, but he headed the mathematics section of the Institute of France, served on several government committees, and received several titles from Napoléon. When Lagrange died his eulogy was given by Laplace, his only equal.

See also Berlin Academy; French Revolution; Laplace, Pierre-Simon de; Mathematics; Mechanics; Royal Academy of Sciences.

Reference

North, John. *The Norton History of Astronomy and Cosmology*. New York: W. W. Norton, 1995.

Lamarck, Jean-Baptiste-Pierre-Antoine de Monet de (1744–1829)

Jean-Baptiste-Pierre-Antoine de Monet de Lamarck was an early champion of the idea that biological species evolve. A young son of a noble family from southern France, Lamarck began his scientific studies in botany. After an injury forced him to retire from the army in 1768, he tried various careers, then went to Paris where he became part of the circle surrounding Georges-Louis Leclerc de Buffon at the Royal Botanical Garden. Lamarck's great energy and ability, combined with his rejection of the Linnaean system of botanical classification, won him admission to the Royal Academy of Sciences after he published a three-volume guide to the plants of France, *French Flora* (1779). His other early interests included mineralogy and meteorology, but after 1780 he focused entirely on botany, producing the eight-volume botanical section of the *Encyclopédie Methodique,* the expanded and topically arranged reworking of the original *Encyclopédie.* He progressed through the ranks of the academy, reaching the highest level of pensionary in 1790, and continued an alliance with Buffon at the Royal Botanical Garden. In 1781 Buffon appointed Lamarck to the newly created unsalaried position of correspondent, but this was actually a mixed blessing. Buffon expected Lamarck to guide Buffon's loutish son, whom he was grooming to take over the garden, on a grand tour of Europe. On Buffon's death, Lamarck received a new salaried position as botanist to the king and keeper of the herbaria, but then had to defend his position in the storm of the French Revolution.

Despite suspicion caused by his noble ancestry, Lamarck managed the transition from monarchy to republic and from Royal Botan-

Jean-Baptiste-Pierre-Antoine de Monet de Lamarck was a botanist and naturalist who proposed one of the first theories of evolution. Though Lamarck's evolutionary theory is now considered invalid, Charles Darwin, the father of modern evolutionary theory, credited Lamarck with laying the foundation for the new understanding of the evolution of the species. (Library of Congress)

ical Garden to Museum of Natural History quite smoothly, being appointed professor of insects and worms on the museum's inauguration in 1794. Despite the fact that Lamarck's publications to this point had been botanical, he was known to have a large shell collection and thus was befitted for the post, at which it was hoped he would produce a treatise on conchology. However, his time in the first half of the decade was principally devoted not to his new discipline, but to the exposition of an original physical and chemical theory. Lamarck's lack of training in the subject, and his rejection of modern physics and chemistry, meant that his theory, based on the universality of fire, was widely ignored. His venture into meteorology, in which he took a great interest, also led to a rebuff when he suggested that the Moon had an influence on the weather. These events contributed to the sense of intellectual isolation and even paranoia that played an increasingly important role in Lamarck's career.

The year 1801 saw the fruit of Lamarck's work connected with his professorship, the *System of Invertebrate Animals; or, General Table of Classes, Orders, and Genera of Such Animals.* Lamarck's zoological studies led him to formulate a new philosophy of life, for which he used the term *biology,* just coming into use in the early eighteenth century. Lamarck's 1809 *Zoological Philosophy* expounded his biology. He was a strong believer in the mutability of species, which set him at odds with the ruler of French natural history, Georges Cuvier, an equally firm believer in the fixity of species. Lamarck and Cuvier were also at odds on the question of species extinction. Cuvier, looking at the fossil record, argued that species extinction was common, whereas Lamarck, invoking the vast areas of the world, particularly under the sea, unknown to European science, denied the possibility of extinction. Species mutability served as an alternative form of explanation of the difference between living and fossil forms. Rather than some species becoming extinct, they had simply changed slightly in the time between the deposit of the fossil and observation of the present-day living creature.

Lamarck's vision of species change (he did not use the word *evolution*) was not random but progressive. He even supported the idea of the spontaneous generation of the simplest living forms from nonliving material. The drive of nature was to create a continuous series of ever more complex living forms, although environmental constraints caused gaps in the series of forms. Lamarck did not regard nature as an autonomous force, nor was he a vitalist who ascribed this drive to complexity to a "life force" not explainable in mechanical terms. Instead, he saw the drive to complexity as a function of the circulation of the "subtle fluids," electricity and caloric, which combined with the material structure to form living things. His classification schemes for invertebrates were initially arranged in a single series of increasing complexity, from the simplest infusorians at the bottom to the mollusks at the top, leading to the vertebrates. By the publication of the seven-volume *Natural History of the Invertebrates* (1815–1822) Lamarck modified this scheme to admit of a second series, beginning with spontaneously generated intestinal worms and ending with cirripedes. He would later admit of other branches, abandoning his initial idea of a single series. The *Natural History of the Invertebrates* was a highly successful work of zoological classification, and was treated as an authority by many who did not accept or were not even interested in Lamarck's theories of species transformation.

Lamarck is often identified with the doctrine of the inheritance of acquired characteristics, overthrown by modern genetics. This idea was not original with Lamarck, and, although important, did not play the central role in his thought. His discussion of this topic occurs in the context of environmental factors modifying the series of organic beings. The concept Lamarck used was that of habit. The most famous example is the giraffe, which develops a long neck as the result of the habit of straining to reach the higher branches of trees to eat the leaves. Conscious intention plays no role. On the interior of the living being, habit forms channels for the more easy circulation of the subtle fluids along certain paths. Conversely, lack of use, as in the case of the eyes of the mole, causes the subtle fluids to stop circulating along a certain path, and eventually to the decay of an organ or faculty. He was willing to extend this theory to the origins of man, speculating on how an arboreal species could become bipedal when forced to relocate to the ground. Variation for Lamarck was purposeful, not random as it would be for Charles Darwin (1809–1882) in his later theory of evolution by natural selection.

Lamarck's theory attracted few followers and many opponents. Lamarck had invoked little evidence of its operation, and much evidence, such as the lack of transitional forms in the fossil record, was against it. Some, in France and elsewhere, opposed what they saw as the theory's materialism and its denial of the necessity of a divine designer and creator. More immediately damaging was the opposition of Cuvier, who simply refused to take Lamarck's theory seriously. Lamarck, who was blind for the last few years of his life, grew even more isolated from the mainstream of the scientific community. In his increasing bitterness he made one of the earliest prophecies of global environmental collapse caused by human action. Despite Cuvier's ridicule, Lamarck's ideas were not forgotten and contributed to the stock of pre-Darwinian evolutionary theory.

See also Botany; Buffon, Georges-Louis Leclerc de; Cuvier, Georges; Encyclopedias; Museum of Natural History; Zoology.

References

Burkhardt, Richard W., Jr. *The Spirit of System: Lamarck and Evolutionary Biology*. Cambridge: Harvard University Press, 1977.

Corsi, Pietro. *The Age of Lamarck: Evolutionary Theories in France, 1790–1830*. Translated by Jonathan Mandelbaum. Berkeley: University of California Press, 1988.

Laplace, Pierre-Simon de (1749–1827)

Pierre-Simon de Laplace was France and Europe's leading physical scientist for decades, and his work represents the culmination of the Newtonian tradition of mathematical physics. He was also a mathematician of genius. From an obscure family of Norman landowners and minor officials, Laplace was originally intended for the church. He entered the University of Caen in 1766, but left for Paris to make a career as a mathematician in 1768. One of his Caen professors gave him a letter of introduction to Jean Le Rond d'Alembert, who was impressed by the young man's mathematical skills and got him a job as an instructor in France's Military

School. This was not a challenging teaching post, but it enabled Laplace to stay in Paris, the center of French and European mathematics. The next order of business was to get elected to the Royal Academy of Sciences, and Laplace presented an impressive number and variety of mathematical papers from 1770 to 1773. He was passed over for more senior candidates for admission to the academy in 1771 and 1772, but was successful on the third try, in 1773, still at a very young age for admission.

Laplace's early efforts focused on two of the most active areas in contemporary mathematics: the theory of probability and celestial mechanics (a term he originated). His most dramatic innovation in probability was in the definition of inverse, or "Bayesian," probability, the determination of the probabilities of causes where the event is known. (Laplace's analysis of inverse probability came after that of the English clergyman Thomas Bayes [1702–1761], but Bayes's work was not immediately followed up on, and the field built on Laplace.) For Laplace, probability had nothing to do with genuine randomness, an idea he rejected. Rather, probability was a matter of unknown causes. A philosophical determinist, Laplace stated in a memoir submitted to the academy in 1776 that an intelligence that comprehended the relation of every component of the universe at one instant could determine its state at any moment of the past or future.

Laplace's early work on celestial mechanics, the subject that would eventually bring him his greatest fame, tried to explain irregularities in the motion of the planets by the hypothesis that the gravitational force of a body was not propagated instantly, as was the standard Newtonian theory, but took time to reach the other body. He also applied probability theory to the distribution of comets.

By 1780 Laplace was undoubtedly France's leading mathematician and mathematical physicist (he did not conceal his awareness of this fact). His capacity for work was prodigious. Although burdened by public responsibilities—in 1784 he was appointed an examiner of cadets for the Royal Artillery, serving as the examiner of Napoléon Bonaparte among many others, and he served on a royal commission to investigate Paris hospitals—he found time to further extend his studies of probability and celestial mechanics, and even to branch into the completely new areas of demography and experimental science. Laplace's experimental work was carried out as an assistant to Antoine-Laurent Lavoisier, France's leading chemist. Their joint project, for which Laplace invented the ice calorimeter, was an investigation of the properties of heat.

By the middle of the decade Laplace's attention was again focused on the project that dominated his next twenty years, the mechanics of the solar system. He abandoned his earlier theory of the noninstantaneous propagation of gravity. His "Memoir on the Secular Irregularities of the Planets and Satellites," read before the academy in 1787, contains two breakthroughs on problems in the field. Laplace demonstrated that the long-term ("secular") acceleration of Jupiter and the deceleration of Saturn were aspects of the same phenomenon, and that over the course of time they would reverse themselves. This removed one objection to the idea that the solar system was stable. The other innovation was a demonstration of the stability of the orbits of Jupiter's first three moons. The two were followed by an explanation for a long-standing problem in celestial mechanics, the secular acceleration of the Moon first demonstrated by Edmond Halley (1656–1742). In 1788 Laplace claimed that the Moon's acceleration was caused by a combination of the effects of the Sun's gravity and of variations in Earth's orbit caused by the actions of the planets. The memoir concluded with a remarkably ill-timed peroration, given the imminence of the French Revolution, comparing the harmony and stability of the solar system to that of human society.

Laplace was largely indifferent to politics, save as it affected his own career. Unlike his friend and colleague Lavoisier he survived the Revolution, although he prudently left Paris before the height of the Terror, returning only after the overthrow of the revolutionary extremists in 1794. The most important contribution Laplace made to the revolutionary restructuring of France was his leading role on the commission that invented the metric system. In 1796 he was elected president of the First Section of the Institute of France, the replacement for the Royal Academy of Sciences. He also helped set up the scientific and mathematical curriculum at France's new educational institutions, including the Polytechnic School. Laplace composed an advanced textbook of mathematical physics in two volumes, *Exposition of the System of the World* (1796). This work has attained notoriety for its tentative putting forth of what has come to be known as the "Kant-Laplace hypothesis" of the origin of the solar system. Laplace was not concerned with the history of the solar system so much as he was with explaining why all the planets orbited the Sun in the same direction, as well as rotating in the same direction, a question not answerable by Newtonian physics. He argued that this uniformity was caused by the fact that the celestial bodies had all condensed from the solar atmosphere.

The Napoleonic period was the height of Laplace's power and influence, although not his scientific creativity. A brief tenure as minister of the interior proved a fiasco, but Laplace accumulated a number of honorary posts and distinctions, including ennoblement as a count of the empire in 1806, and the pensions he received made him for the first time a rich man. Laplace dominated the physical science activities of the First Class of the Institute, and Georges Cuvier, the leader of the life sciences, was a friend and ally. As a working scientist Laplace published his four-volume mathematical explanation of the solar system, *Treatise on Celestial Mechanics* (1799–1805). It is in connection with this work that the most famous story of Laplace arises. Supposedly, Napoléon commented unfavorably on the work's failure to mention God, and Laplace replied that he had no need of that hypothesis. The evidence for this story is not overwhelming, but it is consistent with Laplace's general religious attitude, which did not deny the existence of God, but exiled him from the affairs of the cosmos.

Laplace led, along with his friend the chemist Claude-Louis Berthollet, a working group on physical sciences based at Arcueil, a village outside Paris where Berthollet and Laplace owned adjoining houses. The remainder of the group was composed of talented young scientists beginning their careers, such as Laplace's disciples Jean-Baptiste Biot (1774–1862) and Siméon-Denis Poisson (1781–1840). Laplace's main physical interests moved from the gravitational interaction of vast planetary bodies to explaining physical phenomena by short-range forces of attraction operating between small particles. He and his Arcueil followers applied this method with some success to such outstanding physical problems as the speed of sound, capillary action, and optical refraction. He also published two outstanding works on probability during the Napoleonic period: the *Analytical Theory of Probabilities* (1812) and the less technical *Philosophical Essay on Probabilities* (1814). These studies were the foundation of nineteenth-century studies of probability and statistics.

Age, and the political embarrassment of Napoléon's defeat, caused Laplace to lose much of his preeminence in French science after 1815, although he transferred his political loyalties to the new regime adeptly. (His vote in the French Senate in favor of Napoléon's exile to St. Helena, after the fulsome flattery he addressed to Napoléon in power, attracted unfavorable comment.) The growing popularity of the wave theory of light and chemical atomism in France caused the Laplacian program in physics to look rather old-fashioned. Much of the work of the last decade of his life was oriented to providing a mathematical treatment of the caloric theory of heat, an intellectual dead end, considering the decline of the caloric theory. Laplace kept working to the end, however, publishing a fifth volume of *Treatise on Celestial Mechanics,* mostly consisting of refinements of his previous work, in 1825. His death in 1827, one hundred years after his hero Isaac Newton's, symbolized the end of French scientific preeminence.

See also Alembert; Jean Le Rond d'; Arcueil, Society of; Astronomy; Calorimeters; French Revolution; Lavoisier, Antoine-Laurent; Mathematics; Mechanics; Metric System; Napoleonic Science; Newtonianism; Physics; Probability; Religion; Royal Academy of Sciences.

References

Crosland, Maurice. *The Society of Arcueil: A View of French Science at the Time of Napoleon I.* Cambridge: Harvard University Press, 1967.

Gillispie, Charles Coulston, with the assistance of Robert Fox and Ivor Grattan-Guiness. *Pierre-Simon Laplace, 1749–1827: A Life in Exact Science.* Princeton: Princeton University Press, 1997.

Lavoisier, Antoine-Laurent (1743–1794)

Antoine-Laurent Lavoisier was the protagonist of the chemical revolution, the person most responsible for the creation of modern chemistry. From a family of lawyers, Lavoisier could trace his interest in science to his youth as a student at the College of Four Nations in Paris. (Lavoisier was among the few scientists of the time Paris born and bred.) He initially learned chemistry from a popular demonstrator at the Royal Botanical Garden, Guillame-François Rouelle (1703–1770), a leader in the introduction of phlogiston chemistry to France. Lavoisier attended Rouelle's demonstrations in 1762. The following year he began to accompany a distinguished geologist and old family friend, Jean-Étienne Guettard (1715–1786), on his field trips. This work led to Lavoisier's first paper before the Royal Academy of Sciences, on gypsum, delivered in 1765. He had also entered a Royal Academy contest on the best means to illuminate the streets of Paris at night, and although he did not win, his treatment was impressive enough to merit a special gold medal from the French government. He narrowly missed entry into the chemistry section of the Royal Academy of Sciences in 1768, when he won the academicians' vote but the government preferred his opponent, the metallurgist Antoine-Gabriel Jars (1732–1769). Jars's subsequent death cleared Lavoisier's way for entrance into the academy, where he was very active.

Unlike many aspiring scientists, Lavoisier was not driven by financial ambition. The only surviving child in his family, he was already very wealthy. He invested in the Farmers-General, a group that subcontracted tax collection from the government. The Farmers-General were wealthy, but unpopular. The work was hard, absorbing much of Lavoisier's energy. As an official of the Farmers-General, he won even more unpopularity with his championing of a smuggler-proof wall encircling Paris. The most important contribution the Farmers-General made to Lavoisier's scientific life was the young daughter of a Farmer, Marie-Anne Pierrette Paulze (1758–1836), who became Madame Lavoisier in 1771. Madame Lavoisier was very useful to her husband's career as an assistant, as an engraver who produced illustrations for his books, as a translator—despite his keen interest in the work of British chemists, Lavoisier never learned English—and as the hostess of a scientific salon. Lavoisier's wealth significantly aided him as a chemist, as he could afford the best equipment. He used a very sensitive balance in his first important series of experiments in 1768 and 1769, which disproved the idea—inherited from alchemy and still upheld by some chemists—that water transmuted to the earth by demonstrating that the earth found in the distilled water was dissolved glass from that apparatus. Lavoisier admired the precision of contemporary physicists and sought to introduce it into chemistry, although his weighings also drew on previous chemical practices.

Lavoisier's most famous chemical experiments, undertaken in 1772, were on the question of combustion. The dominant phlogiston theory held that combustion was the release of the phlogiston combined with the combusted substance. Lavoisier's future disciple Louis-Bernard Guyton de Morveau (1737–1816), in a recent series of experiments, however, demonstrated that metals actually gained weight when they were combusted to form calxes. Lavoisier's experiments in 1772 demonstrated that sulfur and phosphorus also gained weight on combustion. Lavoisier theorized that this was because burning substances "fixed" air from the atmosphere. He had already formulated the idea of the gaseous state, that a greater or lesser amount of the "matter of fire" could determine whether a body was solid, liquid, or gas. At the time when Lavoisier began his revolutionary experiments on combustion, the only pneumatic chemist whose work he knew was Stephen Hales, and he was unaware of the overthrow of the notion of an undifferentiated "air." He quickly realized the importance of the British pneumatic research for his own work and studied what was available of the writings of Joseph Black and Joseph Priestley. After abandoning the idea that what was being fixed was simply air, Lavoisier's first theory was that the gas that was being fixed was Black's "fixed air," now known as carbon dioxide. He came to realize that the gas was actually what Priestley had discovered and named "dephlogisticated air." For Lavoisier, in combustion this air lost its heat, which he identified as a substance called caloric.

A gifted scientist and the founder of modern chemistry, Antoine-Laurent Lavoisier also played an active role in public life, which brought him to the guillotine during the French Revolution. (Nancy Carter/North Wind Picture Archive)

A series of experiments in the early 1780s involving the newly invented ice calorimeter led Lavoisier and his experimental partner, Pierre-Simon de Laplace, to conclude that respiration and combustion were analogous processes. The caloric given off by the reaction provided animals with heat. The last piece of the puzzle fell into place in 1783, when Lavoisier heard of the recent work of Henry Cavendish, who had combined "inflammable air" (hydrogen) with "dephlogisticated air" (oxygen) to produce water, a reaction Cavendish explained in terms of phlogiston theory. Hearing of this experiment enabled Lavoisier to identify water as a compound of the two gases, rather than, as Cavendish would have it, a combination of "dephlogisticated air" and phlogiston. Lavoisier then explained what had been a problem for him—what happened to the oxygen when a metallic calx was treated with an acid. He explained that the hydrogen released by the reaction combined with the oxygen to form water. By 1785 Lavoisier was ready to openly challenge the phlogiston theory, setting forth his new theory of combustion in a paper, "Reflections on Phlogiston," read to the Royal Academy of Sciences.

In addition to his new theory on the role of oxygen in combustion and respiration, Lavoisier's other key innovation was methodological. It was what became known as his "balance-sheet" approach to chemical operations, in which the masses of the substances that participated in the operation were

The arrest and execution of Antoine-Laurent Lavoisier was a heavily mythologized event. In this late-nineteenth-century painting by L. Langenmantel, Lavoisier stands upright and is bathed in light emanating from above, a contrast to the distorted positions of the men who have come to arrest him, some of whom crouch in darkness. (Edgar Fahs Smith Collection, University of Pennsylvania Library)

assumed to be equivalent to the masses produced at the end of the operation. This idea was not new, but Lavoisier was able to make it a central feature of chemistry by weighing gaseous as well as solid and liquid substances. The idea of weighing gases emerged only gradually—at first Lavoisier treated the volume of gas as the key measurement. He also sometimes used an exaggerated degree of precision in reporting his measurements.

Lavoisier set about creating a new set of chemical institutions to carry out what he had referred to as early as 1773 as his "revolution." As director of the Royal Academy of Sciences in 1785, he stacked the chemical section with supporters of the new chemistry while pushing through a massive series of organizational reforms. He also set about creating a new standardized chemical nomenclature—something many chemists, regardless of their position on the phlogiston issue, thought needed. Lavoisier's nomenclature, as introduced in the *Method of Chemical Nomenclature* (1787) that appeared under his and three other French chemists' names, carried with it Lavoisier's ideas. For example, Lavoisier coined the word *oxygen* to describe Priestley's dephlogisticated air. The term does not refer to oxygen's role in combustion, but to what Lavoisier saw as its other important property, its role in forming acids, *oxygen* being Greek for "acid former." Lavoisier believed that all acids were compounds of oxygen (although not that all compounds of oxygen were acids). This theory lasted in the French chemical world to the early nineteenth century.

In 1789 Lavoisier published his *Elementary Treatise of Chemistry* setting forth his theories. An English translation by Robert Kerr (1755–1813) appeared the next year and a German one by Sigismund Friedrich

Hermstaedt (1760–1833) in 1792. The work set forth the concept of a chemical element as a substance that could not be broken down, but Lavoisier's concept of an element differed fundamentally from that of modern chemists—for example, he included light and caloric as chemical elements. The *Elementary Treatise* not only set out Lavoisier's chemical theory, but also included a substantial discussion of laboratory technique and equipment. Also in 1789 Lavoisier and a young follower, Pierre Adet (1763– 1834), started a new journal, the still-existing *Annales de Chémie,* devoted to the new chemistry. Lavoisier's actual experimental agenda in these years focused on the composition of organic substances, where he pioneered combustion analysis.

Lavoisier's capacity for work was astounding. In addition to his experiments and many tasks for the academy and the Farmers-General, he was appointed to the Gunpowder Commission in 1775, served as secretary to the Agricultural Commission from 1785, and ran his estate at Frechines as an experimental farm. Lavoisier originally supported the French Revolution and served the revolutionary government with his customary indefatigability on such matters as the commission that created the metric system. He fought long and hard, but ultimately in vain, to save the Royal Academy of Sciences from the revolutionary government. He was executed not for his scientific activities, but for his role in the Farmers-General, which the Revolution had abolished.

See also Agriculture; Chemical Nomenclature; Chemistry; French Revolution; Laplace, Pierre-Simon de; Nationalism; Phlogiston; War.

References

Brock, William H. *The Norton History of Chemistry.* New York: W. W. Norton, 1993.

Gillispie, Charles Coulston. *Science and Polity in France at the End of the Old Regime.* Princeton: Princeton University Press, 1980.

Holmes, Frederic Lawrence. *Antoine Lavoisier—the Next Crucial Year; or, The Sources of His Quantitative Method in Chemistry.* Princeton: Princeton University Press, 1998.

———. *Lavoisier and the Chemistry of Life: An Exploration of Scientific Creativity.* Madison: University of Wisconsin Press, 1985.

McKie, Douglas. *Antoine Lavoisier: Scientist, Economist, Social Reformer.* New York: Harper and Row, 1952.

Leiden Jars

The Leiden jar was the first condenser, or device for the accumulation and storage of electrical charge. It originated from experiments performed by two electrical experimenters working independently, the Prussian Ewald Georg von Kleist (1700–1748) in 1745 and the Dutch physicist Pieter van Musschenbroek (1692–1761) early in 1746. Kleist did not realize the importance of what he had created, so credit for inventing the jar is usually given to the Leiden-based Musschenbroek, hence the term *Leiden jar.* Musschenbroek was trying to duplicate recent electrical experiments of the German Georg Mathias Bose (1710–1761) on the electrification of water in a jar, by running a wire from an electrical generator into a glass vessel containing water. Following standard procedure for electricians, the jar was insulated. The actual discoverer of the Leiden jar was a visitor to Musschenbroek's laboratory, the novice experimenter Andreas Cunaeus (1712–1788). Not knowing that the jar must be insulated, Cunaeus picked it up with one hand while attempting to draw a spark from the conductor with the other. He received a severe shock, far greater than any produced by normal electrical experimentation. On hearing of this, Musschenbroek attempted to duplicate the experience, and succeeded. Experimenters with the Leiden jar were a hardy lot, but Musschenbroek claimed that he would not undergo the experience again for the whole kingdom of France.

The Leiden jar was quickly introduced into the major centers of electrical experiment in France, England, and Germany as well as the remote outpost of Benjamin Franklin's Philadelphia, and it captured the imagination of educated people as had no piece of experimental apparatus since the heyday of the air pump in the late seventeenth century. Exaggerated stories of the damage the shocks caused to experimenters circulated. Despite the pain, being shocked became fashionable, and it was quickly discovered that not just one person, but many, holding hands in a circle, could receive the electrical discharge. Jean-Antoine Nollet, who introduced the Leiden jar to France, shocked more than 180 soldiers in front of the king, and repeated the experiment with more than 200 Carthusian monks. This was spectacular, but the jar presented serious intellectual problems to Nollet and other electricians. Contemporary electrical theory, based on effluvia, could not explain the jar, and Musschenbroek's letter announcing it admits that he had reached the point in his study of electricity when he knew and understood nothing. Franklin's revolutionary theory of "positive" and "negative" electricity derived much of its success from its ability to explain the behavior of the Leiden jar. Franklin claimed that the inside and outside of the jar were positively and negatively charged, respectively, and the glass itself was impermeable to electricity. The discharge of the jar occurred when the inside and outside were connected.

Later in the eighteenth century the jar itself was refined with the substitution of metal for the water on the inside of the vessel and eventually the development by English experimenters of the parallel plate condenser, or "Franklin square."

See also Electricity; Franklin, Benjamin; Nollet, Jean-Antoine.

Reference

Heilbron, J. L. *Electricity in the 17th and 18th Centuries: A Study in Early Modern Physics.* 2d ed. Mineola, NY: Dover, 1999.

Lewis and Clark Expedition

The expedition of Meriwether Lewis (1774–1809) and William Clark (1770–1838) into western North America from 1804 to 1806 was fundamentally one of scientific and geographical exploration. Its chief sponsor was Thomas Jefferson (1743–1826), third president of the United States, whose deep and wide interest in science is unparalleled among presidents. The exploration of western North America and scientific understanding of its geography, Native American population, fauna, and flora were longtime goals of Jefferson, connected with his hopes for westward territorial expansion. In 1793 Jefferson and the American Philosophical Society had sponsored an abortive expedition to be carried out by André Michaux. Elected president in 1800, Jefferson longed to distinguish his presidency with a great scientific accomplishment. Lewis, his secretary, consulted and trained with members of the Philosophical Society to prepare for leading the expedition. Lewis was expected to make natural-historical observations and collections (Jefferson hoped that mammoths would be found in the American interior) and exact astronomical observations for the purpose of mapping.

Jefferson took personal care of the first batch of specimens sent from the expedition, including sixty carefully labeled plant specimens as well as the skins of pronghorn antelopes, a coyote skeleton, a jackrabbit skeleton, and parts of many other North American animals. Live specimens included magpies, a prairie dog, and a sharp-tailed grouse. The bulk of the expedition's scientific gains entered the American learned community on the return of Lewis, Clark, and the rest of their party in 1806. The range of information available in American natural history was immensely expanded. The German botanist Frederick Pursh (1774–1820), author of *Florae Americae Septentrionalis* (1814), included 124 specimens from the expedition's collection in his descriptions, naming genera after both Lewis and Clark. (It was galling for American science, though, that not only was the book the work of a European rather than an American botanist, but it was also published in London!) The Lewis and Clark zoological specimens, mostly in the collection of the American Philosophical Society, contributed to a series of classic early works in American natural history, including *American Ornithology* (1808–1814) by Alexander Wilson (1766–1813), *Fauna Americana* (1825) by Richard Harlan (1796–1843), and *American Natural History* (1826–1828) by John Godman (1794–1830).

The geographical information proved even more important, although the difficulty of precise astronomical observations to establish latitude and longitude meant that the maps the expedition made were not completely accurate. Clark's map, included in the official account of the expedition, *History of the Expedition under the Command of Captains Lewis and Clark, to the Sources of the Missouri, Thence across the Rocky Mountains and down the River Columbia to the Pacific Ocean, Performed during the Years 1804–5–6* (1814), established the principal features of the American West, disabusing geographers of many accepted ideas such as the southwestern rather than northwestern origin of the Missouri River. The Lewis and Clark expedition also set an American precedent for government-sponsored scientific expeditions.

See also American Philosophical Society; Botany; Exploration, Discovery, and Colonization; Michaux Family; Nationalism; Zoology.

Reference

Greene, John C. *American Science in the Age of Jefferson*. Ames: Iowa State University Press, 1984.

Lichtenberg, Georg Christoph (1742–1799)

Georg Christoph Lichtenberg was both a great German Enlightenment writer and a working scientist and science teacher. The child of a Lutheran minister and amateur astronomer, Lichtenberg was educated at the University of Göttingen, where he later became a professor of mathematics. He participated as an astronomer in an effort to produce an accurate map of Hanover, the German principality where Göttingen was located. He also edited the astronomical manuscripts of Johann Tobias Mayer. After the death of his colleague, the experimental physicist Johann Christian Polykarp Erxleben (1744–1777), Lichtenberg took over the Göttingen courses in experimental physics. He became increasingly fascinated by electricity, importing electrical equipment from England as well as commissioning pieces from German artisans. As was the custom in German Protestant universities, Lichtenberg built his collection of experimental equipment himself, although the university bought it from him in 1787. His most important electrical discovery was connected with a large electrophorus, a device recently invented by Alessandro Volta, in his laboratory. In 1777 Lichtenberg noticed that dust had settled on the electrified surface of the resinous "cake" of the electrophorus in patterns resembling stars, the "figures of Lichtenberg."

Lichtenberg was more important as a disseminator of scientific ideas than as an original scientist. In addition to the lectures and demonstrations in his classes, he brought out an improved edition of Erxleben's textbook of physics in 1784, brought up to date with his own notes, that was quickly recognized as the best textbook available. He was also a ruthless critic of work he considered inferior, gaining particular notoriety for his attacks on the "physiognomic" claims of Johann Kaspar Lavater (1741–1801), who claimed that character could be read in the outlines of the face. Lichtenberg lacked the nationalism of many late-eighteenth-century German scientists, and was one of the earliest to accept Antoine-Laurent Lavoisier's chemistry and to incorporate Italian work in electrical physics. A passionate lifelong Anglophile (Hanover was a possession of the king of Great Britain as elector of Hanover, and English students were common at Göttingen), Lichtenberg was admitted as a fellow of the Royal Society in 1793.

See also Electricity; Literature; Mayer, Johann Tobias; Physics.

References

Brinitzer, Carl. *A Reasonable Rebel: Georg Christoph Lichtenberg*. Translated by Bernard Smith. New York: Macmillan, 1960.

Heilbron, J. L. *Electricity in the 17th and 18th Centuries: A Study in Early Modern Physics.* 2d ed. Mineola, NY: Dover, 1999.

Lightning Rods

Comparisons between electrical sparks and lightning grew more common in the eighteenth century largely as a reaction to the greater quantity of electricity generated and perceived, particularly after the invention of the Leiden jar. The English electrician Stephen Gray (1666–1736) suggested that lightning was electric, and Albrecht von Haller, in a report on recent German discoveries in electricity published in the Dutch journal *Bibliotheque Raisonée* in 1745, asserted that electricity and lightning displayed basically the same qualities, producing light and flame and being conducted by metals. This comparison was greatly strengthened by the powerful shocks administered by the Leiden jar. As early as the description by Pieter van Musschenbroek (1692–1761) of his first encounter with the jar, the experience of being shocked by it was compared with the stroke of lightning. When Jean-Antoine Nollet used the spark of a Leiden jar to kill a sparrow, dissection revealed that the sparrow was in the same condition as a man killed by lightning. Like Haller, Nollet listed the similarities between lightning and electricity.

The identification of lightning as an electrical phenomenon, however, is credited to Benjamin Franklin, who was, if not the originator of the idea, its most avid promoter and the one most alert to its implications. Franklin's famous kite experiment of 1752 (actually first performed by a French experimenter named Jacques de Romas [1713–1776]) used a wire attached to a kite to draw electricity from the clouds onto a metal key. Even more influential was an experiment Franklin himself only suggested but never carried out, the sentry-box experiment set forth in his *Experiments and Observations on Electricity* (1751). A man would stand in a sentry box located on a high tower during a thunderstorm. A pointed metal rod would stick out from the top of the box for twenty or thirty feet. The rod would gather electricity from the passing clouds, and the man would touch it and receive the sparks. The idea was not that the rod would "catch" the lightning, but that by absorbing the electrical fluid from the cloud, it would prevent lightning from forming. The experiment was actually tried in France by the rival natural philosophers Nollet and Georges-Louis Leclerc de Buffon, both of whom reported success. This type of experiment could be extremely dangerous—in 1753 a German experimenter working in Russia, Georg Wilhelm Richmann (1711–1753), tried the experiment and was instantly killed when lightning struck the rod.

Franklin believed that sharp objects caused the electrical fire to dissipate rather than concentrating in the form of a lightning bolt, and his suggestion for lightning rods in *Experiments and Observations on Electricity* laid great stress on their pointedness. The success of the sentry-box experiment seemed to prove Franklin's theory of lightning, and lightning rods quickly spread as protectors of large structures, despite the misgivings of some religious people who saw them as an impious attempt to thwart the will of God. Despite the fact that church towers, often the highest structures in an area, were frequent targets of lightning strikes, there was reluctance to protect them. From a technical viewpoint the termination of the rod was the great sticking point, as Franklinist defenders of pointed rods clashed with the English electrician Benjamin Wilson (1721–1788), a champion of short, rounded rods. Wilson argued that long, pointed rods might attract electricity that otherwise would pass harmlessly overhead. (Nollet had his doubts about the whole project, wondering why anyone would want

This illustration shows how Luigi Galvani used a lightning rod to "galvanize" a frog's legs. (National Library of Medicine)

to attract lightning anyway.) Wilson and Franklin clashed on a committee of the Royal Society established in 1772 and chaired by Henry Cavendish with the purpose of deciding how the British gunpowder magazines at Purfleet should be protected from lightning. The committee eventually entangled itself with disputes between Franklin's radical London associates and Wilson's aristocratic patrons. Franklin triumphed, but when the Purfleet magazines, equipped with points, were damaged by lightning in 1777, the pointed rods were taken down and replaced by rounded ones. By that time Franklin, in Paris promoting the American Revolution, had lost interest in the dispute, and openly hoped that George III (r. 1760–1820) would have no lightning rods at all! Despite this temporary setback, pointed rods triumphed in the nineteenth century, although modern students of lightning protection find little difference between the two styles and some experiments support the rounded model. The lightning rod became a standard example of the actual, practical benefits of science during the late Enlightenment.

See also Electricity; Franklin, Benjamin; Technology and Engineering.

References

Heilbron, J. L. *Electricity in the 17th and 18th Centuries: A Study in Early Modern Physics.* 2d ed. Mineola, NY: Dover, 1999.

Sutton, Geoffrey V. *Science for a Polite Society: Gender, Culture, and the Demonstration of Enlightenment.* Boulder: Westview, 1995.

Linnaeus, Carolus (1707–1778)

The Swedish botanical and zoological classifier Carl von Linné is better known by the Latin version of his name as Carolus Linnaeus. The eldest son of a Lutheran minister, Linnaeus was originally intended to follow in his father's profession. Botany attracted his interest from an early age, and against his family's wishes he planned a medical career. In 1727 he entered the University of Lund, which proved unsatisfactory, and he transferred to Uppsala, Sweden's leading medical

Carolus Linnaeus, an eighteenth-century botanist and scientist, created a system to catalog every plant and animal by genus and species. Though much modified, his schema forms the basis of the modern system of classification. (Perry-Castaneda Library)

school, the following year. There he continued to cultivate botany, and became particularly fascinated with the sexuality of plants. The idea that plants reproduced sexually, established experimentally by Rudolf Jakob Camerarius (1665–1721) in 1694, had only recently reached Sweden. An avid collector of plants, Linnaeus was struck by the chaos in which European botany had been stranded by the description of thousands of plants by hundreds of botanists, with no standard names or ways of identification. He was dissatisfied by the dominant classification system available, that of the French botanist Joseph Pitton de Tournefort (1656–1708). In 1730 Linnaeus started classifying plants by the number and arrangements of their sexual parts, the pistils and stamens. This was the beginning of the "sexual system," to which he retained an unswerving devotion for the rest of his life.

In 1732 Linnaeus departed his home in southern Sweden for a prolonged journey to the far north of the country, Lapland, home of the nomadic Lapps. His trip was sponsored by the Royal Swedish Academy of Science, and he was expected to identify resources for economic development. Linnaeus's journal, published thirty years after his death, contains much information on the customs of the Lapps as well as the natural history of the area. Like many eighteenth-century people, Linnaeus idealized remote and pastoral people like the Lapps, who he speculated had been spared the worst effects of the Fall of man. A Lapp drum he acquired became a prize possession, and Linnaeus sometimes dressed up in a Lapp costume and demonstrated the drum. He even had his portrait painted in a Lapp costume.

In 1735 Linnaeus went on a tour of Europe, seeing some of the great natural-history collections and taking a medical degree at a Dutch diploma mill, the University of Harderwijk. He hoped to publish some of his manuscripts in the Dutch Republic, the center of European learned printing. His ambition was more than fulfilled, and the next few years saw a remarkable burst of publication. The first edition of Linnaeus's *System of Nature,* a great rarity, appeared in 1735 in Leiden. The next year he published *The Botanical Library* and *Foundations of Botany.* Linnaeus's Lapland botanical observations were the basis of his *Flora Lapponica* (1737), which classified the Lapp plants according to the new sexual system. The same year he published *Critica Botanica* and *Genera of Plants,* which defined all plant genera then known to European botanists. Linnaeus's ability to finance these publications is testimony to his remarkable skill at finding patrons. The most important patrons at this stage in his career included the Dutch physician and botanist Johann Gronovius (1690–1762) and the rich merchant Georg Clifford (1685–1760). Linnaeus's tremendous botanical knowledge and skill at identifying plants won Linnaeus, a shrewd and imaginative self-promoter, a great reputation among European savants, particularly after he managed to coax a banana into flower, a feat never before accomplished in the Netherlands. He would later duplicate the feat in Sweden.

Despite the wide respect Linnaeus earned as a botanist, the botanical world was skeptical of his classification system, and some even advised Linnaeus to drop the classification project in favor of descriptive botany. Many were skeptical of the extreme emphasis Linnaeus put on sex in classification, and to some it even seemed somewhat improper. In England there was some hesitance in allowing women to learn the Linnaean system, although this was overcome. Throughout Linnaeus's career, French botanists in particular would resist his classification scheme, preferring to classify plants on the basis of a broad range of natural resemblances rather than following what they viewed as Linnaeus's reductionistic emphasis on sex. Georges-Louis Leclerc, the Comte de Buffon, head of the Royal Botanical Garden, in particular opposed Linnaeus's system, but it was adopted in France by royal decree in 1774. In the late eighteenth and early nineteenth centuries, the Linnaean system became universally known.

Linnaeus returned to Sweden and set up medical practice in Stockholm, receiving a medical chair at Uppsala in 1741. His native land attracted most of his interest—he went on several journeys of observation and collection to remote Swedish provinces, and *Swedish Flora* appeared in 1745 and *Swedish Fauna* in 1746. The acquisition of a collection of dried plants and drawings from Ceylon led to the publication of *Ceylonese Flora* in 1747. He also built up the Uppsala Botanical Garden and started a collection of exotic animals. Linnaeus used his central intellectual position as the great classifier to build up an enormous collection of natural-historical specimens that had been sent to him, and acquired an unenviable reputation for accepting gifts of specimens but never returning the favor.

Linnaeus was a gifted teacher, and much of his system's success rested on his ability to

A portrait of Linnaeus in Lapland dress (National Library of Medicine)

attract disciples, mostly from Sweden and Germany. This somewhat made up for Linnaeus's inability to speak any modern language other than Swedish, which few spoke outside Sweden. He was among the last important scientists for whom Latin was an altogether sufficient means of international communication. From Uppsala were dispatched what Linnaeus called his apostles—young men who covered a great portion of the world in their search for exotic, and economically useful, plants. Pehr Kalm (1716–1779) went to North America, Pehr Osbeck (1723–1805) to China, and Daniel Carl Solander (1733–1782) accompanied James Cook on his voyage around the world in the *Endeavour.* Linnaeus commemorated his disciples by naming plants after them. The prerogative of naming was one he guarded jealously, and field botanists who gave their own names to plants frequently found them changed by Linnaeus in the subsequent editions of his botanical works. His rival,

Albrecht von Haller, referred to him mockingly as the new Adam, because of his claim to name every species.

The principles of Linnaeus's classifications of species and genera—and he regarded classification as by far the most important work of botany—were set forth in *Botanical Philosophy* (1750). The great work of Linnaean botany in practice was *Species of Plants* (1753). Its two volumes contain descriptions of nearly 6,000 plants, arranged in 1,098 genera. It stands as the beginning point of the modern system of botanical nomenclature. The tenth edition of Linnaeus's *System of Nature* (1758) occupies the same role in zoology. Linnaeus's expertise in zoology never matched that which he had acquired in botany, but he was eager to classify animals as well as plants. His animal classifications were based on a broader range of characteristics than his plant classifications, and established many of the basic categories used today—for example, he named the class of mammals, and, an ardent supporter of maternal breast-feeding, he permanently removed cetaceans from the class of fish by insisting on the production of breast milk by females as the defining and eponymous characteristic of the class. (His identification of rhinoceroses as rodents was less successful.) In both botany and zoology Linnaeus systematized and made universal the use of binomial nomenclature, the identification of each species by two Latin words, one for the genus and the other for the species itself. In medicine Linnaeus published *Materia Medica,* a popular reference book describing hundreds of plants and their medicinal uses, in 1749, and the short *Double Key to Medicine* in 1766. The *Double Key,* like *Foundations of Botany,* consisted of collections of gnomic aphorisms. Linnaeus's philosophy of medicine was basically mechanical, although with a number of features all his own, such as a division of diseases into feminine and masculine.

Linnaeus was a remarkably keen observer who did not conceal the delight he took in the phenomena of nature. His writings in Swedish, particularly his rapturous descriptions of Sweden's short summers, have become a permanent part of the Swedish literary heritage.

A practicing natural theologian, Linnaeus combined a strong, although increasingly unorthodox, piety with tremendous egotism. He believed himself chosen by God to set knowledge of the natural world on a firm footing. His opponents were not merely intellectual rivals, but heretics. The plethora of honors he received, both from the international world of science and from the Swedish state, even further enhanced his ego. At his own desire, his tomb was inscribed with the words *Princeps Botanorum* (Prince of Botanists).

See also Botany; Exploration, Discovery, and Colonization; Nationalism; Natural Theology; Religion; Royal Swedish Academy of Science; Sexual Difference.

References

Blunt, Wilfrid. *The Compleat Naturalist: A Life of Linnaeus.* New York: Viking, 1971.

Frangsmayr, Tore, ed. *Linnaeus: The Man and His Work.* Berkeley: University of California Press, 1983.

Linnean Society

The Linnean Society was the most successful eighteenth-century society devoted entirely to natural history. It was formally founded in London on 8 April 1788, although there had been informal meetings before that. The first president and guiding light of the society was the physician and botanist James Edward Smith (1759–1828), author of *English Botany* (1790) and subsequent works on the flora of the British Isles. Smith was an avid supporter of the Linnaean system of botanical classification who had purchased Linnaeus's collections, manuscripts, and books for which the society provided an institutional home. Smith's sway over the society had elements of tyranny, as those eminent British natural historians who had crossed him in some way were excluded, and the Linnaeus collection continued to be his personal property, with items given away or sold. When Smith married and moved to Norwich in 1796, he continued to hold the office of president until his death, although he attended few society meetings. After his death the society purchased Linnaeus's collections from his estate, incurring a massive debt that would not be paid off until 1861.

The Linnean Society began publishing *Transactions* in 1791, providing a much needed outlet for natural-history papers. The society remained dominated by botanists, and rising interest in zoology and entomology led to the formation of the Zoological Club of the Linnean Society in 1820. This was part of the overall early-nineteenth-century trend toward greater specialization in scientific societies. The Linnean Society continues to exist to the present.

There was also a French Linnean Society, founded in Paris at the end of 1787 by a group of young natural historians, including Pierre-Auguste Broussonet (1761–1807), the secretary to the Agricultural Society. Broussonet had spent several years in England and was excited by the Linnaean system. The promotion of Linnaeus's system in France was a direct challenge to the rival system of the powerful Georges-Louis Leclerc de Buffon. Buffon, along with his assistant at the Garden of Plants, Antoine-Laurent de Jussieu (1748–1836), quickly crushed the French Linnean Society by letting it be known that any young naturalist who joined it would be excluded from the Royal Academy of Sciences and other centers of scientific prestige. The members provided the nucleus for the Society of Natural History that was founded after the Revolution. Other Linnean Societies were founded in Philadelphia (1806), Boston (1814), and Lyon (1822).

See also Academies and Scientific Societies; Botany; Buffon, Georges-Louis Leclerc de; Linnaeus, Carolus; Zoology.

References

Allen, David Elliston. *The Naturalist in Britain: A Social History.* Princeton: Princeton University Press, 1994.

Gillispie, Charles Coulston. *Science and Polity in France at the End of the Old Regime.* Princeton: Princeton University Press, 1980.

Literature

Literary writers showed an increasing awareness of science in the eighteenth century and incorporated it into their work. The most obvious way this occurred was in works specifically on scientific topics. Science was a surprisingly popular topic for poetry. Several elaborate treatments of the Newtonian system in verse appeared in eighteenth-century England, such as *Creation, a Philosophical Poem in Seven Books* (1712) by Sir Richard Blackmore (1654–1729). Blackmore presented science as a way of refuting "Epicurean" atheism. The death of Isaac Newton (1642–1727) was greeted by an outpouring of poems in his praise, which continued for decades. The most influential was "To the Memory of Newton," by James Thomson (1700–1748), written the same year as Newton's death. Thomson would go on to write the extremely popular *The Seasons,* which incorporated much science. The tradition of expository scientific poetry in English was extended to natural history by Erasmus Darwin, in his popular botanical epic *The Loves of the Plants* (1789) and *The Temple of Nature* (1803). Science also appeared in prose narratives, such as Denis Diderot's dialogue *D'Alembert's Dream* (1769) and Voltaire's romance *Micromégas* (1752).

A more subtle way science affected literature was the use of science to provide image, metaphor, and even plot structure. The Newtonian conception of light as split into colors attracted poets, many of whom incorporated Newtonian optics into their metaphors and descriptions, particularly of the rainbow. The concept of "attraction," used by scientists to describe gravity or electricity, was also frequently used in poems and prose narratives to describe relations between people. One of the most elaborate literary uses of science was Johann Wolfgang von Goethe's novel *Elective Affinities* (1809) in which the arrangements and rearrangements of two pairs of lovers were structured according to Pierre Joseph Macquer's version of the chemical

English author Mary Shelley is remembered for what she called her "hideous progeny," Frankenstein; or, the Modern Prometheus *(1818). (Corbis)*

theory of "elective affinities" (a theory somewhat old-fashioned by 1809).

Not all writers supported the science of their time. Even in the heyday of English Newtonianism, scientists were satirized (although Newton himself was immune), sometimes as triviality-obsessed fools, and sometimes as impious materialists bent on exiling God from the universe. Antiscience writing increased during the romantic period (although many romantics, particularly in Germany, were avid followers or even practitioners of the science of their time) and was more serious than satirical. The most antiscientific poet of the entire period was William Blake (1757–1827), who denounced Newton himself (along with Francis Bacon [1561–1626] and John Locke [1632–1704], the two other intellectual heroes of the English Enlightenment) as one who would reduce the wonder of the universe and the glory of God to mere mathematics and mechanics. John Keats (1795–1821) shared Blake's attitude, although somewhat less passionately. The scientist sometimes appeared as an evil, manipulative character, as in the tales of the German romantic Ernst Theodor Wilhelm Hoffman (1776–1822).

The most influential—and one of the most ambivalent—romantic books dealing with science was *Frankenstein* (1818) by Mary Wollstonecraft Shelley (1797–1851). Shelley was aware of current developments in science and made her tale consistent with recent scientific claims, particularly in the field of electricity. Victor Frankenstein, the creator of the nameless monster, shares some of the character traits of the evil scientist, such as alienation from human society and arrogance in attempting to duplicate the work of God, but is also presented as a sympathetic character. *Frankenstein* is often regarded as founding the modern tradition of science fiction.

See also Diderot, Denis; Goethe, Johann Wolfgang von; Lichtenberg, Georg Christoph; Lomonosov, Mikhail Vasilyevich; Romanticism; Voltaire.

References

Hayes, Roslynn D. *From Faust to Strangelove: Representation of the Scientist in Western Literature.* Baltimore: Johns Hopkins University Press, 1994.

Nicolson, Marjorie Hope. *Newton Demands the Muse: Newton's "Opticks" and the Eighteenth-Century Poets.* Princeton: Princeton University Press, 1946.

Shaffer, Elinor S., ed. *The Third Culture: Literature and Science.* Berlin: Walter de Gruyter, 1998.

Lomonosov, Mikhail Vasilyevich (1711–1765)

The versatile Mikhail Vasilyevich Lomonosov was Russia's first outstanding scientist. From a prosperous background, he was educated at various Russian institutions and at the University of Marburg under Christian Wolff. Lomonosov's mechanistic, corpuscular non-Newtonian physics owed much to Wolff. Lomonosov was admitted as an adjunct at the Imperial Academy of Sciences of St. Petersburg in 1742, and assigned the task of cataloging the mineral and fossil collection. In addition to experimental physics, his many interests included chemistry—he became the professor of chemistry at the St. Petersburg academy and set about the creation of an advanced chemical laboratory. In 1758 he became head of geography at the academy and worked on mapping the vast territories of Russia. He also practiced observational astronomy and meteorology. Lomonosov's membership in the academy was fraught with political tension, and he was even imprisoned for several months. One reason for this was his resentment of the foreign, particularly German, scientists who filled so many positions in the academy. (One exception to this animosity was Leonhard Euler, who introduced some of Lomonosov's work to Western scientists.) Lomonosov was a proud Russian who used his considerable poetic skills to encourage young Russians to emulate the great scientists of the European past. He also published a grammar of the Russian language, and pioneered the development of Russian for scientific purposes. Lomonosov mainly published in the *Transactions* of the St. Petersburg academy. The fact that much of his work was left in manuscript and relegated to obscurity limited his influence on his scientific contemporaries.

See also Imperial Academy of Sciences of St. Petersburg.

References

Dictionary of Scientific Biography. S.v. "Lomonosov, Mikhail Vasilievich."

Heilbron, J. L. *Electricity in the 17th and 18th Centuries: A Study in Early Modern Physics*. 2d ed. Mineola, NY: Dover, 1999.

Longitude Problem

The problem of the longitude, locating the east-west position of a ship on the open sea, was the classic technological problem of the early modern period, assaulted by many of the greatest scientists of the scientific revolution, including Galileo Galilei (1564–1642), Christiaan Huygens (1629–1695), and Edmond Halley (1656–1742). The great astronomical observatories founded in the period, most notably the Paris Observatory and the English Royal Observatory at Greenwich, had the solution of the longitude problem high on their agendas. All failed, leaving the problem for the eighteenth century. If anything, it was of increasing urgency, given the expansion of the territory covered by European vessels. As the celestial bodies seemed to rotate around Earth from east to west, they did not seem to offer a way to know one's position on it. Existing methods, based on observation of the Moon, or simply estimating the speed one had been traveling for a given time, were maddeningly and even dangerously imprecise. Most approaches to the longitude reduced the problem to that of finding the difference between the time on the ship, set by observation of the Sun's meridian at noon, and the time at a fixed point, usually that of the home port. The difference in time could be translated into spatial terms as the difference in longitude between the two points. There were all sorts of bizarre schemes for this, but the two main approaches were using astronomical events to give the correct time and creating a clock able to give accurate time on a ship. If the home-port time of a celestial occurrence were known, all that would be necessary would be to compare the ship's own time on observation of the occurrence. For example, if the time when an eclipse would occur at a fixed point were known, all that would be necessary would be to compare the time that the ship's navigator saw the eclipse, and the distance between the two points would be known. This method was limited in its uses, however, as eclipses were quite rare. Galileo's idea of using the frequent eclipses of the moons of Jupiter became dominant in geography and cartography on land, but the difficulty of observing Jupiter's moons from a moving ship made it difficult if not impossible at sea.

The greatest eighteenth-century sea power, Great Britain took the lead in most eighteenth-century longitude schemes, although its colonial rival, France, was not far behind. Two unsuccessful longitude solvers, William Whiston and Humphrey Ditton (1675–1715), set forth a project in 1713 for the creation of a network of stationary ships over the seas, whose crews would fire guns at designated times, enabling passing ships to set their distances by factoring in their knowledge of the speed of sound. This idea was impractical on many levels, and never seriously considered. Whiston and Ditton's lobbying of the British Parliament for a more active approach to the problem along with London's maritime leaders whom they had organized resulted in the Longitude Act of 1714. This act established a prize of 20,000 pounds for a solution accurate to half a degree of a great circle around Earth; 15,000 pounds for a solution accurate within two-thirds of a degree; and 10,000 for a solution accurate within a degree. It also set up the Longitude Board whose ex officio members included the astronomer royal, the president of the Royal Society, and the first lord of the Admiralty, among others. The board disposed of funds to encourage promising ideas and was the first great institutional patron of science. It was deluged with solutions, most of them crackpot, and for the first decade and a half of its existence never met and concerned itself with little beyond sending out rejection letters. The French Royal Academy of Sciences meanwhile had used a bequest from the magistrate Rouille de Meslay to set up a prize of 125,000 livres for the longitude and other improvements in navigation, and were considerably more active, awarding 2,000 livres in 1720.

Serious eighteenth-century longitude ideas divided into two categories: the creation of an accurate shipboard clock and the astronomical method known as "lunar distances." Lunar distances rested on the invention of a new astronomical instrument, the octant. This happened twice in 1731, with the independent work of the Englishman John Hadley (1682–1744) and the Philadelphian Thomas Godfrey (1704–1749). An arrangement of mirrors enabled a navigator to hold the distances between two celestial objects steady, even on the deck of a rolling ship. By observing the angular separation of the Moon and a given star, then comparing the time of observation with a table giving the times when that angular separation would appear from a fixed point such as London or Paris, the navigator could get the time differential and thus the longitude. All this plan required were accurate, mathematically skilled navigators and accurate tables of the extremely complex lunar motion, and legions of astronomers all over Europe set to work to provide the latter. Although the English and French scientific establishments poured effort and money into the project, the most accurate tables were the work of a German, Johann Tobias Mayer. Mayer's death prevented him from claiming the prize, although his widow received 3,000 pounds from the board.

By comparison, the clock idea was somewhat old-fashioned. The leader in the creation of a navigational clock was a self-taught English clock maker of genius named John Harrison (1693–1776), who worked outside the London-based English clock-making establishment. Harrison had contacted Astronomer Royal Edmond Halley early in the project and enjoyed some support from the Longitude Board and the Royal Society. But he also faced opposition from a series of royal astronomers, including James Bradley and Nevil Maskelyne, who strongly favored lunar distances and were ex officio members of the Longitude Board, often supervising the trials. Harrison received several thousand pounds from the Longitude Board, at one point benefiting from the personal intervention of King George III (r. 1760–1820), but never won the prize he sought. The French, meanwhile, were also investigating the possibility of an accurate watch, led by the royal clock maker Ferdinand Berthoud (d. 1807). After shipboard watches were tested on voyages to Saint Domingue in 1769 and 1771, their use became common in the French marine.

In England Harrison was pitted against Maskelyne, the greatest exponent of lunar distances, who Harrison believed applied unnecessarily stringent conditions to the tests of the clocks and did not care for them properly when they were in his custody. Maskelyne's annual *Nautical Almanac and Nautical Ephemeris,* first published in 1767, with its associated lunar tables, was the best available and put the lunar-distance method on a sound footing. This idea originated in the work of the Frenchman Nicolas-Louis de Lacaille (1713–1762) in the 1750s, but the French had never followed up Lacaille's work. They did publish a French translation of Maskelyne's almanac, beginning in 1772, a project with which Maskelyne cooperated even while the two countries were at war. The British navy required its navigators to be certified as proficient in Maskelyne's method, although this was not consistently enforced at first. Updated, Maskelyne's works served the international navigational community into the early twentieth century. It is due to Maskelyne's lunar tables that the meridian of the Royal Observatory at Greenwich became the determining point for world time.

The lunar-distance method had the disadvantages of not being possible on moonless nights, and of requiring several observations and much tedious and difficult calculation. The chronometric method using timepieces eventually became the most common way to find the longitude. The problem was not the accuracy of the watches, particularly after Captain James Cook used a timekeeper based on Harrison's on his second voyage, from 1772 to 1775, and enthusiastically testified to its merits (although he also praised Maskelyne's almanacs). The difficulty was the cost of reproducing accurate timepieces. Late-eighteenth-century London watchmakers, most notably John Arnold (1736–1799) and Thomas Earnshaw (1749–1829), simplified Harrison's designs and began mass production of accurate shipboard watches, which became the dominant way of finding the longitude by the 1820s. The Longitude Board itself was disbanded in the new Longitude Act of 1828. Its greatest prize was never awarded.

See also Astronomy; Bradley, James; Cook, James; Exploration, Discovery, and Colonization; Instrument Making; Maskelyne, Nevil; Mayer, Johann Tobias; Whiston, William.

References

Howse, Derek. *Nevil Maskelyne: The Seaman's Astronomer.* Cambridge: Cambridge University Press, 1989.

Sobel, Dava. *Longitude: The True Story of a Lone Genius Who Solved the Greatest Scientific Problem of His Time.* New York: Walker, 1995.

Lunar Society of Birmingham

The Lunar Society, so-called because it met on a day near a full moon to enable its members to walk home after sunset, was a highly informal organization of men interested in new developments in science and technology. Its informality makes its history as an institution difficult to trace. Beginning in the mid-1760s there were gatherings in Birmingham whose membership eventually included Erasmus Darwin; the chemical industrialist James Keir (1735–1820); the ceramic manufacturer Josiah Wedgwood (1730–1795); the engineer James Watt (1736–1819) and his business partner, Matthew Boulton (1728–1809), the founder and leader of the group; and the Scottish professor and physician William Small (1734–1775), its leading spirit. On Small's death, more formal arrangements seem to have been thought necessary, and the society's first "official" meeting was 31 December 1775.

Meetings were theoretically monthly but in practice intermittent, and since the society never published its proceedings it is not known what was said there. The society was altered greatly by Joseph Priestley's arrival in Birmingham in 1780. It organized a subscription to provide him with a chemical laboratory and changed its meeting day from

Sunday to Monday in deference to Priestley's professional duties as a minister. Priestley's researches seem to have dominated the society's business, with a corresponding rapid decline after his laboratory was sacked by a mob hostile to his political views in 1791. The Lunar Society's exact ending date is unknown, although there is evidence of meetings into the early nineteenth century.

See also Academies and Scientific Societies; Darwin, Erasmus; Industrialization; Priestley, Joseph.

Reference

Schofield, Robert E. *The Lunar Society of Birmingham: A Social History of Provincial Science and Industry in Eighteenth-Century England.* Oxford: Clarendon, 1963.

M

Macquer, Pierre Joseph (1718–1784)

Pierre Joseph Macquer was the leading French chemist before Antoine-Laurent Lavoisier. Like many eighteenth-century chemists, he first trained as a physician, receiving an M.D. from Paris in 1742. He entered the chemistry section of the Royal Academy of Sciences as an adjunct in 1745 and became chemistry lecturer at the Royal Botanical Garden.

Although Macquer kept up his medical connections and was a member of the Royal Society of Medicine, most of his practical work dealt with industrial chemistry. He was head of both the French porcelain manufactory at Sèvres and the dye works at the Gobelins. His numerous experiments on porcelain resulted in a theory of its manufacture that went beyond a recipe into a description of the chemical processes involved. It also improved the quality of French porcelain. As a dyer, Macquer's greatest achievement was the chemical analysis of the painter's color Prussian blue, and its adaptation into a dye. He also researched the chemical properties of arsenic, and how to improve wine making, among many other subjects. He was also active in the academy, rising to the highest rank of pensionary in 1772, sitting on several commissions, and serving as director in 1774, in which capacity he firmly and successfully resisted an attempt by the government to foist on the academy a candidate the academicians considered unworthy.

Macquer wrote textbooks called *Elements of Theoretical Chemistry* (1749) and *Elements of Practical Chemistry* (1751) that became standard, replacing the much simpler and more pharmaceutically oriented *Course of Chemistry* (1675) by Nicolas Lémery (1645–1715). He endorsed the phlogiston theory, and also used tables of chemical affinity. His most influential work was his *Dictionary of Chemistry* (1766), which was translated into English, German, Danish, and Italian. This was the first chemical dictionary, and an early attempt to reform and standardize chemical language, a project that Macquer had begun in his earlier writings. Macquer insisted that chemical terms be used with a precise meaning. For example, it was inappropriate to refer to the substance now called silver nitrate as "vitriol of silver" as it contained no vitriolic acid. The second edition of the *Dictionary,* in 1778, acknowledged the new discoveries of the pneumatic chemists. Macquer kept abreast of the new chemistry of Lavoisier, but remained a phlogistonist.

See also Chemical Nomenclature; Chemistry; Industrialization; Royal Academy of Sciences.

References
Coleby, L. J. M. *The Chemical Studies of P. J. Macquer.* London: Allen & Unwin, 1938.
Hahn, Roger. *The Anatomy of a Scientific Institution: The Paris Academy of Sciences, 1666–1803.* Berkeley: University of California Press, 1971.

Madness

Neither theories of madness nor treatment of the mad changed greatly for most of the eighteenth century, but a series of rapid changes around the beginning of the nineteenth century led to the founding of "psychiatry." Traditional explanations of mental disturbance, demonic possession, humoral imbalance, or the "wandering womb" in women were declining during the Enlightenment. No single medical theory replaced the older ideas, but many, including William Cullen (1710–1790) and his disciples at the University of Edinburgh, speculated that madness was caused by poorly functioning nerves. After the mid-eighteenth century the care of the mad was more likely to be a medical specialty, rather than being handled as one part of a general medical practice. Institutions for the mad, whether private, as was usual in Britain, or affiliated with the state or church, acquired a more separate identity from caregiving institutions generally and began to be seen as medical rather than charitable institutions. Madhouses proliferated, although the majority of the insane remained outside them. In Great Britain madness acquired a particularly high public profile with the madness of George III (r. 1760–1820), first a spell in 1788, and then permanently in 1810. The 1788 episode saw the royal household flooded with subjects' suggestions for cures, and the king passed from the court physicians, unspecialized practitioners, to a specialized "mad doctor" and asylum owner, the Reverend Dr. Francis Willis (1718–1807), who was eventually credited with the king's recovery.

Medical treatments for the mad in the eighteenth century were sometimes punitive, with mad people being retained in shackles and kept in filthy conditions in asylums such as England's notorious Bridewell. Less dramatic medical treatments included bleeding and purging, traditionally seen as curative, with a variety of quack nostrums and "secret" formulas, mesmerism, and the use of opiates growing in popularity during this period. Although opiates seemed to be effective temporarily, no treatment seemed consistently effective in producing cures.

The late eighteenth century saw a series of changes in the care of the mad, which eventually led to the emergence of the discipline of "psychiatry," a word coined at that time. Champions of "moral treatment" advocated a less punitive and restraint-based environment for the mad. In France the key—and heavily mythologized—moment was the so-called striking off of the chains of the madmen of the Paris asylum, the Bicêtre, in 1793 by the physician Philippe Pinel (1745–1826). Pinel's actual activities in the Bicêtre and the women's madhouse, the Saltpêtrière, were less dramatic than the legend, but he was responsible for a move from an emphasis on restraint to attempts at cures, or at least management, through moral treatment. Pinel attributed most cases of madness to specifically mental conditions, rather than organic and irreparable damage to the nerves, and believed that kind, but authoritative, treatment could cure or at least alleviate the sufferings of many victims. His *Treatise Medical-Philosophical on the Treatment of Mental Alienation* (1801) was translated into Spanish, German, and English. Similar movements away from punitive treatment occurred in Britain, where the lead was taken by a Quaker charitable institution, the York Retreat, founded by the tea merchant William Tuke (1732–1822) in 1796, and in Florence under the leadership of Vincenzo Chiarugi (1759–1820).

The York Retreat, publicized in *Description of the Retreat* (1813) by Samuel Tuke (1784–1857), had many imitators in England and America. However, many physicians were also troubled by the only minor role afforded medicine in the "moral treatment" model,

Philippe Pinel's removal of the chains from the mad people of Bicêtre was heavily mythologized, as can be seen in this painting (artist unknown). (National Library of Medicine)

and attempted to combine a moral with a medical approach. The American physician Benjamin Rush (1745–1813), in his *Medical Inquiries and Observations upon the Diseases of the Mind* (1812), combined moral therapy in the Pinel tradition, restraint, and medical treatment. Rush, one of medicine's great bleeders, believed that his favorite therapy could also benefit the insane, recommending that an attack of madness be immediately treated with the letting of twenty to forty ounces of blood. Another who combined moral and medical treatment was Johann Christian Reil, author of *Rhapsodies on the Use of Psychological Treatment Methods in Mental Breakdown* (1803). Reil, the coiner of the word *psychiatry,* also founded the field's first journal, *Journal of Psychological Therapy,* in 1805. He was followed by other German romantics who emphasized mental rather than physical causes of insanity, including J. C. A. Heinroth (1773–1843), author of *Textbook of Mental Disturbances* (1818) and one of the first to teach psychiatric medicine in a university.

See also Hospitals; Medicine; Psychology; Reil, Johann Christian.

References

Porter, Roy. *The Greatest Benefit to Mankind: A Medical History of Humanity.* New York: W. W. Norton, 1998.

———. *Mind-Forg'd Manacles: A History of Madness in England from the Restoration to the Regency.* Cambridge: Harvard University Press, 1987.

Manchester Literary and Philosophical Society

The Manchester Literary and Philosophical Society, founded in 1781, went through several phases in its development as northern England's premier scientific association. Initially, it was dominated by the physicians and apothecaries of the Manchester Infirmary (founded 1752) and had a pronounced Unitarian (and to a lesser extent Quaker) tinge. Its dominant ideologies were faith in progressive change through intellectual effort and suspicion of wealth and hereditary power. Joseph Priestley was the hero of many early members. Despite the society's presence in the heart of the English Industrial Revolution, it had little interest in questions of economic development, and few manufacturers

John Dalton, the chemist, was a leader in the Manchester Literary and Philosophical Society. (Library of Congress)

were members. The society's main activity was the reading of papers by its members, and unlike other northern English scientific societies like the Lunar Society of Birmingham, it published transactions. These were referred to as the *Manchester Memoirs,* and began in 1785. The society also provided its members with an extensive library.

By the early nineteenth century the society like many British institutions had become politically more conservative, had acquired more manufacturers as members (they formed an absolute majority by 1810), and had a scientific star, the chemical atomist John Dalton, who served as its president from 1817 to his death in 1844. Many of Dalton's theories and ideas were first put forth in papers presented to the society and published in its *Memoirs.* The society had become the north's leading scientific group, providing an alternative scientific culture to that of London and the English university towns, dominated by the Royal Society—one friendlier to trade and religious dissent. Rather than egalitarianism and social reform, it now emphasized the potential of natural science to provide rational diversion for young men, to adorn the social elite, and to function as a solvent of class tensions. The "Lit and Phil" also helped spawn a number of other Manchester scientific organizations, such as the Manchester Natural History Society in 1821 and the Royal Manchester Institution in 1823. Although the Manchester Literary and Philosophical Society's interests were not originally technological, engineers began to join in the 1820s and to introduce a greater interest in technological problems. In addition to Dalton, other leading scientists who were active in the Manchester society were the chemist, industrialist, and experimental physicist William Henry (1774–1836), a second-generation member, and Dalton's pupil James Prescott Joule (1818–1889). The society remains active to the present day.

See also Academies and Scientific Societies; Dalton, John.

Reference
Thackray, Arnold. "Natural Knowledge in Cultural Context: The Manchester Model." *American Historical Review* 79 (1974): 672–709.

Marggraf, Andreas Sigismund (1709–1782)

Andreas Sigismund Marggraf was mid-eighteenth-century Germany's leading chemist. The son of a Berlin grocer, Marggraf entered the Court Apothecary shop in 1726. He also attended the chemical lectures of Johann Heinrich Pott (1692–1777), a student of Georg Ernst Stahl, at the Medical-Surgical College in Berlin. He pursued his interest in chemistry as a medical student at the University of Halle in 1733 without taking a degree. The following year he studied at the mining academy of Freiberg, returning to Berlin in 1735. He was admitted as a member of the Berlin Academy in 1738 and became director of its new chemical laboratory in 1753. This was a central position of power and patronage in German chemistry. Lorenz Florens Friedrich von Crell recognized Marggraf's leadership by dedicating the first volume of his *Chemical Journal* to him in 1778. Marggraf's reputation extended outside Germany. In France, where his work was frequently cited in chemical publications, he was named one of the six foreign members of the Royal Academy of Sciences in 1777.

Marggraf's chemical achievements included the identification of beet sugar as the same substance as cane sugar, laying the foundation for the beet-sugar industry, and the isolation and identification of zinc by the heating of calamine and charcoal. Marggraf also obtained formic acid by the distillation of red ants. He was known for the small quantities he worked with, the mark of a skillful chemical analyst. In addition to his work at the academy, Marggraf taught private chemistry lessons. His pupils included the pharmacist Jacob Reinbold Spielmann (1722–1783); the distiller Johann Carl Friedrich Meyer (1739–1811); Franz Karl Achard (1753–1821), who tried to build on Marggraf's work by developing a commercially viable process for extracting beet sugar; and Martin Heinrich Klaproth.

See also Berlin Academy; Chemistry; Crell, Lorenz Florens Friedrich von; Klaproth, Martin Heinrich; Royal Academy of Sciences.

Reference

Hufbauer, Karl. *The Formation of the German Chemical Community, 1720–1795*. Berkeley: University of California Press, 1982.

Maskelyne, Nevil (1732–1811)

Nevil Maskelyne, the fifth astronomer royal of Great Britain, perfected the astronomical solution to the longitude problem after centuries of effort had been devoted to the question. However, his support of the lunar method led him into bitter conflict with rivals who supported solutions based on accurate timekeepers.

Astronomy was his obsession since his school days. After graduating from Cambridge University and entering the clergy, Maskelyne was admitted as a fellow of the Royal Society in 1758. He went to St. Helena as part of the team sent to observe the 1761 transit of Venus, although the observers were frustrated by the passage of a cloud. He remained on the island for several months to observe the passage of Sirius.

On his return Maskelyne began his involvement with the chief task of his career, the longitude. His *British Mariner's Guide* (1763) provided instruction for using the lunar-distance method—the observation of the Moon's movement against the background of the stars—using the new and accurate tables of Johann Tobias Mayer. He also set out for Jamaica that year as part of the Longitude Board's team for testing the chronometer of the Yorkshire clock maker John Harrison (1693–1776), who became his bitterest enemy. Harrison suspected Maskelyne, who had hopes of winning the Longitude Board's prize with an improved lunar-distance method, of sabotage. Maskelyne's reputation as the arch-opponent of the chronometric method for finding the longitude persisted to the end of his life, making him a frequent target of vitriolic attack.

In 1765 Maskelyne was appointed astronomer royal on the death of Nathaniel Bliss (1700–1764). His base was now the Royal Observatory at Greenwich, which he devoted considerable energy to refurbishing. Maskelyne's greatest work as astronomer royal was the annual *Nautical Almanac and Nautical Ephemeris* with the associated tables, first published in 1767. The *Almanac* was a remarkable feat of organization, with Maskelyne collecting a body of observations from astronomers in many parts of the world (in addition to his own extremely accurate observations from Greenwich) and overseeing the efforts of calculators, individual contractors who did the actual number crunching. The *Almanac* was published several years in advance, for the benefit of those undertaking long voyages. Maskelyne's observations were used by students of celestial mechanics, including Pierre-Simon de Laplace, as well as navigators, and his work is the reason the Royal Observatory is the starting point of world time.

Besides his navigational work, Maskelyne also coordinated British observations of the second transit of Venus in 1769. In 1774 he went to the mountain of Schiehallion in Scotland to measure the gravitational attraction exerted by a mountain on a plumb bob. This "attraction of mountains" was a long-standing problem for astronomers using precise instruments near mountains and hills. The experiment was also the first attempt to directly measure Newton's gravitational constant, and relevant for the question of Earth's density, although that was not Maskelyne's concern. For his work at Schiehallion, Maskelyne received the Copley Medal of the Royal Society for 1774.

See also Astronomy; Longitude Problem; Mayer, Johann Tobias; Observatories; Transits.

Reference

Howse, Derek. *Nevil Maskelyne: The Seaman's Astronomer.* Cambridge: Cambridge University Press, 1989.

Masons

See Freemasonry.

Masturbation

The eighteenth century saw the beginning of the great medical crusade against masturbation, which lasted into the twentieth century. Although masturbation had long been considered a sin, it was made a medical issue during the Enlightenment and moved to the forefront of medical and pedagogic concern. This began with a pamphlet by an anonymous English quack, *Onania,* which probably first appeared in 1715, although the earliest surviving edition is the fourth, from 1718. (The reference is to the biblical character Onan. The word *onanism* first appears in 1719.) The author blamed masturbation for ulcers, epilepsy, consumption, impotence, sterility, overall weakening, and an early grave. As treatment he recommended repentance, cold baths, and a drug available from the bookseller for a fee—a standard feature of quack pamphlets. *Onania* became one of the great best-sellers of the eighteenth century, going through fifteen editions by 1730 and many more thereafter, in addition to a German translation in 1736. The editions swelled with letters purportedly sent by masturbators suffering the consequences of their depraved habit and begging the author's assistance.

The second best-seller on the subject was the work of the Swiss physician Samuel-August Tissot (1728–1797), a popular and voluminous writer on health. He first published on masturbation in 1758, in an appendix to a Latin work on bilious fevers. In 1760 he published a French work, *Onanism.* Influenced by *Onania* and his own experience with patients, Tissot painted a gruesome picture of the masturbator, male or female, as subject to hideous and crippling diseases. Tissot also provided the campaign against masturbation with medical legitimacy. He claimed that physicians throughout history had condemned it, often distorting the writings of ancient and early modern medical writers by applying their warnings about the danger of sexual overindulgence in general to masturbation in particular. He also provided a medical theory, claiming that semen, or to a lesser degree an analogous fluid in women, was the highest perfection of bodily fluid, and a substance necessary to life. Masturbators wasted this substance, thereby weakening their constitutions and shortening their lives. This required some agility to explain why heterosexual intercourse, entailing a similar loss of fluid, was not debilitating, or why bloodletting, which Tissot recommended as a treatment for many diseases, actually strengthened the body. Tissot recommended cold baths and a restricted diet as a cure for masturbation.

Whatever its weaknesses, Tissot's thesis won general acceptance. Both the *Encyclopédie* and the *Encyclopedia Britannica* followed Tissot in their discussions of masturbation, and such leaders of European thought as Voltaire and Immanuel Kant found his arguments convincing. Physicians advised parents to closely supervise their children, and schoolmasters increased their surveillance of their charges. Specially designed nightclothes to prevent children from masturbating appeared by 1785, and by the end of the century some German surgeons were drawing and fixing the foreskin over the glans of the penis, a procedure called infibulation and claimed to cure masturbation. Some masturbators internalized the Tissot thesis and expected their systems to collapse if they continued the practice.

See also Medicine; Popularization.

Reference

Stengers, Jean, and Anne van Neck. *Masturbation: The History of a Great Terror.* Translated by Kathryn A. Hoffman. New York: Palgrave, 2001.

Materialism

The denial of spiritual reality and the reduction of the universe to matter was widespread among philosophers and scientists of the Enlightenment, although it was vigorously opposed by religious organizations and thinkers. Materialism had many roots in early modern thinking. There was a revival of interest in the philosophy of the ancient Epicureans, particularly the Roman poet Lucretius (c. 100 to 90–c. 55 to 53 B.C.). In Britain and France the philosopher John Locke (1632–1704) remarked that God could, if he wished, endow matter with the capacity for thought, setting off a philosophical debate that lasted for most of the century. In France the Cartesian tradition of mechanistic thinking and the restriction of spirit to God and the human soul could easily be taken a step further into pure materialism. The Cartesian description of animals as purely material and mechanical beings was widely accepted in eighteenth-century science, even by non-Cartesians. Physicians, many of whom in the early eighteenth century followed the doctrine of "iatromechanism," could easily extend the idea of the body as a mechanical and material system to explain all of human actions, without the necessity of a "soul." This was the path taken by the French physician and materialist Julien Offroy de La Mettrie. Materialists sometimes carried the denial of the spiritual world to the point of atheism (to the extent that they could openly express this) and almost always connected materialism with determinism.

Belief in dead, inert matter was more likely to be held by antimaterialists, who claimed that spirit was necessary to explain matter's activity. Materialists believed that matter was inherently active. The rise of vitalism from the mid-eighteenth century gave a tremendous boost to materialism, particularly in France. Given the greater strength of anticlericalism and atheism in French culture, French materialists were far more radical and confrontational than British. (In Germany the strength of the church and the Leibnizian tradition prevented materialism from getting much of a foothold at all during the Enlightenment and romantic periods.) Some of the most important French materialists were associated with the *Encylopédie,* notably Denis Diderot, its editor, and Baron Paul-Henri-Dietrich d'Holbach (1723–1789), who contributed many articles. Diderot gave literary form to his "vital materialism" in *D'Alembert's Dream,* written in 1769 but considered too daring to be published and circulating clandestinely in manuscript. The militantly atheistic d'Holbach published the most famous materialist work of the century, *The System of Nature* (1770). Although familiar with the science of his day, d'Holbach, a militant atheist, was less interested in the scientific aspects of atheism than in the philosophical and moral case against theism. Like many materialists, he believed that accepting the world's and humanity's material nature would lead to social reform. His science was basically Newtonian, incorporating an active matter. The French materialist tradition was carried on after the

French Revolution by a group of thinkers known as the ideologues, led by Antoine-Louis-Claude Destutt de Tracy (1754–1836) and the physician Pierre-Jean-Georges Cabanis (1757–1808). Cabanis became particularly notorious for his claim that the brain secreted thought in the same way that the liver secreted bile, and, like d'Holbach, connected a materialist analysis of humanity with social reform.

British materialism was less influenced by physiology than by physics and was much less antitheistic than the French tradition. Joseph Priestley attempted to transcend the division between materialism and religion by setting forth a version of materialism that he claimed was more conducive to religion than was the belief in spirits. Drawing on Ruggiero Giuseppe Boscovich's atomistic physics, Priestley argued in *Disquisitions Relating to Matter and Spirit* (1777) and subsequent works that matter was neither inert nor defined principally by extension, but was more accurately a set of forces. He claimed that matter-spirit dualists had no coherent explanation for how they interacted, and that there was no direct evidence of the existence of a separate nonmaterial soul. Priestley claimed that the immaterial soul was unscriptural, and even asserted that God was in some sense material. He also accepted the link between materialism and determinism. Priestley's religious materialism was not accepted by the defenders of religion and had little appeal to materialists who had absorbed the radical atheist materialism of the French Enlightenment.

Materialism was intellectually marginalized in the early nineteenth century. The reaction against the Enlightenment and the French Revolution that took different forms in the old regime states and Napoleonic France condemned it as atheistic, while romantics viewed materialism as soulless and mechanistic.

See also Diderot, Denis; La Mettrie, Julien Offroy de; Natural Theology; Priestley, Joseph; Vitalism.

References

Tapper, Alan. "The Beginnings of Priestley's Materialism." *Enlightenment and Dissent* 1 (1982): 73–82.

Vartanian, Aram. *Diderot and Descartes: A Study of Scientific Naturalism in the Enlightenment.* Princeton: Princeton University Press, 1953.

Vizthum, Richard C. *Materialism: An Affirmative History and Definition.* Amherst, NY: Prometheus, 1995.

Wellman, Kathleen. *La Mettrie: Medicine, Philosophy, and Enlightenment.* Durham: Duke University Press, 1992.

Yolton, John. *Thinking Matter: Materialism in Eighteenth-Century Britain.* Minneapolis: University of Minnesota Press, 1983.

Mathematics

Mathematics during the Enlightenment was closely bound up with physical science, and many of the most important contributors to physics, such as Jean Le Rond d'Alembert and Leonhard Euler, considered themselves to be primarily mathematicians. This was not merely a question of applying mathematics, but of a belief that large sectors of physics—notably mechanics and optics—were branches of mathematics. The program of "rational mechanics," for example, was to reduce mechanics to a deductive science modeled on Euclidean geometry. Physical problems led to many developments in mathematics itself. The debate between Euler and d'Alembert on the nature of a mathematical function began with d'Alembert's discovery of an equation describing the motions of a vibrating string. Mathematics possessed the prestige associated with Isaac Newton (1642–1727) and was considered as presenting the model for the physical sciences. Those sciences less amenable to mathematization, such as botany, were often considered less prestigious than fully mathematized ones, such as astronomy. It was even hoped to mathematize the human sciences, and much of the popularity of probability theory, which advanced greatly during this period, was based on the hope that it would provide a mathematical system for analyzing and solving social problems.

Mathematics itself moved toward abstraction and formalization. Diagrams and geometrical drawings disappeared from the most advanced mathematical texts, replaced by strings of symbols, many of whose meanings originated in this period. For most of the Enlightenment geometry was definitely of minor interest to the most advanced mathematicians. Even Newtonian celestial mechanics, which Newton himself had presented geometrically, was more and more presented in series of equations.

The foremost mathematical issue in the early eighteenth century was the contest of the two rival systems of the calculus, the Leibnizian based on differentials and the Newtonian based on fluxions. Partly due to the support of the influential Bernoulli dynasty of Swiss mathematicians, the Leibnizian calculus won out on the European continent, while the British mathematical community, dominated by Newton and his allies, went the way of fluxions. The isolation of British mathematics meant that it was increasingly backward compared to the Continent, only rejoining the main stream of European mathematics in the nineteenth century. (The most important British mathematician after Newton's time was Colin Maclaurin [1698–1746], whose most important work concentrated on planar curves.)

The Swiss and French were the intellectual leaders of mathematics in the early eighteenth century. Since mathematics jobs in Switzerland were not plentiful, the Swiss fanned out over continental Europe. The greatest Swiss mathematician, Euler, spent his productive career in Berlin and St. Petersburg. There he laid the foundation for subsequent mathematical development in many fields. Euler was particularly important for establishing "analysis," the understanding of mathematical or physical phenomena by reduction to equations, as mathematicians' dominant method. Euler's analysis drew principally on the Leibnizian calculus. French mathematicians, such as the child prodigy Alexis-Claude Clairaut (1713–1765), were less footloose. Most were closely associated with the Royal Academy of Sciences. Mathematics played little role in most university curricula in France during the eighteenth century. It was more strongly represented in scientific academies and societies, and in technical and military schools.

Analytical mathematics as developed by Euler, d'Alembert, and their colleagues was enshrined in a standard textbook, the French *Course of Mathematics for the Use of the Marine Guard* by Étienne Bezout (1730–1783), published in four volumes from 1764 to 1767 and an expanded six-volume edition from 1770 to 1782. Bezout drew on his experience as an examiner of the French marine guards and artillerists. His work was widely circulated, republished, and translated. Part of it was adopted for the mathematical use of the U.S. military academy at West Point, and it was also taught at Harvard University.

The tradition of analysis was carried on in France by a group of mathematicians, many of whom had been d'Alembert's protégés. These included Joseph-Louis Lagrange, Pierre-Simon de Laplace, and the Marquis de Condorcet, Marie-Jean-Antoine-Nicolas de Caritat. By 1781 it looked to Lagrange as if mathematics might have passed its fruitfulness as a discipline. In a letter to d'Alembert, he compared mathematics to a mined-out mine. This pessimistic prophecy was belied by Lagrange's own work—his *Analytical Mechanics* (1787) was the crowning glory of the intellectual program of rational mechanics—and by the work of his peers. Laplace's work in celestial mechanics was the equivalent of Lagrange's in rational mechanics, although Laplace was less gifted as a pure mathematician. French mathematics continued to progress after the French Revolution, although the Marquis de Condorcet was killed for political reasons. Important French mathematicians whose careers prospered after the Revolution were Adrien-Marie Legendre (1752–1833), Gaspard Monge (1746–1818), and Lazare-Nicolas-

Marguerite Carnot (1753–1823). All came up through the French system of military and technical education, and all were supporters of the Revolution. Carnot's most important contribution was not in the area of theoretical mathematics but as the "organizer of victory," indispensable in the survival of the French Republic. All three contributed to a range of mathematical fields, and Monge was active in chemistry and experimental physics as well. They also shared in the revival of descriptive geometry as an active field of research. Legendre's *Elements of Geometry* (1794) replaced Euclid's (c. 325 B.C.–c. 265 B.C.) *Elements* as the basic text of Euclidean geometry, holding that position for the next century. It also contains the first proof that *pi* squared is irrational. Legendre's devotion to Euclid was such that he never acknowledged the possibility of a non-Euclidean geometry and wasted much effort in attempts to prove Euclid's parallel postulate. Legendre also made contributions in calculus, number theory, and celestial mechanics.

Monge and Carnot were involved in the founding of what became Europe's leading center of mathematical excellence, the Polytechnic School, founded in 1795. Monge was also involved in politics, serving briefly and unsuccessfully as minister of the navy, and accompanying Napoléon Bonaparte on the expedition to Egypt. Napoléon admired mathematicians, and Monge returned his high regard. (It was said that such was his personal devotion that he grew sick on those rare occasions when Napoléon lost a battle.) Monge led the revival of solid, three-dimensional geometry, which had roots in Euler's work. He also advanced the study of analytic geometry, which expanded rapidly in the opening of the nineteenth century. Another professor at the Polytechnic School was Jean-Baptiste-Joseph Fourier (1768–1830) who contributed to the mathematical study of heat and made a controversial but very fruitful mathematical innovation in his trigonometric expansion of functions, now known as Fourier series.

The revolutionary era saw French dominance of the mathematical world reach a very high point, but this started to change in the early nineteenth century. This is partly due to one man, Carl Friedrich Gauss, a German and the greatest and most versatile mathematician since Euler. Gauss was a university professor, and his rise to mathematical stardom indicates the larger role the German universities, particularly Gauss's Göttingen, would play in the nineteenth century. Along with Fourier and Siméon-Denis Poisson (1781–1840), a Polytechnic graduate, Gauss was a leader in applying mathematical rigor to areas of physics, such as acoustics, electricity, heat, and magnetism, previously dominated by experiment. Chemistry was also becoming more quantitatively exact, with the rise of atomism, optics, and with the revival of the wave theory by Augustin-Jean Fresnel (1788–1827).

Britain began to reenter the mathematical mainstream with the founding of the Analytical Society in 1813. The society's founders were three young Cambridge men, George Peacock (1791–1858), John Herschel, and Charles Babbage (1792–1871). (Continental mathematics was introduced to Ireland in the same period, beginning with the appointment of Bartholomew Lloyd [1772–1837] to the chair of mathematics at Trinity College Dublin in 1812.) The society's efforts bore fruit in 1817, when Peacock was appointed mathematical examiner at Cambridge, England's leading mathematical university, and procured the use of continental differentials in the place of Newtonian fluxions.

See also Alembert, Jean Le Rond d'; Bernoulli Family; Euler, Leonhard; Gauss, Carl Friedrich; Germain, Sophie; Lagrange, Joseph-Louis; Laplace, Pierre-Simon de; Mechanics; Newtonianism; Probability.

References

Boyer, Carl B. *A History of Mathematics.* 2d ed. Revised by Uta C. Merzbach. New York: John Wiley and Sons, 1991.

Hankins, Thomas L. *Science and the Enlightenment.* Cambridge: Cambridge University Press, 1985.

Maupertuis, Pierre-Louis Moreau de (1698–1759)

Pierre-Louis Moreau de Maupertuis was a leader in the introduction of Newtonianism to France, as well as an innovative student of life. From a family of provincial French nobility, Maupertuis became involved in science as a member of a group that met at the Parisian Café Procope in the early 1720s. He was admitted to the Royal Academy of Sciences as an adjunct in geometry in 1723, and his early papers, beginning with one in 1724 on the shapes of musical instruments, dealt principally with mathematics. He also did experimental work on salamanders and scorpions, and in 1728, on a short visit to London, was admitted as a fellow of the Royal Society.

In 1729 and 1730 Maupertuis studied with Johann Bernoulli in Basel, Switzerland. There he learned the Leibniz-derived mathematical physics associated with the Bernoullis. Johann Bernoulli may have hoped that Maupertuis would promote this physics when he returned to Paris, still dominated by Cartesianism. Maupertuis published in mathematics after he returned and was elected pensionary, or full member, of the Royal Academy of Sciences in 1731. Shortly afterward his work on the problem of the shape of Earth led him in the direction of Newtonianism. This led to a break with Bernoulli, and relations between them were never entirely restored (although Maupertuis remained friendly to younger members of the Bernoulli family). In 1732 Maupertuis published *Figures of the Stars,* which, while not explicitly Newtonian, defended the Newtonian idea of the oblate, or flattened, shape of Earth against the Cartesian idea of its prolate, or oblong, shape. Maupertuis spent the next few years as a Newtonian missionary in France, publishing widely, tutoring his lover, Gabrielle-Émilie du Châtelet, in Newtonian physics, and feuding bitterly with the cartographer and astronomer Jacques Cassini, whose measurements of France seemed to support the prolate theory. Maupertuis was a good hater, and quarrels and feuds accompanied him throughout his career.

In 1736 Maupertuis, along with a small party including the French mathematician Alexis-Claude Clairaut (1713–1765) and the Swedish astronomer Anders Celsius (1701–1744), went to establish Earth's shape by performing measurements in Lapland. Maupertuis's triumphant return, with evidence for the Newtonian theory of Earth's shape, was marred by his quarrel with Clairaut and mockery he received from Parisian wits for bringing back with him two Lapp girls. Maupertuis continued to write both learned works and popularizations, such as the 1742 *Letter on the Comet,* which demonstrated the superiority of Newtonian to Cartesian explanations of the paths of comets. However, he was also growing tired of the debate, and by the middle of the decade his interests were broadening to include biological questions. He also relocated from Paris, accepting the invitation of Frederick the Great (r. 1740–1786) to head the new Berlin Academy in 1746.

Maupertuis's studies of living things are principally known for his theory of conception, which was epigenesist rather than preformationist. As set forth in *Venus Physique* (1745) and *System of Nature* (1751), it traced the conception of a new living organism to a mixture of male and female seminal fluid, drawn from different parts of the body. The particles, endowed with a "memory" of their origin, arranged themselves to form the new being. Monstrosities occurred when this process had somehow been disturbed. This endowment of material particles with the quality of memory was attractive to materialists, and Maupertuis, a devout Catholic, had to spend much energy distinguishing his position from that of materialists like Julien Offroy de La Mettrie and Denis Diderot.

Maupertuis's other great scientific interest during the Berlin years was mechanics, specifically the principle of least action, the idea that nature always acts in such as way as to minimize action, defined as the product of the mass, velocity, and distance. Maupertuis had originally derived this principle from optics, where light travels in the most direct fashion between two points. However, his formulation was marred by lack of clarity and consistency in the definition of "action," and had to be rescued by his Berlin colleague Leonhard Euler, who formulated it much more clearly and powerfully. Maupertuis did not limit the importance of his new principle to mechanics. He wanted to use it to demonstrate the existence of an all-powerful God, who provided the only explanation of how the principle could apply in all cases.

The principle of least action led to a disaster for Maupertuis, his controversy with the German Leibnizian Samuel Konig (1712–1757), who asserted that the principle had been plagiarized by Maupertuis from an unpublished letter of Gottried Wilhelm Leibniz (1646–1716). The controversy rippled through the Berlin Academy, and Maupertuis eventually forced it to sit in judgment over Konig's claim, finding that Konig had forged the letter. Konig resigned from the academy, but Maupertuis's use of it in his feud with Konig attracted much denunciation. The most notable example was Voltaire's *Diatribe of Doctor Akakia* (1752), a vicious and highly amusing satire.

In addition to the hostility engendered by the Konig affair, Maupertuis's last years were marred by ill health, which forced him to permanently leave the harsh climate of Berlin in 1756, and by the Seven Years' War (1756–1763), in which France and Prussia were on opposite sides. He died in the house of Johann II Bernoulli in Basel.

See also Berlin Academy; Châtelet, Gabrielle-Émilie du; Embryology; Materialism; Mechanics; Newtonianism; Voltaire.

References

Beeson, David. *Maupertuis: An Intellectual Biography.* Studies in Voltaire and the Eighteenth Century, no. 299. Oxford: Voltaire Foundation at the Taylor Institution, 1992.

Sutton, Geoffrey V. *Science for a Polite Society: Gender, Culture, and the Demonstration of Enlightenment.* Boulder: Westview, 1995.

Mayer, Johann Tobias (1723–1762)

The German astronomer Johann Tobias Mayer was the most important observer and predictor of the Moon's motion in the mid-eighteenth century. Frustrated in his early ambition to become an artillery officer, the young, mathematically gifted Mayer became a commercial cartographer and then an astronomer through a project to accurately map the Moon's surface. His achievements led in 1751 to an invitation to Göttingen, which the government of Hanover was trying to make a major scientific center with a university, scientific society, and observatory, of which Mayer became director. He taught applied mathematics, physics, and astronomy at the university and was a leading member of the Göttingen Scientific Society. Mayer's most important publication was the 1753 *New Tables of the Motion of the Sun and Moon,* which like many of his writings (including improved tables in 1754) were published in the *Commentaries of the Royal Society of Sciences of Göttingen.* Mayer's tables, the most accurate to date, required the solution of several problems, including atmospheric refraction, corrections for the spheroidal shape of Earth, and the lunar parallax.

The success of Mayer's lunar tables led him, with the encouragement of his correspondent and collaborator Leonhard Euler, to apply for the prize offered by the British Longitude Board for a method of finding longitude by sea. His entry consisted of his tables and a new navigational instrument that he claimed would make accurate observation aboard a moving ship possible, the repeating circle. Mayer's method proved in practice not quite accurate enough to win the full prize of 20,000 pounds, and his receiving any award was complicated first by the Seven Years' War (1756–1763) and then by his own death. Eventually, 3,000 pounds was shared by his widow and surviving family; the Göttingen Scientific Society; the society's president, Johann David Michaelis (1717–1791); and Euler. Mayer's tables were eventually the foundation of Nevil

Maskelyne's successful lunar-distances method of finding the longitude.

Much of Mayer's time in the closing years of his life was taken up by magnetism. His *Magnetic Theory* (1760) discussed the classic problems of the magnetic dip and variation and also established that magnetic attraction follows a Newtonian inverse-square law. He also published on the theory of color mixing.

See also Astronomy; Lichtenberg, Georg Christoph; Longitude Problem; Observatories; Universities.

Reference

Forbes, Eric G. *Tobias Mayer (1723–62): Pioneer of Enlightened Science in Germany.* Göttingen: Vandenhoeck and Rupprecht, 1980.

Mechanics

The discipline of classical mechanics was formed in the eighteenth century. It derived from a number of sources, including *Mathematical Principles of Natural Philosophy* (1687) by Isaac Newton (1642–1727), the development of the Leibnizian calculus, and the practical work of engineers. The leaders in developing "rational mechanics" were Swiss and French mathematicians, including the Bernoulli family, Leonhard Euler, Jean Le Rond d'Alembert, and Joseph-Louis Lagrange. The goal of rational mechanics was to fully describe mechanics in terms of equations (rather than the geometrical diagrams Newton had used) and to make it a deductive science where conclusions could be drawn from premises, with little or no use of experiment. Every important contributor to rational mechanics in the eighteenth century was a skilled mathematician, and mechanics was often seen as a branch of mathematics rather than physics.

Newton's mechanics, summed up in his famous three laws of motion, was powerful, but limited in its application. It essentially described the behavior of idealized mass points subjected to forces, although Newton himself spoke of "bodies," and the concept of mass points was introduced later by Euler. Newton's work was less useful in understanding the behavior of rigid or elastic bodies or fluids.

One of the controversies swirling around Newtonianism and mechanics in the mid-eighteenth century concerned the doctrine of Newton's great rival, Gottfried Wilhelm Leibniz (1646–1716). In 1686 Leibniz had put forth the theory of the *vis viva,* "living force." The living force was the product of the mass of a moving body times the velocity squared. Leibniz and his followers, notably the Bernoullis, believed that this force was the quantity whose total amount was always conserved in the universe, thus preventing the universe from collapsing. Newton and subsequent strict Newtonians regarded *vis viva* as lacking any real physical importance, and that the universe was preserved from collapse by occasional divine intervention rather than by the operation of its own laws. The question of *vis viva* was one of the few mechanical questions in the eighteenth century involving extensive appeals to experimental evidence. The Italian Giovanni Poleni (1683–1761) and the Dutch Newtonian Willem Jakob 's Gravesande (1688–1742) dropped balls of different weights from different heights onto clay, and measured the impressions. The results seemed to endorse Leibniz's concept of the conservation of *vis viva.* The doctrine was also mathematically expanded and refined by d'Alembert, Johann Bernoulli, and Daniel Bernoulli. The conversion of the Newtonians 's Gravesande and Gabrielle-Émilie du Châtelet eventually led to the integration of the *vis viva* into Newtonian mechanics.

Another non-Newtonian concept introduced into mechanics in the eighteenth century was "action." This idea, which is that things in nature are accomplished with the minimum effort, can be traced to discussions between Euler and Daniel Bernoulli in the 1740s. The principle was generalized, losing much of its precision and explanatory power, by Pierre-Louis Moreau de Maupertuis. The principle also led to a vicious priority dispute between Maupertuis and Samuel Konig

(1712–1757), who argued that it had originated in the work of Leibniz. Maupertuis's version of the principle of least action, in addition to being handicapped by his own grandiose cosmological claims, had the disadvantage of not rigorously defining the term *action*. However, more mathematically rigorous formulations of the principle, such as those of Lagrange, eventually carried the day. The non-Newtonian concepts of *vis viva,* which eventually developed into the concept of energy, and action eventually displaced the Newtonian concept of force from the center of mechanics. Other fundamental ideas, linear momentum and the moment of momentum, were first fully formulated mathematically by Euler.

There was much research in the application of mechanics to specific fields, much of it driven by practical concerns. Benjamin Robins (1707–1751), in *New Principles of Gunnery* (1742), investigated how the flight path of projectiles was affected by the resistance of the medium. Daniel and James Bernoulli were leaders in the development of hydrodynamics. Hydraulic problems were of particular importance in the eighteenth century due to the development of engines worked by moving water. Euler's work on the operations of water-powered turbines, published by the Berlin Academy in the 1750s, was inspired by the recent invention of a more effective waterwheel. A more experimental and less mathematical approach to hydraulics and hydrodynamics was the work of some French savants, including d'Alembert, the Marquis de Condorcet, and the abbé Charles Bossut (1730–1814). The three, led by Bossut, performed a long series of experiments recounted in Bossut's *New Experiments on the Resistance of Fluids* (1777). The experiments involved the towing of boats across a lake on the grounds of the Military School and measuring their mean velocities, and led to conclusions concerning the effect of the swell formed in front of the boat and the depression formed in its wake. Bossut was eventually appointed to the new position of professor of hydrodynamics at the Royal School of Architecture.

Another connection between practical engineering and theoretical mechanics was the study of stress and strain, which was carried on both by mathematicians and by engineers such as Charles-Augustin de Coulomb. This work culminated in the formulation of the concept of the stress tensor by Augustin-Louis Cauchy (1789–1857) in 1822. Eighteenth-century mechanists also very actively investigated vibrating or oscillating bodies. Leaders in this area were d'Alembert, whose work on vibrating strings led to the first full solution of a partial differential equation; Daniel Bernoulli; and above all Euler. Euler was active in all these areas, and in most of them it was his work that had the most influence on the future.

Lagrange's *Analytical Mechanics* (1787) summed up much eighteenth-century mechanics, drawing mostly from Euler's work and that of Lagrange himself. Lagrange boasted that no geometrical diagrams appeared in the work, which presented mechanics solely in terms of analytical equations. He presented mechanics in highly formal terms as a closed system, and his work, unlike Newton's a century earlier, presented no agenda for future research. Its publication was followed by something of a lull in many areas of mechanics. Lagrange also did not thoroughly deal with some active areas of eighteenth-century mechanical research, such as vibrating-string problems or the motions of fluids in tubes.

See also Astronomy; Bernoulli Family; Châtelet, Gabrielle-Émilie du; Euler, Leonhard; Lagrange, Joseph-Louis; Laplace, Pierre-Simon de; Mathematics; Maupertuis, Pierre-Louis Moreau de; Physics.

References

Hankins, Thomas L. *Science and the Enlightenment.* Cambridge: Cambridge University Press, 1985.

Truesdell, Clifford. *Essays in the History of Mechanics.* New York: Springer-Verlag, 1968.

Wolf, Abraham. *A History of Science, Technology, and Philosophy in the Eighteenth Century.* 2d ed. Revised by D. McKie. Gloucester, MA: Peter Smith, 1968.

Medicine

Physicians, surgeons, and other medical practitioners continued to add to their scientific knowledge in the eighteenth century, although the impact on patient care was still minor, and sometimes negative. The leading intellectual figures in early eighteenth-century medicine were university professors, Hermann Boerhaave of the University of Leiden and Georg Ernst Stahl of Halle. Boerhaave was a supporter of the doctrine of "iatromechanism," which held that the human body was best understood as a mechanical system of fluids. He understood medical mechanics in a Newtonian way. Stahl was a vitalist who believed that mechanism (or chemistry) could never fully explain the body, and that living matter fundamentally differed from that which was not living. Boerhaave's influence dominated the first half of the century, spread through his popular textbooks and a network of his students and disciples extending throughout the university world. His influence made Leiden the most highly regarded school of medicine, part of a process whereby French and Protestant institutions moved ahead of the formerly leading Italian universities.

The Leiden influence was particularly strong in the Scottish universities, which became Britain's leading institutions for medical education. They attracted English Protestant Dissenters, barred from England's universities, as well as Scotsmen. Scotsmen and Scottish university graduates dominated the British medical scene and spread out over the British Empire. Medical education took place in England in private schools such as William Hunter's London anatomy school or in hospitals rather than the universities. German and Italian medicine continued to be a university discipline, although Germany at the end of the eighteenth century also saw a substantial increase in the hospital sector. The North American colonies and the United States were medicine's wild frontier, where physicians were few and regulation virtually nonexistent.

Medicine, particularly medicine of the school of Boerhaave, played an important role in the culture of the Enlightenment. Boerhaave's student Julien Offroy de La Mettrie was the only major medical or philosophical thinker to take Boerhaave's system to its logical extreme of a completely materialistic theory of humanity, but many other philosophes were physicians themselves, or saw medicine as fundamental to the reform of society, which they sought. Some, such as Voltaire, were suspicious of physicians and their claims to authority or mocked the constant jurisdictional quarrels of physicians and surgeons and their myriad institutions and colleges, but many thought Europe's increasing population was a result of improved medicine.

Vitalism took on a new lease of life in mid-century, as pure mechanism seemed unable to explain certain biological actions, such as the re-formation of chopped-up polyps and the regeneration of a lobster's severed claws. The studies of the nervous system made by scientists such as Albrecht von Haller were also harder to fit into a mechanical framework. A modified vitalism was disseminated at the medical school of the University of Montpellier in France, a traditional center of medical excellence and one of the few French scientific institutions of the first rank outside Paris.

Meanwhile, the stock of empirical knowledge of the human body was steadily accumulating, embodied in more detailed anatomical atlases. Medical researchers like John Hunter and Pieter Camper (1722–1789) were also in the forefront of comparative anatomy, which in the eighteenth century was a science of examining animal bodies to better understand humans. The study and classification of disease, nosology, was another area of medicine undergoing steady, if not spectacular, progress in the eighteenth century. The Scottish professor William Cullen (1710–1790) and the Montpellier professor Boissier de Sauvages (1706–1767) both produced elaborate classification schemes incorporating thousands of diseases.

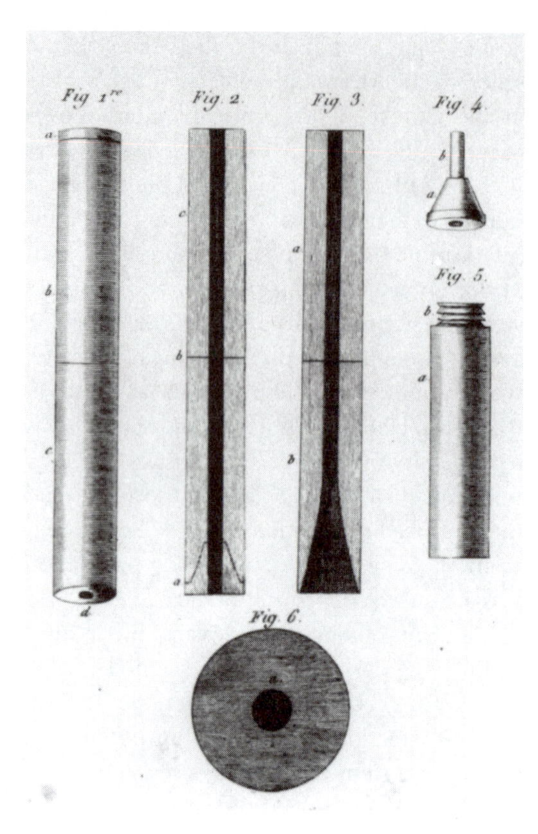

Illustrations of René-Théophile-Hyacinthe Laënnec's early stethoscopes (National Library of Medicine)

(Cullen's student John Brown [c. 1735–1788], founder of Brunonianism, turned the process on its head by claiming that there was only one disease, caused by differing degrees of nervous excitement.) The principal theories of infection were that of contagion, where sickness was transferred from sick to healthy individuals, and miasma, where sickness was the result of environmental conditions. Even the old humoral theory survived in a changed form, as disease was ascribed to a buildup of "morbific" matter. The purge remained a favorite treatment of the eighteenth-century sick.

In the eighteenth century the tradition of learned physicians was growing closer to that of surgeons, whom physicians had traditionally looked down upon as craft specialists. Edinburgh medical training made little of the distinction of physicians' and surgeons' knowledge, and surgeons like the Hunter brothers and Marie-François-Xavier Bichat were at the forefront of medical knowledge. English surgeons were also branching into the field of midwifery, setting off fierce disputes with the established female midwives, but this movement had not yet had much impact on the European continent.

Although the great medical school at the University of Padua, which had dominated European medicine in the sixteenth century, had yielded much of its primacy, it contributed one important study to eighteenth-century medicine. This was the work of the anatomy professor Giovanni Battista Morgagni (1682–1771) on pathology, published as *On the Sites and Causes of Diseases* (1761). This incorporated the results of more than 700 autopsies, linked to case histories. Morgagni gave detailed descriptions of the anatomical results of degenerative diseases of the heart and established that the cause of blueness of the skin was the narrowing of blood vessels, in addition to many other innovations. His work was followed up in England by Matthew Baillie (1761–1823) and greatly improved upon in France by Bichat.

The French Revolution worked a vast change in French medicine, whose impact would be felt in the European and American medical world for decades. The corporate bodies that ruled French medicine, such as the medical faculties of the French universities and the Paris Academy of Surgeons, were dissolved as aristocratic remnants. The three new "Schools of Health" established at Paris, Montpellier, and Strasbourg emphasized bedside learning and made no distinction between learned medicine and surgery. The intellectual prestige of medicine gained from the inclusion of a division devoted to the healing arts on the founding of the Institute of France in 1795. France's hospital system, until the Revolution under the purview of the Catholic Church, was taken over by the state and incorporated into the system of medical education. This in practice meant that the hospitals were taken over by young French physicians and surgeons who made them over from places primarily dedicated to tending the sick and dying to institutions aimed at close clinical observation and the advance of medical science. The traditional balance between book learning and clinical work was now tilted heavily in favor of the clinic, and the new medicine drew from the surgical tradition as much or more than from that of physicians. The distinction between surgeons and physicians disappeared in France, and the Paris hospitals made the city the medical capital of Europe.

Bichat, the most important medical innovator of the period, never held a major hospital position, teaching private medical courses. His books *Treatise on Membranes* (1800) and *General Anatomy* (1801) changed the focus on organs to one on tissues, a concept he originated. Bichat distinguished between twenty-one different kinds of tissues, including nerve and muscle tissues. Diseases were located not in organs but in individual tissues. The ability to identify damaged tissue and diagnose disease became the mark of the Paris-trained doctor, and some, including the Parisian François-Joseph-Victor Broussais (1772–1838), complained that medicine had become too obsessed with diagnosis and too pessimistic about the possibility of cure.

The most important Paris doctor after Bichat was René-Théophile-Hyacinthe Laënnec (1781–1826.) Laënnec is principally known for his invention of the stethoscope in 1816, although this did not at all resemble the modern instrument, being at first simply a rolled-up sheet of paper. The instrument described in Laënnec's *Treatise on Mediated Auscultation* (1819) was made out of wood, and attached to only one ear. The stethoscope enabled physicians, particularly Laënnec, who became a renowned diagnostician, to bypass the patient's own words, which physicians since the ancient world had believed to be their main source of information on the patient's case, and listen directly to the thoracic cavity. This partially depersonalized medical treatment and the discussion of case histories. It also made the stethoscope a common symbol of the scientific physician and Laënnec one of Europe's most popular medical instructors. His pupils, like the Englishman Thomas Hodgkin (1798–1866), brought back stethoscopes and Paris medicine to their native lands. Laënnec also identified tuberculosis as one disease, wherever the tuberculous nodule was located in the body. Ironically, it was tuberculosis that killed him. Another Parisian, François Magendie (1783–1855), pioneered drug experimentation.

Advances in medical knowledge had little impact on patient care. The period saw major disasters in care such as the development of the lying-in hospital, in which childbed fever raged virtually without restraint. There were also major improvements in the period, notably smallpox immunization, William Withering's (1741–1799) identification of foxglove as a heart stimulant, and the experimental demonstration by Scottish naval surgeon James Lind (1716–1794) that oranges and lemons were cures for scurvy. (Lind's experiment in 1754, in which he separated groups of sailors and fed them different foods, is sometimes identified as the first clinical trial.) However, these had little to do with medical theory.

See also Anatomy; Beddoes, Thomas; Bichat, Marie-François-Xavier; Boerhaave, Hermann; Brunonianism; Cheyne, George; Coudray, Angelique Marguerite Le Boursier du; Darwin, Erasmus; Embryology; Haller, Albrecht von; Hospitals; Hunter Family; La Mettrie, Julien Offroy de; Madness; Masturbation; Materialism; Mesmerism and Animal Magnetism; Midwives; Physiology; Reil, Johann Christian; Royal Society of Medicine; Sexual Difference; Sloane, Sir Hans; Smallpox Inoculation; Surgeons and Surgery; Toft Case; University of Edinburgh; University of Leiden; Vitalism.

References

Gay, Peter. *The Enlightenment: An Interpretation, Volume II: The Science of Freedom.* New York: Alfred A. Knopf, 1969.

Lesch, John E. *Science and Medicine in France: The Emergence of Experimental Physiology, 1790–1855.* Cambridge: Harvard University Press, 1984.

Porter, Roy. *The Greatest Benefit to Mankind: A Medical History of Humanity.* New York: W. W. Norton, 1998.

Porter, Roy, and W. F. Bynum, eds. *William Hunter and the Eighteenth-Century Medical World.* Cambridge: Cambridge University Press, 1985.

Mesmerism and Animal Magnetism

The Austrian physician Franz Anton Mesmer (1734–1815) invented the influential doctrine and practice of animal magnetism. For him animal magnetism was a method of healing physical illnesses employing the flows of an invisible fluid associated with living things. However, Mesmer quickly lost control of his movement, and "animal magnetism" included a wide variety of doctrines and practices in the late eighteenth and early nineteenth centuries.

Mesmer's dissertation submitted for the medical degree from the University of Vienna in 1766 concerned the influence of the stars on the human body, categorized not in astrological terms but in quasi-Newtonian ones. The influence of the stars on living things was "animal gravity," just as Newtonian gravity by its influence on the Earth was "animal gravity." In the medical practice he set up in Vienna, Mesmer encountered the idea of the curative powers of iron magnets applied to the human body. After success using magnets to cure a sick woman, Mesmer extended the idea of the "magnetic cure" to the use of substances and processes that were not at all "magnetic" in the conventional sense. The human body itself, he claimed, was the supreme animal magnet. The manipulation of its magnetic flows by direct physical contact between the "magnetizer" and the patient could cure a host of ailments, including hemorrhoids, irregular menstruation, and epilepsy. Mesmer applied his techniques in the Holy Roman Empire and Hungary, achieving some spectacular cures. However, he also ran into what was to become a recurring problem for mesmerists—opposition from organized physicians. In 1778 he relocated to the center of European science, Paris.

Mesmerism took Paris by storm. Mesmer worked some dramatic cures and was invited to speak before the Royal Academy of Sciences, the Faculty of Medicine of the University of Paris, and the Royal Society of Medicine. The French government offered Mesmer an institute of his own, which he declined, fearing the loss of independence. A ferocious pamphlet war broke out between opponents and defenders of the Viennese physician. Demand for Mesmer's services was so great that he was forced to expand beyond one-on-one sessions with the invention of the magnetic tub, a complicated device with bottles of magnetized water positioned on the interior of an oaken tub large enough for several people to get in at once. Holding each other's hands, they created a "magnetic circle." Another result of the increasing demand was that, despite Mesmer's claims to a monopoly, other people started practicing mesmerism.

The official test of animal magnetism was done under the auspices of the Royal Academy of Sciences in 1784. The five-person commission was joined by four physicians. The group was extremely impressive. It included Benjamin Franklin in the chair, Antoine-Laurent Lavoisier, and the astronomer Jean-Sylvain Bailly (1736–1793). Although Mesmer himself refused to cooperate, an ingenious series of tests performed on another magnetizer showed that patients could not distinguish between magnetized and unmagnetized objects if they were uninformed of their state. This led the commission to conclude that animal magnetism did not exist. This did not end the controversy, particularly as all those opposed to the hegemony of the Royal Academy of Sciences could now see mesmerism as another of its victims. Many of the future political radicals who would become prominent during the French Revolution, such as Jacques-Pierre Brissot de Warville (1754–1793), were mesmerist supporters and pamphleteers. But radicals were not the only mesmerists. Mesmer's version of mesmerism spread throughout France as far as the French Caribbean colony of Saint Domingue (now Haiti) with the formation of a series of "Lodges of Universal Harmony," set up on the model of Freemasonry (and in some places coextensive with it) to spread Mesmer's doctrines and practices. Mesmer himself became quite wealthy.

The most important of Mesmer's followers was a nobleman, the Marquis de Puységur (1751–1825). In 1784, while mesmerizing Victor Race, a peasant who lived on his estate and suffered from congestion of the lungs, Puységur discovered that Race had entered a condition he called "magnetic sleep," which we would now call a hypnotic state (the word *hypnotism* was introduced in 1842). Puységur was clearly amazed by the change between Race's normal condition and his condition under magnetic sleep. He ascribed various paranormal abilities, such as clairvoyance and telepathy, to the magnetized patient. Puységur turned the original physically based doctrine of Mesmer based on flows of the magnetic fluid to one that placed vastly more emphasis on the mental conditions of both the magnetizer and the patient and the mental and spiritual rapport between them. The relation between women patients and male mesmerizers was clearly open to suspicion, and most reputable male mesmerizers insisted on having a witness present during mesmerizing sessions. Some argued that women in particular were too fragile and too emotional to be involved in animal magnetism either as practitioners or as patients, but there were a number of women practitioners. Despite these misgivings, and Puységur's departure from the theory and practice of Mesmer, magnetic sleep itself now became a craze and the subject of several treatises.

Mesmerism had definitively evolved far beyond the original doctrine of Mesmer. It was now often practiced as a matter of healing the patient's mind, rather than, as Mesmer himself had practiced, the body. It also became allied to a variety of esoteric and occultist movements, including Swedenborgianism. Mesmerism spread to Germany in a spiritualistic and often Swedenborgian form oriented to magnetic sleep. Disheartened by splits in the mesmerist movement and no longer attracting large numbers of patients, Mesmer left Paris in 1785. He traveled extensively throughout Europe and published several mesmeric works over the following decades, but was not a leader of the movement. There were a number of other mesmeric writers, including writers of mesmeric textbooks and editors of mesmeric journals, setting themselves up as authorities, often in contradiction to Mesmer. By the 1810s mesmerists such as the Portuguese-French priest José Custodio de Faria (1755–1819) and the historian Étienne Felix Henin de Cuvillers (1755–1841) were denying both Mesmer's magnetic fluid and Puységur's emphasis on the spiritual rapport of magnetizer and patient, and treating the magnetic sleep purely as a psychological state attainable by the application of particular techniques. Mesmeric sleep was thus applicable to a number of medical problems, including those curable by the power of suggestion, and was also useful for the alleviation of pain, such as that of surgery or dental extraction.

See also Medicine; Psychology; Royal Academy of Sciences.

References

Crabtree, Adam. *From Mesmer to Freud: Magnetic Sleep and the Roots of Psychological Healing.* New Haven: Yale University Press, 1993.

Darnton, Robert. *Mesmerism and the End of the Enlightenment in France.* Cambridge: Harvard University Press, 1968.

Meteorology

Although no dramatic breakthroughs in the science of weather occurred during the Enlightenment, it was the site of much intellectual, technological, and organizational activity. The systematic collections of weather observations over long periods of time had begun in the seventeenth century and was continued and expanded in the eighteenth. There were two projects led by scientific societies to systematically gather weather observations in the period. The first was the brainchild of the Royal Society secretary James Jurin, who in 1723 invited correspondents to make weather observations using a thermometer, barometer, and rain gauge. The second, which attempted to involve scientific societies rather than individual correspondents, was undertaken by the Meteorological Society of Mannheim, founded in 1780 as an offshoot of the Mannheim Academy. The leaders of the Meteorological Society, Johann Jacob Hemmer (1733–1790) and Stefan Stengel, recruited participants and, aware of the problem of instrumental variation, provided them with standardized thermometers, barometers, rain gauges, and windmills. Observations were to be made three times daily, at 7:00 A.M., 2:00 P.M., and 9:00 P.M. The society established fifty-seven stations in Europe and America (curiously, the Royal Society and Britain's other scientific societies did not respond to the Mannheim offer) and published a series of *Ephemerides* ending in 1795. The project fell apart under the strains of the wars of the French Revolution, and the Meteorological Society of Mannheim itself shut its doors in 1799. The establishment of weather stations and the recording of weather observations went on outside these big projects, and the portion of the globe covered steadily advanced. In the remote and inhospitable country of Siberia, Russian scientists set up several weather stations in the early 1730s. Another project was rooted in the connection many argued existed between weather and disease, particularly epidemics. The French Royal Society of Medicine required each of its 1,000 or more provincial correspondents to take elaborate weather readings three times a day, using advanced Parisian equipment.

The great questions of meteorology in the eighteenth century had to do with the causes of atmospheric phenomena such as clouds, wind, and rain. The dominant theory of clouds was that they were composed of tiny bubbles of water enclosing a matter of fire that made them lighter than air. One difficulty with the theory was in explaining how the bubbles were formed. John Dalton's *Meteorological Observations and Essays* (1793) endorsed a rival theory, that clouds were composed of droplets, but did not succeed in toppling the bubble theory, which lingered well into the nineteenth century. There was little interest in distinguishing between types of clouds until the early nineteenth century, when first Jean-Baptiste-Pierre-Antoine de Monet de Lamarck and then the London Quaker pharmacist and weather observer Luke Howard (1772–1864) set forth classification schemes. Howard's, set forth in a paper delivered in late 1802 or early 1803, "On the Modifications of Clouds," eventually became standard and with modifications is still in use. Howard, a supporter of the droplet theory, divided clouds into cumulus, cirrus, stratus, and nimbus as well as intermediate types.

Abilities to measure the moisture of the air improved with the development of hygrometers. The dominant theory for most of the period was that water vapor dissolved in air. This led to some difficulties, like why water evaporated in a vacuum. The leading late-eighteenth-century meteorologist, a Genevan resident in England, Jean André Deluc (1727–1817), made his reputation with a treatise on using barometers to measure height, *Experiments on the Changes of the Atmosphere* (1772). Deluc wrote a massive treatise, *Ideas on Meteorology* (1786–1787), claiming that water vapor and air were different forms of the same thing, and that water vapor rose, turned to air, and then turned back to water to fall as rain. Deluc's fellow Genevan and great rival Horace-Bénédict de Saussure (1740–1799) argued that water was borne upward in columns of hot air rising by convection. The idea that atmospheric phenomena could be explained by the rising of hot air was originally difficult to reconcile with the fact that higher places, like mountains, were cooler. Johann Heinrich Lambert (1728–1777), in *Pyrometrie* (1779), explained the cooling of hot air as it rose in terms of heat as a physical substance bonding with matter. De Saussure explained it by the rarefication of the air and its greater distance from the Earth. The cooling of rising air was mathematically demonstrated by Siméon-Denis Poisson (1781–1840) in 1823.

Edmond Halley (1656–1742) based his original theory of wind on the movement of hot and cold air masses. George Hadley (1685–1768) refined Halley's theory to take into account the rotation of the Earth, publishing his findings in an article in *Philosophical Transactions* in 1735. In 1746 Jean Le Rond d'Alembert took a completely different approach, in a prizewinning paper submitted to the Berlin Academy. Although d'Alembert's theory that the movements of the wind were influenced by the attractions of the Sun and Moon found few followers, his was by far the most mathematically sophisticated treatment of winds to that date, opening the door to applying fluid dynamics to meteorology. Measurement of wind remained subjective. What eventually became the most common wind scale was devised by a British admiral, Sir Francis Beaufort (1774–1857). He originally devised his wind scale using twelve gradations for his own records in 1805, but it was adopted by the Admiralty in 1838.

See also Hygrometers; Rain Gauges; Thermometers.

References

Frisinger, H. Howard. *The History of Meteorology: To 1800*. New York: Science History Publications, 1997.

Khrgian, Aleksandr Khrsioforovich. *Meteorology: A Historical Survey*. 2d ed., revised. Edited by Kh. P. Pogosyan. Jerusalem: Israel Program for Scientific Translations, 1970.

Middleton, W. E. Knowles. *A History of the Theories of Rain and Other Forms of Precipitation*. London: Oldbourne, 1966.

Meteors and Meteorites

At the beginning of the eighteenth century, the dominant theory of meteors, or "shooting stars," was still that put forth by Aristotle (384–322 B.C.E.): that they were vaporous atmospheric phenomena. Newtonians, including Isaac Newton (1642–1727) himself, assumed that the space between Earth's atmosphere and the planets was completely empty of solid matter. Edmond Halley (1656–1742) argued in 1714 that meteors were actually matter that collected in the ether, although he later abandoned the position. The stones claimed to have fallen from heaven (meteorites) were also viewed as having earthly origins. Those who claimed to have seen stones fall from the sky, and even produced the stones themselves, were viewed by most scientists as superstitious or credulous, on the same level as those who claimed to have witnessed rains of blood or frogs. European scientists explained meteorites as earth stones that had been fused together by lightning or as material ejected from volcanoes.

Thinking about meteors changed in the mid-eighteenth century, particularly after Benjamin Franklin's kite experiment in 1752. Franklin's demonstration of the electrical nature of lightning led many to suspect that meteors were also electrical. Another theory was put forth by the president of Yale, Thomas Clap (1703–1767), in his posthumously published *Conjectures upon the Nature and Motion of Meteors* (1781). Clap suggested that meteors were comets that orbited Earth. Scientists also continued to investigate falling stones. The most impressive of these was the enormous 1,600-pound meteorite found in Siberia in 1749 and investigated by the German Peter Simon Pallas (1741–1811) in 1772. Pallas established that the iron meteorite had no connection with local iron deposits but stopped short of declaring it of extraterrestrial origin.

Conventional thinking was challenged in the 1790s by dramatic meteor showers, particularly one at Siena in 1794 and at Wold

Newton in England the next year and by the publication of *Concerning the Origin of the Mass of Iron Discovered by Pallas and Others Similar, and Concerning a Few Natural Phenomena Connected Therewith* (1794) by Ernst Florens Friedrich Chladni (1756–1827). Chladni firmly identified meteors as the source of meteorites and claimed an extraterrestrial origin for both, although he later wavered on this question. He also established one of the first important meteorite collections. The most important public collection was begun in 1806 in Vienna, by Karl von Schreibers and the Imperial Cabinet of Natural History.

Although the identification of meteors and meteorites convinced many, further investigation was ambiguous on the question of extraterrestrial origin. Prompted by Georg Christoph Lichtenberg, two German students, Heinrich Wilhelm Braudes (1777–1834) and Johann Friedrich Benzenberg (1777–1846), plotted the path of several shooting stars observed during a meteor storm and claimed their evidence suggested terrestrial origin. Chemical analyses of newly fallen meteorites carried out in England in 1802 by Edward Charles Howard (1774–1816) and Jacques-Louis Bournon (1751–1825) and in France the following year by Jean-Baptiste Biot (1774–1862) established that meteorites' chemical composition was different from terrestrial rocks.

The question of the origins of meteors and meteorites remained unresolved in the early nineteenth century. The idea that interplanetary space was not a void and meteors could originate there was strengthened by the discovery of the asteroids. Atmospheric formation was another possible explanation. Another, popular in France although originated by the German astronomer Heinrich Wilhelm Matthäus Olbers (1758–1840), was that meteors were ejected by lunar volcanoes. At Yale Clap's intellectual heirs argued that meteorites were the debris of orbiting comets.

See also Volcanoes.

References

Burke, John G. *Cosmic Debris: Meteorites in History.* Berkeley: University of California Press, 1986.

Hughes, David W. "The History of Meteors and Meteor Showers." *Vistas in Astronomy* 26 (1982): 325–345.

Metric System

The creation of the metric system, which involved many of revolutionary France's leading scientists, was the culmination of decades of projects for the reform of the chaos of French measurement. French measurements varied by province, by what was being measured, and even by economic function—for example, it was not unusual for the same commodity to be sold wholesale and retail by different measures. This variation was widely perceived as having held France back economically, particularly by comparison with Britain, which had a greater degree of uniformity in its measures. Lack of uniform measures was one of the many complaints made to the new government in the early days of the Revolution. Creation of a new standardized system of weights and measures was both a necessary reform from the point of view of the government and a way for the Royal Academy of Sciences to demonstrate its usefulness to the new order.

Like many reform ideas, measurement reform had been previously attempted in the ministry of Anne-Robert-Jacques Turgot (1727–1781). It fit with Turgot's technocratic approach that the measurement be based on a scientific phenomenon, the beating of a pendulum, rather than being simply the standardization of an already existing measure. The attempt to find a standard measurement of length, from which other measures could be derived, came to an end with the end of Turgot's ministry. Although early in the Revolution the idea of simply forcing the rest of the country to adopt the Paris measures had some support, the revolutionaries had the far more ambitious and time-consuming idea of establishing a universal measure–based science. The fundamental characteristics of the new measure were set forth by the National Assembly in the spring of 1790. The new measure would be "objective," as opposed to the Paris measure based on a standard iron bar, and would potentially command universal assent. (Initial hopes for cooperation with Britain and the United States were not fulfilled.) The system would be interrelated, as weight and volume measures would be defined in terms of the distance measurement, and decimalized for ease of calculation. (The French revolutionaries carried decimalization to a higher pitch than anyone before or since, instituting a ten-day week and even proposing a ten-hour day.) The original pendulum idea was dropped the following year in favor of one based on the size of Earth on the recommendation of a commission of the Royal Academy of Sciences chaired by Jean-Charles de Borda (1733–1799), inventor of the Borda repeating circle, and including the academy's leading scientists and mathematicians such as Pierre-Simon de Laplace, Antoine-Laurent Lavoisier, and Charles-Augustin de Coulomb. One-ten-millionth of the length from the North Pole to the equator—a quarter of Earth's circumference—was fixed as the length of the new meter.

Anne-Robert-Jacques Turgot, French administrator, suggested reform of the system of weights and measures as part of an overall Enlightened program. (Perry-Castaneda Library)

The difficulty lay in accurately measuring Earth to the degree of precision required. France led the world in geodesy, the science of Earth measurement, and the Borda circle was the world's most accurate portable instrument for triangulation. The actual measuring was to take place between Dunkirk and Barcelona through Paris, the "meridian of Paris," and then extrapolated for the entire distance. The actual measuring was entrusted to two astronomers, both clients of Joseph-Jérôme Le Français de Lalande (1732–1807). Jean-Baptiste-Joseph Delambre (1749–1822) would take the northern leg, and Pierre-Françoise-André Mechain (1744–1804) the southern. The two astronomers faced enormous political and technical difficulties, particularly with the vagaries of both domestic and international politics (Mechain's measurements were handicapped by the fact that France and Spain were at war). The project also faced difficulties at the center, with repeated political purges and the abolition of the Academy of Sciences.

With many interruptions the measurements of the two astronomers would take from 1792 to late 1798, and would be the most accurate to that date. However, France could not wait that long for its new meter. It is a measure of the speed of events that the law making the use of the meter mandatory, passed on 1 August 1793 and supposedly taking effect the next year, long preceded the definition of the meter. It established a "provisional" meter to be used until the actual meter could be defined.

The completion of Delambre's and Mechain's measurements were marked by an international gathering at Paris of scientists from countries dominated by or allied with France. (The British, at war with France, were not invited.) The data were presented to the scientists on 2 February 1799. After some confusion caused by the measurements' revelation that Earth was more irregular in shape than previously thought, the new standard was enshrined in a platinum "meter bar." Subsequent measurements followed. The chemist Louis Lefèvre-Gineau in 1799 defined the gram as the weight of a cubic centimeter of rainwater in a vacuum at the temperature of maximum density. However, the new measures were only grudgingly adopted in France, and in 1812 the government abandoned the whole project, switching to a standardized version of the old Paris measurements. France returned to the metric system only in 1837. Delambre assembled the records of the metric expeditions in the three-volume *The Basis of the Decimal Metric System* (1806–1810). In doing so he discovered that Mechain had fudged his data, although in such a way as not to throw off the final calculation.

See also Astronomy; French Revolution; Laplace, Pierre-Simon de; Nationalism.

References

Alder, Ken. *The Measure of All Things: The Seven-Year Odyssey and Hidden Error That Transformed the World.* New York: Free Press, 2002.

Gillispie, Charles Coulston. *Science and Polity in France at the End of the Old Regime.* Princeton: Princeton University Press, 1980.

Michaux Family

The natural historians André (1746–1802) and François-André (1770–1855) Michaux, father and son, were prominent botanical explorers of the early United States, personally linking the scientific communities of France and the young republic. Interested in plants from an early age, André Michaux studied under Bernard de Jussieu (1699–1777) at Trianon and at the Royal Botanical Garden in Paris. From 1782 to 1785 Michaux traveled as a botanist with a French expedition to Persia, sending back, along with plants and seeds, the first sample of cuneiform writing to arrive in Europe. Back in Paris, in 1785 Michaux was commissioned royal botanist with the mission of finding useful plants for France in America. Originally landing in New York, he arrived in Charleston in 1786. The city, where he established a 111-acre botanical garden, was his base of operations as he ranged over North America as far south as Florida and as far north as the Hudson Bay region. Plants Michaux first described or collected include the swamp chestnut oak, the Oconee bell, the big-leaf magnolia, native cane, blue-eyed grass, and the Carolina willow. A member of the Agricultural Society of South Carolina, Michaux also worked to acclimate foreign plants, mostly from Asia, in America. Notable examples include the camellia, the mimosa from Persia, the gingko tree from China, and the crape myrtle from India.

In 1790 Michaux was trapped in Charleston by the French Revolution and the freezing of his funding. He bore no grudge against the French Republic, becoming a strong supporter of the Revolution. In 1792, while on a visit to Philadelphia, Michaux proposed that the American Philosophical Society sponsor a natural-historical expedition to the Pacific Coast. Despite the support of Thomas Jefferson (1743–1826), this scheme eventually came to nothing, but the instructions drawn up for Michaux were one source for those later given to the Lewis and Clark expedition. Michaux also became mixed up in a scheme by French revolutionary diplomats to recruit Americans for an attack on Spain's New World possessions.

Returning to France in 1796, Michaux wrote two books on American plants based on his personal observations, *Oaks of North America* (1801) and the first systematic botanical description of eastern North America, *Flora boreali-americana* (1803). Despite Michaux's early training by de Jussieu, *Flora boreali-americana* employed Carolus Linnaeus's rival classification system. Michaux died on Madagascar in November 1802, while accompanying a French expedition to the South Seas as a botanist.

As a young man François-André Michaux accompanied his father on many of his early explorations. As an adult he returned to Charleston, arriving in October 1801, unfortunately during a yellow fever epidemic. He caught the disease, but survived. While in Charleston François-André Michaux arranged the liquidation of his father's botanical garden, sending plants and seeds back to France and transferring the land and the plants that were left from the property of France to the Agricultural Society of South Carolina, of which he, like his father, was a member. He explored and collected plants in the Alleghenies and down the Ohio River and in Tennessee and Kentucky. His book on his American travels, *Travels to the West of the Allegheny Mountains* (1804), contains observations on American society as well as botany. After returning to France Michaux came to America again while he was engaged in a project to find North American trees with useful lumber that could be acclimated in France from 1806 to 1808. His findings on the trip were reported in the three-volume *North American Sylva* (1810–1813), the first scientific catalog of America's trees. On his return to France Michaux was a principal supplier of scientific books and journals to America until the 1820s, when poor eyesight and failing health forced him out of Paris. American recipients of continental scientific works from Michaux during this period include Jefferson, the American Philosophical Society, the South Carolina Medical Society, and the Charleston botanist Stephen Elliott (1771–1830). Although Michaux's provincial existence removed him from the active scientific community of Paris, his intellectual life was not over. In 1834 he was appointed superintendent of an experimental royal forest at Harcourt, where he established France's first arboretum in 1853.

See also American Philosophical Society; Botanical Gardens; Botany; Exploration, Discovery, and Colonization; French Revolution.

Reference

Savage, Henry, Jr., and Elizabeth Savage. *André and François-André Michaux.* Charlottesville: University Press of Virginia, 1986.

Microscopes

The great seventeenth-century microscopists found few immediate successors in the eighteenth century, although Anton van Leeuwenhoek (1632–1723) soldiered on, publishing his observations to his death. Unlike the telescope, which was used by an established scientific community of astronomers, the microscope lacked a professional constituency in the early eighteenth century. (It was not yet commonly used in medicine.) The principal users of microscopes were hobbyists and amateurs. The London community of instrument makers, such as John Marshall and James Wilson, popularizer of the easily portable "screw-barrel" single-lens microscope, dominated the industry throughout this period. The improvements they made were largely in the mounts and lighting of microscopes rather than in the magnification itself.

Of the two main types of early microscopes, the single and compound lens, the single lens was the more serious scientific tool. (Another type of microscope was the solar microscope, used to project microscopic images to be seen by a large audience. It was principally used for entertainment and demonstration purposes, as its poor resolving power made it of little use in scientific research.) Most important discoveries of the period, such as the discovery of polyp regeneration by Abraham Trembley (1700–1787), were made with single-lens microscopes. One of the few pieces of original scientific work done in the eighteenth century with a compound microscope was the work on wood by the apothecary John Hill (c. 1707–1775), published in 1770 as *The*

Construction of Timber, From Its Early Growth; Explained by the Microscope and Proved from Experiments in a Great Variety of Kinds. By the late eighteenth century microscopes had come into increasingly common use in medical research. Wilhelm Friedrich von Gleichen-Russwurm (1717–1783) introduced phagocytic staining to improve the visibility of tissues in 1778. The early dermatologist Robert Willan (1757–1812) made extensive use of microscopic imagery in his *On Cutaneous Diseases* (1798). Not all physicians trusted microscopes, however. Marie-François-Xavier Bichat avoided them due to their propensity for error.

Compound microscopes were handicapped by chromatic aberration, the breaking of white into colored light. The solution of this problem for refracting telescopes was not immediately followed by the solution for microscopes. Leonhard Euler sent several microscope makers down a false trail by suggesting that using multiple lenses in the eyepiece would solve the problem. Although different arrangements of lenses cut down considerably on chromatic aberration, reliable achromatic microscopes became widely available only in the 1820s.

The main problem with the image revealed by single-lens microscopes was spherical aberration. In 1811 Sir David Brewster (1781–1868) suggested that this problem could be minimized if lenses were ground from substances with a higher refractive index than glass. Andrew Pritchard eventually created a diamond lens in 1824, but the expense and difficulty of creating lenses from precious stones meant that they never became common. The achromatic compound microscope eventually took over scientific microscopy. This did not happen overnight, however. Three of the most important microscopic discoveries of the early nineteenth century—the mammalian ovum in 1827 by Karl Ernst von Baer (1792–1876) and "Brownian movement" in 1827 and the cell nucleus in 1831 by Robert Brown—were made with single-lens microscopes.

See also Brown, Robert; Instrument Making; Telescope.

References

Bradbury, S. *The Evolution of the Microscope.* Oxford: Pergamon Press, 1967.

Stafford, Barbara. "Images of Ambiguity: Eighteenth-Century Microscopy and the Neither/Nor." In *Visions of Empire: Voyages, Botany, and Representations of Nature,* edited by David Phillip Miller and Peter Hans Reill, 230–257. Cambridge: Cambridge University Press, 1996.

Midwives

The Enlightenment period saw the beginnings of an effective male challenge to the traditional female monopoly of obstetrical care. This challenge was as yet limited to the English-speaking world, where the surgical community developed the new specialty of the "man-midwife." The mark of the new specialist was the obstetrical forceps, at first the secret of the Chamberlen dynasty of accoucheurs but generally available after around 1730. The center of man-midwife activity was London, where such specialists as William Smellie (1697–1763) practiced. The female midwives put up a fierce resistance, spearheaded by Elizabeth Nihell, author of the passionately feminist *Treatise on the Art of Midwifery: Setting Forth Various Abuses Therein, Especially As the Practice with Instruments* (1760). Nihell attacked man-midwives as bunglers who mutilated women and children with their instruments and lacked the sympathy for women that midwives needed to possess. Not all man-midwives fitted Nihell's caricature, but the introduction of men into middle- and upper-class midwifery in Britain (and America, where man-midwifery was promoted by the London-trained surgeon William Shippen [1736–1808]) did not noticeably improve care.

Nihell benefited from training at the Paris Hôtel-Dieu, France's leading center for training midwives. On the Continent improved training for women midwives helped keep the profession female-dominated (although male

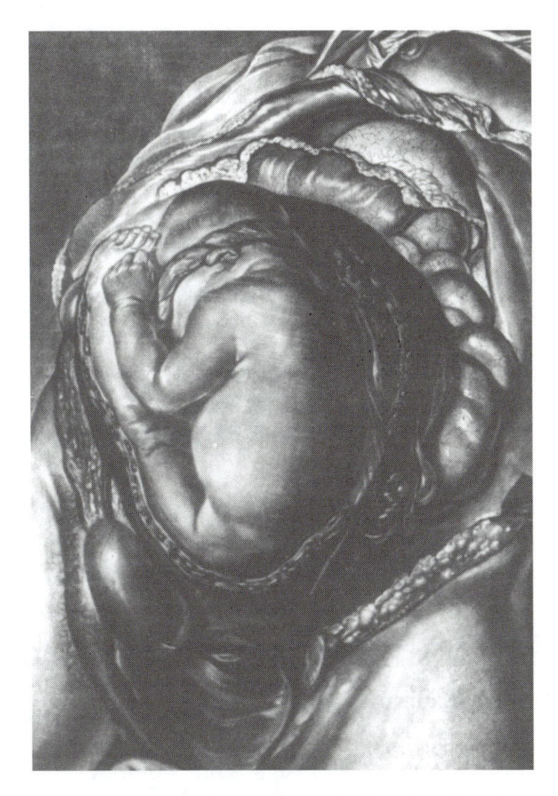

A uterine pregnancy illustrated in a book on the subject by Charles Jenty, published in 1757 (National Library of Medicine)

practitioners were not unknown). Continental midwives who published original work on obstetrics included the Frenchwoman Marie-Louise Lachapelle (1769–1821), author of the three-volume *Practice of Obstetrics* (1821–1825), and her student Marie-Anne Victorine Boivin (1773–1847), inventor of a vaginal speculum. The most influential of all Enlightenment midwives, Madame Angelique Marguerite Le Boursier du Coudray, worked by teaching rather than publishing.

The real disaster for pregnant women in the eighteenth and nineteenth centuries was the spread of the lying-in hospital, where women near to delivery were housed rather than delivering at home. Due to poor sanitation, the attendance on the women of medical students fresh from dissection, and lack of understanding of infection, mortality rates from puerperal fever, or "childbed fever," at these institutions were horrendous.

See also Coudray, Angelique Marguerite Le Boursier du; Medicine; Women.

References

Alic, Margaret. *Hypatia's Heirs: A History of Women in Science from Antiquity to the Late Nineteenth Century.* London: Women's Press, 1986.

Gelbart, Nina Rattner. *The King's Midwife: A History and Mystery of Madame du Coudray.* Berkeley: University of California Press, 1998.

Porter, Roy. *The Greatest Benefit to Mankind: A Medical History of Humanity.* New York: W. W. Norton, 1998.

Museum of Natural History (Paris)

The Museum of Natural History in Paris was established by the revolutionary government in 1793 as the successor of the old Royal Botanical Garden. The revolutionaries, working with reform forces in the Royal Botanical Garden itself, abolished the old position of intendant, from which Georges-Louis Leclerc de Buffon had been the unchallenged ruler until his death. Instead of three full professors, the new museum had twelve who met as a council to make decisions and elect a director for a term of one year. The chairs were divided into three for botany, two for zoology, two for chemistry, two for anatomy, and one each for mineralogy (taken by the leader of the new institution, Buffon's old assistant Louis-Jean-Marie Daubenton [1716– 1800]), geology, and illustration (which was abolished in 1822). Professors received lodgings at the museum, and an annual income of 5,000 francs, which most supplemented with work elsewhere. They were required to give forty lectures a year on a course in their subject. Many avoided this by hiring a lesser savant to actually give the lectures. Often the lesser savant would receive the professorship on the professor's retirement.

Although the botanical mission of the old garden continued, and even expanded with the establishment of subgardens such as one for tropical plants, the principal mark of the new institution was made in zoology, in which it was the world leader by the early nineteenth century. The original zoology professors were Jean-Baptiste-Pierre-Antoine de Monet de Lamarck and Étienne Geoffroy Saint-Hilaire, joined in 1795 by Georges Cuvier who was hired as the lecturer for the professor of comparative anatomy. (Cuvier himself became professor in 1802.) The museum added a menagerie of live animals in 1794, drawing from the royal collection at Versailles in the aftermath of the abolition of the French monarchy. The menagerie became a popular destination for Parisians and helped the museum's image. The museum's collections and its professors benefited immeasurably from the triumph of the Revolution at home and French arms in Europe. In 1795 the Dutch stadtholder's collection became one of the largest of the many natural-history collections that would be sent back to the museum by the conquering French armies. Rather than all being held in a central collection, items acquired by the museum were held in collections attached to professorial chairs and curated by persons hired by the professors. Cuvier's comparative anatomy collection, not opened to the public until 1806, was particularly famous. It contained 11,486 items in 1822, filled several rooms (in an arrangement based on Cuvier's own system of natural classification), and was the basis of much of his research. The museum also founded a journal, *Annals,* in 1802, which became the principal forum for natural history in France.

The fact that each professorship was independent and theoretically equal made it impossible for anyone at the Museum of Natural History to establish the supremacy that Buffon had had over the garden, although Cuvier had become dominant by the end of the Napoleonic era. The fifty or so families resident at the museum formed a conflicted and faction-ridden community. The peculiar situation of the museum community also contributed to a high degree of intermarriage, nepotism, and the handing down of professorial chairs and other positions among families.

See also Botanical Gardens; Cuvier, Georges; French Revolution; Geoffroy Saint-Hilaire, Étienne; Lamarck, Jean-Baptiste-Pierre-Antoine de Monet de; Museums; Napoleonic Science; Zoology.

References

Appel, Toby A. *The Cuvier-Geoffroy Debate: French Biology in the Decades before Darwin.* New York: Oxford University Press, 1987.

Limoges, Camille. "The Development of the Muséum d'Histoire Naturelle of Paris, c. 1800–1914." In *The Organization of Science and Technology in France, 1808–1914,* edited by Robert Fox and George Weisz, 211–240. Cambridge: Cambridge University Press, 1980.

Outram, Dorinda. *Georges Cuvier: Vocation, Science, and Authority in Post-Revolutionary France.* Manchester: Manchester University Press, 1984.

Museums

Museums in the eighteenth century were both increasing in size and number and changing their function. Beginning in the late seventeenth century a movement away from the idea that a museum, or "cabinet of curiosities," should aim at provoking wonder in the beholder reshaped Europe's museum culture. Instead, many museums oriented their collections and displays around the ideal of education. This shift affected many aspects of collecting. In natural history, ingenious creations made by altering or combining animal forms, such as "basilisks," were eliminated from collections in the early eighteenth century. Rarity, although still important, ceased to be the primary consideration in determining an object's worthiness to be collected. Collections began to be organized on the basis of putting similar objects or objects in the same category together, rather than heightening contrasts with unusual juxtapositions. (The great classifier Carolus Linnaeus also worked on organizing some of the Swedish royal collections.) "Natural" objects were increasingly housed in separate rooms, or even separate museums, from "artificial" or human-made ones, as the Florentine Museum of Science was spun off from the Uffizi Gallery in the 1770s.

Museums were rising institutions. The great princely collections were being opened to the public (a narrowly defined public), and a museum was considered to be an asset to a town or even a country. (In the early eighteenth century Peter the Great of Russia [r. 1682–1725] bought several large Dutch collections as part of his project of remaking Russian culture on European lines.) For many collectors, leaving their collections to museums was an attractive alternative to leaving them to heirs, who frequently sold them piecemeal for the cash. Natural-history collections were particularly affected by improved techniques for taxidermy and preservation, such as the arsenical soap invented by the Frenchman Jean-Baptiste Becouer, which for the first time offered the possibility of preserving birds in their natural plumage. Catalogs and guidebooks to Europe's museums appeared at an increasing pace. Although major museums were moving away from wonder and surprise as organizing principles, the gap was filled by smaller private museums, often charging admission and running for profit. The big museums were theoretically public, but this did not mean that anyone could walk in. Admissions were often tightly controlled as by a system of tickets. One reason for limitation in many cases was the expectation that every group of visitors would be shown around the museum by a member of the staff.

The association of museums with scientific institutions such as scientific societies, medical associations, and universities was an old but problematic one. Many institutions lacked the resources or staff to curate collections properly, and several strong collections simply melted away in institutional hands. Probably the most successful collection of those attached to a scientific institution was that of the French Royal Botanical Garden, the "Cabinet of the King." So successful was the Cabinet of the King, particularly under Buffon's rule, that on the reorganization of the garden after the French Revolution it eclipsed the garden itself, and the new institution was called the Museum of Natural History. As such it became one of France's premier institutions for scientific research, an unusual role for a museum to play at the time. Another important Paris museum was the Conservatory of Arts and Trades founded in 1794, one of the earliest museums of technology.

In addition to the Museum of Natural History, the other great museum that can trace its roots to the Enlightenment is the British Museum. Unlike the Museum of Natural History, the British Museum was universal, collecting art, artifacts, books and manuscripts, as well as natural objects. It was founded to house the vast collections of Sir Hans Sloane, who had left his collection to a group of trustees with directions that they attempt to sell the collection to the nation. This was accomplished with some difficulty. Although the founding collections of the museum also included two collections of manuscripts of principally historical and antiquarian interests, there are indications that it was thought of as primarily a scientific institution—its first three principal librarians, administrative heads, were physicians like Sloane himself. Major scientific collections acquired by the British Museum included the fossil collection of Gustav Brander (1720–1757), the Hatchett collection of minerals acquired in 1794, and many natural and artificial objects acquired on Captain Cook's voyages. The museum also acquired Sir Joseph Banks's huge collection of books on natural history at his death.

See also Brown, Robert; Museum of Natural History; Sloane, Sir Hans.

References

Mazzolini, Renato G., ed. *Non-verbal Communication in Science prior to 1900.* Florence: Leo S. Olschki, 1993.

Miller, Edward. *That Noble Cabinet: A History of the British Museum.* London: Andre Deutsch, 1973.

Murray, David. *Museums: Their History and Use with a Bibliography and a List of Museums in the United Kingdom.* 1904. Reprint, Staten Island: Pober, 2000

N

Napoleonic Science

The period of Napoléon Bonaparte's rule over France, first as first consul (1799–1804) and then as emperor (1804–1814), was a great era of French science. In part this was due to Napoléon's continued support of those institutions that traced their roots to the revolutionary period, such as the Polytechnic School, the Museum of Natural History, and the First Class of the Institute of France, as well as Napoléon's own enthusiasm and personal patronage of science. Napoléon was trained in mathematics as part of his schooling as an artillery officer, and he had a genuine interest in science and enjoyment of the company of scientists. He took pride in his membership of the Institute, to which he was elected in 1797, and took an active part in its deliberations, attending meetings and serving on committees. He even signed some of his military dispatches as "Member of the Institute." He was accompanied by scientists, including Claude-Louis Berthollet and Étienne Geoffroy Saint-Hilaire, on the 1798 Egyptian expedition. In 1800, after the seizure of power, he served a six-month term as president of the Institute. However, he stopped attending Institute meetings in 1802, after he attained the position of consul for life.

As a patron of science, Napoléon worked through and expanded the traditional eighteenth-century method of prize competitions, announcing a huge prize of 60,000 francs for discoveries in electricity comparable to those of Benjamin Franklin and Alessandro Volta, whom he tried to lure to Paris. Napoléon was particularly interested in electricity, which he correctly believed was a field in which enormous discoveries remained to be made. This prize was never awarded, but a smaller one was awarded to Sir Humphry Davy. This award to a foreigner was not exceptional, as Napoléon, like previous French governments, believed that one effective way to advance French science was luring outstanding foreign scientists to Paris. (Napoléon also ransacked scientific collections belonging to defeated foes to find choice items to take back to Paris.) Foreign scientists like Benjamin Thompson, Count Rumford, were encouraged to settle in Napoléon's capital. Another significant Napoleonic prize was the decennial prizes of 10,000 francs each, to be awarded on the anniversary of Napoléon's seizure of power in 1799. These prizes were mostly for science and technology, and the first—and only—set of winners included such stalwarts of the regime as Pierre-Simon de Laplace and Claude-Louis Berthollet. Napoléon's prizes

Napoleon I was greatly interested in science and its potential contributions to French power and cultural prestige. (Library of Congress)

had the advantage that they were not merely awarded but actually paid, something not always true of scientific prizes. He also extended the degree to which scientists could hope for social recognition by granting titles of nobility or membership in orders of chivalry to scientists.

Napoléon preferred scientists to literary men and philosophers, whom he viewed as politically unreliable. The scientists he liked were apolitical, more concerned with advancing science and technology than with projects of social and cultural reform in the Enlightenment tradition. Napoléon encouraged French scientists to think in nationalistic terms, rather than seeing themselves as part of an international scientific community, a task made considerably easier by the near-constant state of war. Several leading scientists played important roles in his regime, particularly in his vast reorganization of French education. Laplace, after a brief and unsuccessful period as minister of the interior, received the largely honorary but lucrative position of chancellor of the French Senate. Berthollet, probably Napoléon's closest friend among leading French scientists, was a senator, was made a count of the empire, was paid 150,000 francs to clear his debts in 1807, and received a pension enabling him to support himself and his scientific activities. Like previous French regimes, Napoléon hoped that science could contribute to economic development. For example, he planned a system of institutions to train a hundred chemists in improved means of extracting sugar from sugar beets, hoping to make France self-sufficient in sugar—although he fell from power before these institutions were set up, and France went back to importing cane sugar. Napoléon also encouraged the development of a French manufacturing sector that would eclipse that of Britain, currently undergoing its Industrial Revolution.

One of Napoléon's most long-lasting legacies to France was his reorganization of the educational system, in which he employed several scientists. Particularly important was the chemist Antoine-François de Fourcroy (1755–1809) and the naturalist Georges Cuvier. Surprisingly, given the high participation of scientists in Napoléon's reforms, the reformed system was criticized for its traditional curriculum based on the classics and French literature with comparatively little emphasis on science and mathematics. Despite the persistent traditionalism of the French educational system, the density of educational and scientific institutions in Paris and the practice of one scholar holding multiple positions at different institutions enabled the city to support a remarkably large number of both senior and junior scientists. In Napoléon's time, it was the unquestioned capital of European science.

Napoléon's preference for apolitical scientists meant that many he favored and who in return fulsomely flattered him quickly reconciled themselves with the new regime after his fall from power. Even Berthollet signed the Act of Deposition of 1814, and Laplace voted in the Senate for Napoléon's exile to St. Helena.

See also Arcueil, Society of; Berthollet, Claude-Louis; Laplace, Pierre-Simon de; Nationalism; War.

References

Crosland, Maurice. *The Society of Arcueil: A View of French Science at the Time of Napoleon I.* Cambridge: Harvard University Press, 1967.

Daston, Lorraine. "Nationalism and Scientific Neutrality under Napoleon." In *Solomon's House Revisited: The Organization and Institutionalization of Science,* edited by Tore Frangsmyr, 95–119. Canton, MA: Science History Publications, 1990.

Nationalism

Although the ideal and reality of scientific internationalism was strong in the Enlightenment, scientists were also part of national communities, and nationalism and national differences affected both the style and the content of what scientists did. This was particularly marked as science was being practiced in an increasing number of places. Countries like Prussia, Scotland, and Sweden, previously on the fringes of European science, developed strong national-science institutions and traditions. New nations emerging in the Americas, such as the United States, also sought to participate independently in the European-centered world of science. Among the factors contributing to the independence of national-scientific traditions were the rise of scientific societies organized on a national or regional basis, increasing use of vernacular languages rather than Latin in books and the increasing number of scientific periodicals, and the intertwining of science and government. Scientific projects were often harnessed to projects of national self-definition, such as exact mapping of the national territory. Scientists were sought after by national governments to discover new economic resources in national or colonial territories.

Scientists could become virtual symbols of their countries' cultural claims, the two most striking examples in the eighteenth century being the cult of Sir Isaac Newton (1642–1727) in Britain and the talismanic role of Benjamin Franklin in America. Carolus Linnaeus came to occupy a similar role in Sweden. France did not have such an iconic individual, but the leading role of France, the French language, and the Royal Academy of Sciences were used as evidence of the greatness of French culture. Scotland, coming from scientific nonentity in the seventeenth century to a leading role in the eighteenth, found in science some compensation for the loss of its national political institutions by the union with England in 1707.

The two nations with the greatest amount of scientific activity in the eighteenth century, Britain and France, developed strikingly different scientific identities. Britain isolated itself from the most advanced mathematics by its refusal to adopt the Leibnizian calculus, a refusal connected with the power of the cult of Newton, the rival of Gottfried Wilhelm Leibniz (1646–1716). British mathematicians began to remedy their backwardness only in the early nineteenth century. British isolation not only crippled British mathematics, but also rendered British science unable to absorb many developments in mechanics, mathematical physics, and to a lesser degree astronomy. French science, by contrast, was highly mathematized, and French scientists viewed mathematization as a necessary step in the formation of any physical science discipline. Britain's advantage for most of the eighteenth century lay in its superior instruments, partly a product of its more egalitarian social system, which did not feature as large a gap between artisanal instrument makers and gentlemen of science as did France. Although British science carried a pronounced empirical cast, there were differences between England and Scotland. The Scots were more likely to produce "system builders" such as James Hutton in geology or William Cullen (1710–1790) in medicine.

Germany mostly followed France in mathematics and science, but developed an increasingly independent scientific identity by the late eighteenth century. Although the closest thing to a German central government, the Holy Roman Empire, was of diminishing importance in the period, and German states engaged in wars and conflicts with each other, a German identity transcending the borders between states was rising, in science as in other areas. National feeling played a role in the German debate over the new antiphlogistic chemistry of Antoine-Laurent Lavoisier in the 1790s. Lavoisier's chemistry was frequently referred to as "French chemistry," and the chemical claims of the French, always chasing after new ideas, contrasted unfavorably with the alleged technical superiority and intellectual solidity of German chemists. The late eighteenth century saw a rise in the self-estimation of German intellectuals as a more spiritual and metaphysical people than the "materialistic" British and French. The sense of a particular German national mission to overcome "shallow" mechanistic science underlay much romantic science and *Naturphilosophie.* German science, like Dutch and Swiss, was also marked by strong universities.

There were two main reasons a nation would support science: prestige and economic development. Generally, economic development, while always important, was particularly important for smaller and poorer nations that could not hope to compete with the leaders in pure science. State initiative played an enormous role in the diffusion of modern science in countries on Europe's periphery. The Polish Commission on National Education, founded by the Diet in 1773, sponsored the training of teachers and the translation of French mathematical textbooks into Polish. The spectacular rise of Swedish science beginning in the mid-eighteenth century was largely fueled by government hopes that science could lift Sweden into the ranks of Europe's developed countries. State subsidies provided for the manufacture of high-quality scientific instruments, a contrast with Britain, where the London instrument makers worked primarily for the market. The change in government in 1772, when Gustav III (r. 1771–1792) seized power from the Swedish Diet, led to scientific decline, as the new regime withdrew much of its support.

Issues of prestige were not always national in the modern sense. Europe's monarchs were also keenly aware of personal prestige. The reorganization of the Berlin Academy by Frederick II (r. 1740–1786) was meant to raise the prestige not of Prussian science, but of Frederick himself as a patron of the sciences on the model of the French king Louis XIV (r. 1643–1715), founder of the Royal Academy of Sciences on which the new Berlin Academy was modeled. Rather than raising German scientific culture, Frederick sought the best French and Swiss scientists, and the Berlin Academy published in French. Napoléon also sought out foreign scientists to bring to Paris, but wanted to raise both his own glory and the glory of French science by emphasizing the national character of French scientific achievements. Given the near-constant warfare between France and Great Britain during the revolutionary and Napoleonic eras, national identification in both scientific communities increased considerably.

See also Colonial Science; Exploration, Discovery, and Colonization; French Revolution; Japan, Western Science in; Napoleonic Science; War.

References

Daston, Lorraine. "Nationalism and Scientific Neutrality under Napoleon." In *Solomon's House Revisited: The Organization and Institutionalization of Science,* edited by Tore Frangsmyr, 95–119. Canton, MA: Science History Publications, 1990.

Hufbauer, Karl. *The Formation of the German Chemical Community, 1720–1795.* Berkeley: University of California Press, 1982.

Porter, Roy, and Mikulas Teich, eds. *The Scientific Revolution in National Context.* Cambridge: Cambridge University Press, 1992.

Natural Theology

The tradition of a mutually supportive intellectual alliance of science and religion that had been so important during the scientific revolution continued to play a central intellectual role in the eighteenth century. The main intellectual connection between religion and science was the concept of design, and particularly of benevolent design. Natural theology based on the argument from design was strongest in Protestant Europe, particularly but not exclusively the British Isles and its colonies. Although some works of natural theology were produced in Catholic France, the much more anti-Christian attitude of many French scientists and philosophers meant that little original work in natural theology took place there.

An important moment in the institutionalization of natural theology occurred in late-seventeenth-century England with the foundation of the Boyle Lectures in the will of the chemist Robert Boyle (1627–1691). Boyle Lecturers and other English and British American scientists and popularizers linked natural theology with Newtonian physics, a project encouraged by Isaac Newton (1642–1727) himself. Another root of Enlightenment natural theology was the work of Newton's great rival, the German philosopher Gottfried Wilhelm Leibniz (1646–1716). Liebniz's belief that the actual universe was the "best of all possible worlds," meaning that it combined the maximum order with the maximum diversity, would be savagely mocked by Voltaire, among others. Newton and Leibniz differed in that Newton gave a much larger role to God's continuing management of the universe, invoking divine intervention to prevent accumulating irregularities in the movements of the planets. The famous debate between Leibniz and the Newtonian Samuel Clarke (1675–1729) that took place in 1715 and 1716 was one between rival claims to orthodoxy—both argued that their natural philosophy was a better support to belief in the Christian God. (This debate was first published in 1717, and many times thereafter in the eighteenth century.) In addition to its usefulness to believing Christians, such as Newton and Leibniz, natural theology could also be used by Deists to demonstrate the existence of God.

Natural history as well as physics were useful for natural theology. The most influential published Boyle Lectures in natural theology were those delivered by the Reverend William Derham (1657–1735) in 1711 and 1712, published as *Physico-Theology* in 1713. Derham and other British natural theologians brought design down from the heights of the Newtonian divine Lawgiver presiding over the fundamental forces of the universe to natural-historical detail, pointing out living creatures had been perfectly adapted for their own survival and flourishing as well as their usefulness to human beings. God's benevolent design extended into the details of the human realm as well, as Derham pointed out that the fact that God gave human beings distinct handwritings served society by making it more difficult to forge checks. British works of natural theology were translated into Dutch and German, and inspired numerous continental emulators.

Despite its rich tradition, natural theology was increasingly vulnerable to eighteenth-century developments in science, religion, and philosophy. With the increasing mechanization of the universe by later Newtonians, culminating in the celestial mechanics of Pierre-Simon de Laplace, the original Newtonian role of God as a maintainer of the solar system and corrector of irregularities—the "God of the gaps"—was no longer necessary. Voltaire and others attacked the idea of the benevolence of the universe, an idea many Enlightenment thinkers considered incompatible with the prevalence of pain, suffering, and disaster in the universe. On a more fundamental level, the argument from design came under increasing attack from the Enlightenment philosophers David Hume (1711–1776) and Immanuel Kant.

Hume, in his brilliant and posthumously published *Dialogues Concerning Natural Religion* (1779), pointed out that the design argument rested on an irrational extension of the human experience of design and designers to the cosmic realm. He claimed that the universe in many ways seemed to have grown like a tree rather than having been designed and constructed like a house. Hume also questioned the benevolence of what had been "designed." Kant, in addition to agreeing with many aspects of Hume's arguments, suggested that making the question of the existence of God dependent on scientific understanding made it potentially vulnerable to changes in scientific understanding. He viewed moral arguments for the existence and nature of God, rather than scientific ones, as convincing. On the religious side, some devout Christians found natural theology intellectually arrogant in its assumption that the purposes of God in creating things as they were could be known.

Despite these criticisms of natural theology, it remained influential into the early nineteenth century. This was not just on the level of popularizers like Derham; many important scientists saw what they were doing as natural theology. Carolus Linnaeus included an elaborate discussion of the problem of suffering and pain in the later editions of his *System of Nature.* He believed that the experimental method could be applied to theological questions, but his most important treatment of this idea, *Nemesis Divina,* was never published. The classic early-nineteenth-century statement of natural theology was the frequently reprinted work of the Anglican clergyman William Paley (1743–1805), *Natural Theology* (1802), a commonly used textbook in English universities. Paley's emphasis on the complexity of living things, the perfection of their adaptation, and the impossibility of their origin in any process save intelligent design was part of the intellectual milieu of the young Charles Darwin (1809–1882).

See also Bonnet, Charles; Hutchinsonianism; Kant, Immanuel; Linnaeus, Carolus; Newtonianism; Popularization; Religion.

References

Brooke, John Hedley. *Science and Religion: Some Historical Perspectives.* Cambridge: Cambridge University Press, 1991.

Lindberg, David C., and Ronald L. Numbers, eds. *God and Nature: Historical Essays on the Encounter between Christianity and Science.* Berkeley: University of California Press, 1986.

Naturphilosophie

One of the most influential syntheses of German romanticism and German science was *Naturphilosophie,* the "philosophy of nature." Its main founder was professor Friedrich Wilhelm Joseph von Schelling (1775–1854), who published the central document of the movement, *Ideas for a Philosophy of Nature,* in 1797. Another influential champion was the biologist and professor Lorenz Oken (1779–1851). Although strongest in Germany, the movement also greatly influenced Scandinavia, which looked to Germany for intellectual and cultural leadership. For *Naturphilosophs,* the investigation of nature was a spiritual quest with an ultimately spiritual goal. Nature itself was not a material phenomenon existing outside humans, but ultimately a product of the human spirit. *Naturphilosoph* science, unlike much Enlightenment science, was not primarily oriented to technological progress. The mission of *Naturphilosophie* was ultimately to restore, on a higher level, the original unity of man and nature held to exist in the golden age or before the Fall of Man, when the products of the human spirit became separated from the human spirit itself.

Naturphilosophs understood nature itself as moved by transcendent, nonmaterial forces. A nonmechanical drive to organization manifested itself in the crystallization of minerals or the growth of living things. What the scientist should study was not phenomena taken in isolation, but these forces and the systems that they shaped. Only the *Naturphilosoph,* with his spiritual awareness, could truly understand nature—traditional scientists merely worked in the outward-seeming. *Naturphilosophs* and other German romantic scientists were holists, emphasizing entities as wholes that worked in a certain way, rather than collections of parts. Their holism was combined with dualism—*Naturphilosophs* saw the world as governed by pairs of opposed forces and placed great emphasis on symmetry.

Naturphilosophs, despite their somewhat off-putting penchant for mystical language, were not outsiders in the scientific community. They accepted much of the mechanical science of the day, seeing it as incomplete rather than erroneous. When they actually did science, they followed the classic experimental methods of early modern and Enlightenment science. Although *Naturphilosophie*'s mysticism and transcendental orientation led to a partially deserved bad reputation as an impediment to the development of science, it could also be scientifically productive. One example is Johann Wilhelm Ritter's discovery of ultraviolet light. When Ritter heard about William Herschel's discovery of infrared light, he quickly realized that because the universe of *Naturphilosophie* was symmetrical, there must be something at the opposite end of the spectrum. He devised an ingenious series of tests that demonstrated the existence of ultraviolet.

The area on which *Naturphilosophie* had the most impact in the early nineteenth century was biological science. *Naturphilosophs* inverted the traditional hierarchy of the sciences, treating the life sciences as the model for the others, and attempting to apply biological concepts of development to the inanimate universe. *Naturphilosophie* offered a way to raise the status of natural history within the faculties of German universities, offering a justification for its separation from medicine. Vitalism, with its emphasis on the nonmaterial definition of life, was congenial to *Naturphilosophs*. *Naturphilosophs* emphasized the parallels between the anatomy of different creatures, viewing them as a range of geometrical variations on a few basic forms. They were more concerned with abstract variation than the actual functions of the anatomical organs and parts. Oken, for example, held that bones were all modifications of the vertebra and suggested that the skull was created by the fusing of four vertebra. This program was compatible with the "philosophical anatomy" of the French zoologist Étienne Geoffroy Saint-Hilaire, and an alliance was formed between the two movements in the 1820s, but *Naturphilosophs* differed from their French contemporaries in their transcendental and mystical orientation.

After the 1820s *Naturphilosophie* lost most of its vitality. In Sweden it was vigorously and successfully opposed by Jöns Jakob Berzelius. In its German homeland new scientific leaders such as the organic chemist Justus von Liebig (1803–1873), who famously referred to *Naturphilosophie* as the "Black Death," and the physiologist and physicist Hermann Ludwig Ferdinand von Helmholtz (1821–1894) asserted that mechanical and reductionistic science, including mechanical and reductionistic biology, was the only proper science, ridiculing *Naturphilosophie*'s transcendent orientation.

See also Berzelius, Jöns Jakob; Ørsted, Hans Christian; Ritter, Johann Wilhelm; Romanticism; Sexual Difference; Vitalism.

References

Cunningham, Andrew, and Nicholas Jardine, eds. *Romanticism and the Sciences.* Cambridge: Cambridge University Press, 1990.

Jardine, Nicholas. "*Naturphilosophie* and the Kingdoms of Nature." In *Cultures of Natural History,* edited by N. Jardine, J. A. Secord, and E. C. Spary, 230–245. Cambridge: Cambridge University Press, 1996.

Lenoir, Timothy. *The Strategy of Life: Teleology and Mechanics in Nineteenth-Century German Biology.* Dordrecht, Netherlands: D. Reidel, 1982.

Navigation

See Exploration, Discovery, and Colonization; Longitude Problem.

Newtonianism

During the Enlightenment many of Europe's scientists and philosophers declared themselves Newtonians. But Newtonianism had many meanings, far beyond the mere acceptance of Newton's universal gravitation or laws of motion. Isaac Newton himself lived on until 1727, and although the creative period of his science had long since passed, he had a tremendous influence on English science through his position as president of the Royal Society. This influence was not altogether for the good—British mathematics, through its stubborn adherence to the Newtonian calculus and rejection of the calculus of Newton's hated rival Gottfried Wilhelm Leibniz (1646–1716), became isolated from the European mathematical mainstream, and the great British mathematicians of the seventeenth century found no successors.

The most naive, and possibly the most widespread, form of Newtonianism was the simple acknowledgment of Newton as the greatest man, or at the very least the greatest scientist, who ever lived. This was most succinctly summed up in the declaration that Newton was the luckiest man who ever lived, because there was only one universe, and only one man could discover its laws. Scraps of paper with Newton's writing on them were treasured as relics. The Scottish philosopher and historian David Hume (1711–1776), not given to overstatement, described Newton as "the greatest and rarest genius that ever rose for the ornament and instruction of the species." Newton's system was expounded and praised by a small army of lecturers, demonstrators, and popularizers, of whom Voltaire and Francesco Algarotti (1712–1764), author of the widely translated *Newtonianism for Ladies* (1737), were perhaps the most successful. But the Newton of the eighteenth century was a bowdlerized Newton. In the torrent of panegyric many things important to Newton himself—biblical prophecy and alchemy, for example—were simply ignored. The link that Newton himself had seen between his natural philosophy and his religion was dissolved, and Newtonianism became an ideology practiced by those of all religions and none. Extreme anticlericals in France such as Voltaire claimed to be Newtonians, as did the developers of the English tradition of natural theology that climaxed with the publication of *Natural Theology* (1802) by William Paley (1743–1805).

The acceptance of Newtonian physics on the European continent was a slow process, in which its principal rival was Cartesianism. Newtonianism's strongholds were Britain and the Dutch Republic, where it was introduced in the beginning of the eighteenth century. The Dutch Republic produced some of the most influential Newtonians, such as two professors at the University of Leiden, the physician Hermann Boerhaave and the physics professor Willem Jakob 's Gravesande (1688–1742). The Cartesian stronghold was France, where young intellectuals like Voltaire—championing the cause of reform and greater personal freedom—seized upon Newtonianism as evidence of the superiority of the freer intellectual environment of England.

The first scientific issue that tested the difference between the two philosophies was the shape of Earth. Cartesians predicted that the spherical Earth would be slightly elongated at the poles, Newton and subsequent Newtonians that it would be slightly flattened. In 1718 a leading French Cartesian, the astronomer Jacques Cassini, published measurements indicating that the Earth was elongated. Newtonians took up the challenge, and a bitter controversy developed, largely on national lines. In the 1730s the French government took measures to settle the question. An expedition under Charles-Marie de La Condamine (1701–1774) set out for Ecuador near the equator in 1735, and the subsequent year another expedition under Pierre-Louis Moreau de Maupertuis set out for the Arctic Circle. The results of the measurements were a triumphant vindication of the Newtonians. The successful prediction of the return of Halley's comet in 1759 by Alexis-Claude Clairaut (1713–1765) was another victory of Newtonianism. The French by this point accepted the Newtonian system with some modifications and additions, the most important being the addition of Leibniz's concept of "living force," *vis viva,* calculated as the mass times the square of the velocity of a moving object. (Newton had been aware of living force but attached no physical importance to it.) The French further refined Newtonian physics by the use of the more sophisticated Leibnizian calculus. The full perfection of the Newtonian system of gravity as applied to the planets was reached in the *Treatise on Celestial Mechanics* (five volumes, 1799–1825) of Pierre-Simon de Laplace. But this was a paradoxical triumph. Laplace's system denied the need for divine intervention to keep the universe running, which had been central to Newton's own philosophy.

Methodologically, the Newtonian legacy to the eighteenth century was rich and contradictory, ranging from the austere mathematics of *Mathematical Principles of Natural Philosophy* (1687) to the experimentalism of *Opticks* (1704). The concept of Newtonian method was broad enough to include both the arch-experimenter Stephen Hales and Jean Le Rond d'Alembert, who worked strictly from mathematical and logical deduction and despised experiment. The model of Newtonian mechanics, of a science reduced to a small set of mathematical laws, remained a powerful one, but not all sciences could be approached in the same manner, and not all Newtonian scientists did so. Efforts to reduce magnetic or electrical interactions to an equivalent of Newton's inverse-square law of gravitation bore little fruit until late in the eighteenth century with Henry Cavendish, Franz Maria Ulrich Theodor Hoch Aepinus (1724–1802), and Charles-Augustin de Coulomb.

See also Alembert, Jean Le Rond d'; Bassi, Laure Maria Catarina; Boerhaave, Hermann; Cartesianism; Châtelet, Gabrielle-Émilie du; Desaguliers, John Theophilus; Hales, Stephen; Hauksbee, Francis; Literature; Mathematics; Maupertuis, Pierre-Louis Moreau de; Mechanics; Nationalism; Optics; Physics; University of Leiden; Voltaire; Whiston, William.

References

Gay, Peter. *The Enlightenment: An Interpretation, Volume II: The Science of Freedom.* New York: Alfred A. Knopf, 1969.

Hankins, Thomas L. *Science and the Enlightenment.* Cambridge: Cambridge University Press, 1985.

Jacob, Margaret. *Scientific Culture and the Making of the Industrial West.* New York: Oxford University Press, 1997.

Nollet, Jean-Antoine (1700–1770)

Jean-Antoine Nollet was France's premier electrician and natural-philosophical demonstrator in the mid-eighteenth century. A provincial abbé of obscure background, Nollet came to Paris in the 1720s and ingratiated himself with leading students of electricity such as Pierre Poliniere and Charles-François de Cisternay Du Fay (1698–1739). He traveled to the Dutch Republic and Britain to observe the great Newtonian demonstrators John Theophilus Desaguliers, Willem Jakob 's Gravesande (1688–1742), and Pieter van Musschenbroek (1692–1761). By 1735 Nollet was offering a series of natural-philosophical afternoon lectures in intimate surroundings, popular with the men and women of upper-class Paris. He introduced some of the organization and theoretical sophistication of the Anglo-Dutch tradition with the informality of Parisian salon culture. Dramatic electrical demonstrations such as the electrification of a boy hanging from the ceiling were the centerpiece of Nollet's presentations. He also went into business designing and selling his own equipment, and published a six-volume exposition of physical science, *Experimental Lessons in Physics* (1743–1748). In 1739 he won admission to the Royal Academy of Sciences, and he accumulated many other honors in the subsequent decades.

Nollet's natural philosophy was eclectic, incorporating much of Newtonian mechanics into a basically Cartesian picture of a world filled with subtle matters. His preference for Cartesian vortices over Newton's theory of universal gravitation did not preclude friendly personal relations with the Newtonians

Voltaire and Gabrielle-Émilie du Châtelet. Nollet's influential Cartesian theory of electricity, set forth in his *Essay on Electricity* (1746) and known as the "systeme Nollet," ascribed electricity to the flux and reflux of subtle matter squeezed out of the pores of certain objects. Nollet's electrical theory was an immediate success, not only in France but also in Britain, the Dutch Republic, and Germany. His leadership among French electricians was reinforced when he introduced the Leiden jar to France at a meeting of the Royal Academy of Sciences in April 1746 and carried out more spectacular experiments, electrifying circuits of hundreds of people. In 1749 he carried out a triumphal tour of Italy, exposing various charlatans using electricity for medicine.

Nollet's reign over electricity turned out to be brief, however, as he was challenged in 1752 with the publication of a French translation of Benjamin Franklin's work on electricity. The history of electricity prefaced to the work by its French translator, Thomas François Dalibard (1703–1779), did not even mention Nollet, an omission meant as a deliberate insult. The translation was sponsored by Nollet's enemy and Dalibard's friend the Comte de Buffon, Georges-Louis Leclerc, who detested the abbé as an ally of his rival in natural history, René-Antoine Ferchault de Réaumur (1683–1757). Nollet, who first thought Franklin a fictional character invented by his French enemies, fought the Franklinist interpretation vigorously, devising new experiments to demonstrate that glass was not impervious to electricity as Franklin stated. He also attacked Franklin's doctrine of "negative electricity," and pointed out the many electrical phenomena, such as attraction, which Franklin's system failed to explain. The strength of Nollet cut France off from the development of electrical theory for some time, but eventually the younger generation of electricians passed from Nollet's to Franklin's system. By Nollet's death both the study of electricity and physics in general, which had moved to more quantitative approaches, had passed him by.

See also Cartesianism; Electricity; Franklin, Benjamin; Popularization; Volta, Alessandro Giuseppe Antonio Anastasio.

References

Heilbron, J. L. *Electricity in the 17th and 18th Centuries: A Study in Early Modern Physics.* 2d ed. Mineola, NY: Dover, 1999.

Sutton, Geoffrey V. *Science for a Polite Society: Gender, Culture, and the Demonstration of Enlightenment.* Boulder: Westview, 1995.

O

Observatories

The Enlightenment period saw astronomical observatories grow in size and number and shift in their function, with a growing emphasis on the housing of larger and larger telescopes. The two major permanent public observatories founded in the seventeenth century, Britain's Royal Observatory and France's Paris Observatory, faced increasing competition in the eighteenth century. The Paris Observatory was dominated by the Cassini dynasty, making first-rate French astronomers reluctant to work there. It focused narrowly on the Cassini program of geodesy and cartography. Much of its instrumentation and physical plant deteriorated or became outmoded before Jacques-Dominique de Cassini's reorganization in the years immediately preceding the French Revolution. As a general astronomical facility it was marginalized in favor of smaller and less elaborate institutions, such as the French naval observatory at the College of Cluny. Cluny was dominated by the geographer to the navy Joseph-Nicolas Delisle (1688–1768), and was the base from which his student the great observer Charles Messier (1730–1817) made his discoveries of comets and other celestial phenomena. The Royal Observatory at Greenwich, the base of the astronomer royal, had one great triumph in the early eighteenth century with the publication of the star catalog of the first astronomer royal, John Flamsteed (1646–1719), in 1725. It declined somewhat after Flamsteed, although not marginalized to the same degree as the Paris Observatory. By the late eighteenth century Greenwich, with its focus on navigation, was less important in pure astronomy than were the comparatively primitive "household" facilities of William Herschel at Slough.

The new permanent observatories (as opposed to temporary facilities built for specific observations of transits or other events) built in the eighteenth century were of two main types: tower observatories and turret observatories. Tower observatories were dominant in the early part of the century, with new observatories built in Berlin (1706–1711), Bologna (1712), Ingolstadt (1725), Montpellier (1745), Göttingen (1751), Mannheim (1772), and Bogotá (1802). The Bogotá Observatory was the first permanent observatory built in the Americas and the last tower observatory of significance. Tower observatories possessed the advantage of getting the observer high above buildings, trees, and other terrain features. They were also easy to adapt existing buildings for, but did not always provide a stable platform for telescopes and other instruments.

Turret observatories, usually built on the model of a large private house with two wings flanking a central dome where the telescope was kept, began to come into fashion in the second half of the century. Notable examples were built in Stockholm in 1753 and in Richmond in 1769, one of several observatories whose building was inspired by the 1769 transit of Venus. Oxford University's Radcliffe Observatory (1772) was also built on the turret design. A particularly important and influential observatory was built in Ireland at Dunsink outside Dublin in 1785 for Trinity College Dublin in conjunction with the college's appointment of a professor of astronomy. Designed by the architect Graham Moyers, the Dunsink Observatory departed from previous designs in that the telescope and other instruments rested directly on the ground, rather than on top of a series of rooms as had been the case in previous designs. To minimize vibration, telescopes and other instruments were isolated from the floors. The first continental observatory built on similar principles was the Seeburg Observatory (1787–1792). This type of design was dominant in the early nineteenth century.

See also Astronomy; Cassini Family; Maskelyne, Nevil; Mayer, Johann Tobias; Telescopes; Transits.

References

Donnelly, Marian Card. *A Short History of Observatories.* Eugene: University of Oregon Books, 1973.

Gillispie, Charles Coulston. *Science and Polity in France At the End of the Old Regime.* Princeton: Princeton University Press, 1980.

Oceanography

The eighteenth century saw growing interest in gathering data on the world's oceans as well as some attempts to theorize about their nature and behavior. The first book entirely devoted to the science of the ocean (including marine biology) was by the Italian soldier-scientist Luigi Ferdinando Marsigli (1658–1730), *Physical History of the Sea* (1724). The work was based on research Marsigli had done on the Mediterranean coast of France in 1706 and 1707, testing the salinity of water taken from the sea and local rivers and streams. Marsigli also argued that coral was a plant rather than a mineral, citing "flowers" that had emerged on a coral he had removed from the sea.

The most active area of oceanographic investigation in the mid-eighteenth century was measuring the density, salinity, and temperature of the ocean at different locations and depths. (Curiously, the recording of the surface temperature of the ocean began well after the recording of the deep-water temperature.) This was originally carried out by individuals, such as Hugh Campbell, whose observations on the density of seawater at different depths was published in the *Gentleman's Magazine* in 1755. The tendency was for observations to be increasingly carried out by the ship's official personnel as part of their duties, and to be done on a more systematic basis. Such observations were part of the unsuccessful voyage of the British captain Constantine John Phipps (1744–1792) to find a sea route through the Arctic in 1773. By 1817 when the French ship *Uranie* began a circumnavigation of the world, it was recording the surface temperature every two hours.

Recording was also aided by the adaptation of the maximum-minimum recording thermometer for oceanographic use in 1794. It was first used extensively on the Russian circumnavigation of 1802–1806, and remained in use for most of the nineteenth century. Benjamin Thompson, Count Rumford, drew on data gathered by investigators to establish the circulation of warm and cold water in the ocean in his paper "On the Circulation of Heat in Fluids," but his work had surprisingly little impact.

Increasing familiarity with the Arctic and Antarctic raised the question of the source of oceanic ice. Many, including Buffon, claimed that saltwater did not freeze and icebergs were originally composed of freshwater from rivers that had frozen and been carried to sea. Edward Nairne (1726–1806) disproved this assertion by experimentally establishing the freezing point of saltwater. There was also great interest in the exact chemical composition of seawater. Antoine-Laurent Lavoisier published an analysis of sea salt in 1772. Two British chemists, Smithson Tennant (1761–1815) and Alexander Marcet (1770–1822), encouraged by Sir Joseph Banks, carried out a lengthy series of analyses of water from different places brought back by British ships. Marcet published some results in *Philosophical Transactions* in 1819, establishing that the salinity of the ocean varies slightly by latitude but not by depth or longitude.

Increasing knowledge of the world's oceans meant increasing awareness of the importance of currents. The foremost expert in the early nineteenth century was the British army officer and surveyor James Rennell (1742–1830), better known as the founder of the Survey of India. Rennell's efforts collecting and analyzing information on winds and currents culminated in the posthumous *An Investigation of the Currents of the Atlantic Ocean, and of Those Which Prevail between the Indian Ocean and the Atlantic* (1832), the most important treatment to that date.

See also Exploration, Discovery, and Colonization.

References

Deacon, Margaret. *Scientists and the Sea, 1650–1900: A Study of Marine Science.* London: Academic Press, 1971.

Stoye, John. *Marsigli's Europe, 1680–1730: The Life and Times of Luigi Ferdinando Marsigli, Soldier and Virtuoso.* New Haven: Yale University Press, 1994.

Optics

Eighteenth-century optics was dominated by the classic *Opticks* (1704) of Isaac Newton (1642–1727). Although Newton himself made only tentative claims about the nature of light, his successors, particularly in the British Isles, interpreted his optics as based on light particles. These particles naturally followed straight lines, but were deflected by short-range forces. The elements of a wave theory in *Opticks*, conceptualized in terms of an ether, diminished in importance in the early eighteenth century. The difficulty for Newtonian opticians was in demonstrating the existence of the particles experimentally. Efforts to measure the force of impact of the light particles proved fruitless, and opponents of the light-particle theory argued that the enormous speed of the particles should produce some kind of shock. An alternative picture of light as matter was that of a "subtle fluid." This position was held by Hermann Boerhaave, who like other fluid theorists came at the subject of light not through classical optics, but through chemistry. Boerhaave considered light in the context of a theory of the universal element of "fire," rather than through refraction, reflection, and other standard optical phenomena.

Wave theory, which likened light to sound, was upheld by persons influenced by Cartesianism, most importantly Leonhard Euler in his *New Theory of Light and Color* (1746). Euler, like other opponents of the particle theory, argued that the Sun would exhaust itself emitting light particles, and that light particles coming from different sources would interfere with each other, eventually filling the universe. Particle theorists attacked the wave theory by referring to Newton's arguments against it in *Opticks*. Newton and his successors pointed out that opaque objects fully blocked light, producing shadows, whereas if light waves acted like sound waves, they should bend around obstacles.

The optical triumphs of the eighteenth century were not theoretical but practical, in the improvement of devices such as telescopes and microscopes. The great triumph in this area was the elimination of chromatic aberration, solved for telescopes (but not microscopes) by English instrument makers, and explained theoretically by Euler. This was a serious but not fatal blow to Newtonian optics, as Newton had claimed that aberration could not be eliminated. Another instrumental innovation was the invention of a photometer to compare the brightnesses of different sources of light by the Frenchman Pierre Boguer (1698–1758). This invention was announced in his *Optical Treatise on the Gradation of Light* (1729). Photometry was further advanced and systematized by the German physicist Johann Heinrich Lambert (1728–1777) in his Latin *Photometry* (1760).

Particulate optics continued to dominate late in the century, being given a more mathematically sophisticated form by the French Newtonians, led by Pierre-Simon de Laplace. However, several attempts to construct a non-Newtonian theory of light emerged by the late eighteenth century. These attempts, though, involving two persons not generally known as scientists, the German poet Johann Wolfgang von Goethe and the physician and future French revolutionary politician Jean-Paul Marat (1743–1793), proved unsuccessful in replacing Newtonian optics.

The early nineteenth century saw two major changes in optics. One was the extension of the spectrum beyond the visible, with William Herschel's discovery of the infrared in 1800, which inspired Johann Wilhelm Ritter's discovery of the ultraviolet the next year. The other change was the revival of the wave theory of light. This was the work of two men working independently, the English physician Thomas Young (1773–1829) and the French engineer Augustin-Jean Fresnel (1788–1827). Both worked in the margins of their national-scientific communities, still dominated by Newtonians. Young's work combined a revival of Euler's vibratory theory, a rediscovery of Newton's ideas about an "aether," and an experimental and theoretical study of "interference." Young's theory of interference enabled him to explain a number of optical phenomena by the relations of light waves at different phases in their undulations. The most dramatic case occurred when the peak of one light wave corresponded with the valley of another of the same frequency, leading the two to cancel out and produce darkness.

Although Young was the innovator, Fresnel presented a more complete, more influential, and far more mathematical theory and was the more influential of the two. Drawing on analytical mechanics and the work of the seventeenth-century physicist Christiaan Huygens (1629–1695) on wave fronts, Fresnel's system also explained the recently discovered phenomenon of polarized light. Fresnel overcame the resistance of the French Newtonian school to win the prize offered by the Academy of Sciences in 1819 in optics. The wave theory of light as it developed in the nineteenth century was founded on his work.

See also Goethe, Johann Wolfgang von; Microscopes; Newtonianism; Physics; Telescopes.

References

Cantor, G. N. *Optics after Newton: Theories of Light in Britain and Ireland, 1704–1840.* Manchester: Manchester University Press, 1983.

Gillispie, Charles Coulston. *The Edge of Objectivity: An Essay in the History of Scientific Ideas.* Princeton: Princeton University Press, 1960.

Wolf, Abraham. *A History of Science, Technology, and Philosophy in the Eighteenth Century.* 2d ed. Revised by D. McKie. Gloucester, MA: Peter Smith, 1968.

Ørsted, Hans Christian (1777–1851)

The Dane Hans Christian Ørsted was among the most versatile scientists of the early nineteenth century. Like many Danish intellectuals, he was also deeply influenced by contemporary German thought, both the critical philosophy of Immanuel Kant and romantic *Naturphilosophie.* The son of a pharmacist, in 1794 Ørsted entered the University of Copenhagen, where he was first exposed to Immanuel Kant. On a leisurely journey through Europe, he met and became friends with Johann Wilhelm Ritter in 1801. Ritter deeply influenced Ørsted's approach to *Naturphilosophie,* but Ørsted, who also frequented scientific circles in Paris, was more experimentally grounded.

Ørsted's science was premised on the basic *Naturphilosoph* idea of the unity of fundamental forces. Ørsted analyzed scientific phenomena in terms of modification of fundamental forces of attraction and repulsion, which were in themselves modifications of a fundamentally unified force. This philosophy led him to suspect as early as 1805 that electricity and magnetism were related or even identical forces, at a time when orthodox scientific thinking asserted that they were unrelated. He was only able to demonstrate this in 1820, by which time he had become ordinary professor of physics at Copenhagen. While lecturing to advanced students, Ørsted brought a wire with an electric current to a compass and observed a slight deviation of the needle. He was able to get a larger effect by modifying the wire, and announced his discovery to the European scientific world the same year in a short paper in Latin, "Experiments on the Effect of the Electrical Conflict on a Magnetic Needle," one of the last important scientific papers in Latin. (Ørsted had great linguistic facility, publishing in Danish, French, German, English, and Latin.) Ørsted explained "electromagnetism" in terms of the "electrical conflict," although he never precisely defined it. This concept never became widely used, researchers preferring the established idea of the electrical current.

Ørsted's other areas of scientific activity included fluid mechanics, where he performed many studies and experiments on compressibility, and chemistry. His goal was to demonstrate the unlimited compressibility of matter, thus vindicating the antiatomism he shared with Kant. As a chemist he developed the basis of the modern Danish chemical vocabulary and researched several compounds of chlorine. In 1825 he extracted aluminum of greater purity than anyone had before, although still impure. The intellectual effort of his closing decades was mostly devoted to expounding his natural philosophy and its connection to religion.

See also Ampère, André-Marie; Electricity; *Naturphilosophie;* Ritter, Johann Wilhelm; Romanticism.

References

Cunningham, Andrew, and Nicholas Jardine, eds. *Romanticism and the Sciences.* Cambridge: Cambridge University Press, 1990.

Ørsted, Hans Christian. *Selected Scientific Works of Hans Christian Ørsted.* Edited and translated by Karen Jelved, Andrew D. Jackson, and Ole Knudsen. Princeton: Princeton University Press, 1998.

P

Periodicals

The eighteenth century saw growth in the number and importance of scientific periodicals. At the beginning of the century there was only one stable, regularly appearing periodical solely devoted to science, the *Philosophical Transactions of the Royal Society.* By the end of the century there were hundreds. More than half were published in German-speaking Europe, although many of those were ephemeral. The audiences addressed varied. Many journals, particularly in medicine and agriculture, were aimed not at scientists, but at literate people who wanted to know how to live healthy lives or farm effectively.

The most important kind of periodical publication for most of the century were those sponsored by scientific societies, led by *Philosophical Transactions* and the *Memoirs* of the Royal Academy of Sciences. These constituted about a quarter of the scientific journals publishing substantive material, as opposed to reviews and abstracts, and a larger percentage of the ones publishing over a long period. Some sort of serial publication was necessary for every scientific society aiming at a European reputation, and the exchange of publications was an important part of the creation of bonds between societies. The periodical press also served the important function of announcing and disseminating information on prize contests.

The more permanent, larger, and better-funded societies were more regular in their publications, whereas smaller ones published less often and at irregular intervals. The American Philosophical Society, for example, published only when it had enough papers in hand to fill a volume. A problem with some society publications (although not *Philosophical Transactions*) is that they published only the works of members of the society, meaning that other papers it might have received from nonmembers did not have an outlet. The Royal Academy of Sciences tried to remedy this problem by the publication of eleven volumes of *Foreign Savants* between 1751 to 1786. The Royal Academy was always generating more work than its publications could hold. As early as 1703, just a year after the first issue of the *Memoirs,* the mathematician Antoine Parent (1666–1719) started *Researches in Physics and Mathematics,* just to publish his papers that had not been accepted for academy publication. Another limitation of society publication was that most, with the exception of the quarterly *Philosophical Transactions,* published yearly or less frequently. A scientist could wait years after submitting a paper to see it in print.

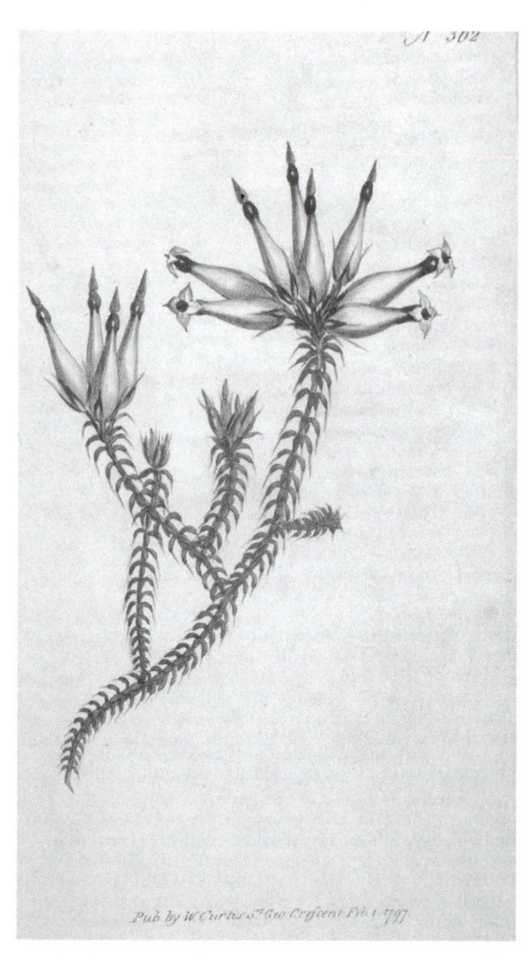

William Curtis's Botanical Magazine, *from which this illustration of Recurve Heath is taken, popularized botany and the Linnean system of classification in late-eighteenth- and early-nineteenth-century England. (Special Collections, National Agriculture Library)*

One of the principal roles of the scientific journal was facilitating communication across the barriers of geography and language. Many journals specialized not in original work, but in translating, abstracting, reviewing, or simply reprinting already published material. The first journal dedicated exclusively to reviews of new scientific writings was the Latin quarterly *Commentaries on Occurrences in Natural Science and Medicine,* published in Leipzig from 1752 to 1798. It acquired a European reputation. (Latin serials were mostly found in Germany and eastern Europe, and were diminishing in the eighteenth century. Most journals were published in the local vernacular.) Journals were also repackaged as books or sets of books, published for those who could not get the originals or needed a translation. The longer-running periodicals also acquired published indexes. The largest number of specialized periodicals was in the field of medicine. The foremost French journal, *General Journal of Medicine, Surgery, and Pharmacy,* founded in 1754, survived the vagaries of politics until 1822. The medical journals were followed at a distance by agricultural journals, largely a phenomenon of the second half of the eighteenth century.

Both the number and the size of scientific periodicals increased rapidly in the second half of the eighteenth century, after a relatively stagnant first half. By the late eighteenth century specialized scientific journals publishing original work had begun to emerge. Three influential examples are the first ongoing chemical journal, Lorenz Florens Friedrich von Crell's *Chemical Journal;* the Jens professor Johann Ernst Immanuel Walch's *Der Naturforscher (The Natural Investigator)* (1725–1778); and, most important of all, the monthly journal first published in 1771 as *Observations on Physics, Natural History, and the Arts,* but usually referred to as "Rozier's journal." Its founder was the abbé François Rozier (1734–1793). Rozier's journal was designed to supplement the publications of the academy and offer scientists something the academy could not: prompt publication. Rozier's journal was capable of publishing papers the next month after they were received, a standard that compares well to modern journals. Antoine-Laurent Lavoisier availed himself of Rozier's journal to publish short versions of his memoirs, before the academy would publish them, several times. The journal covered all areas of science outside mathematics and mathematical physics. Founded somewhat later, in 1795, was Johann Christian Reil's *Archive for Physiology,* which was the first medical journal primarily concerned with theory rather than medical advice or case histories.

A somewhat different case was the French *Annales de Chémie,* founded in 1789 not just as a chemical journal, but one with a specific agenda: to promote Lavoisier's new nomenclature and new theories. A German journal with the same purpose was Alexander Nicholas Scherer's (1771–1824) *Journal of Chemistry,* founded in 1798. It eventually absorbed Crell's phlogistic rival in 1804. Another reason for specialized journals was the rise of specialized scientific societies, many of which published journals. In Britain the Linnean Society and the Royal Geographical Society published journals, which became the most important in their fields and drained some of the best papers from *Philosophical Transactions.*

The form of journal publication was also changing in the late eighteenth and early nineteenth centuries. Experimental descriptions grew more elaborate and were sometimes preceded by reviews of the literature. The epistolary form, commonly used in *Philosophical Transactions,* was being replaced by more formal, third-person narratives.

See also Academies and Scientific Societies; Crell, Lorenz Florens Friedrich von; Popularization; Reil, Johann Christian.

References

Atkinson, Dwight. *Scientific Discourse in Historical Perspective: The Philosophical Transactions of the Royal Society of London, 1675–1975.* Mahwah, NJ: L. Erlbaum Associates, 1999.

Broman, Thomas. "J. C. Reil and the 'Journalization' of Physiology." In *The Literary Structure of Scientific Argument,* edited by Peter Dear, 13–42. Philadelphia: University of Pennsylvania Press, 1991.

Hufbauer, Karl. *The Formation of the German Chemical Community, 1720–1795.* Berkeley: University of California Press, 1982.

Kronick, David A. *A History of Scientific and Technical Periodicals: The Origins and Development of the Scientific and Technical Press, 1665–1790.* 2d ed. Metuchen, NJ: Scarecrow Press, 1976.

Pyenson, Lewis, and Susan Sheets-Pyenson. *Servants of Nature: A History of Scientific Institution, Enterprises, and Sensibilities.* New York: W. W. Norton, 1999.

Phlogiston

The chemical concept of phlogiston was most influentially set forth by Georg Ernst Stahl, building on the work of previous German chemists and alchemists, particularly Johann Joachim Becher (1635–1682). Stahl renamed the substance that Becher had called "fatty earth" as *phlogiston,* a term Becher had coined but used only occasionally. For Stahl, phlogiston was the principle of flammability. He understood combustion as the loss of phlogiston. When a metal was exposed to extreme heat, the resulting calx (now called the oxide) was what was left from the original combination of the metal and phlogiston. When a calx was heated with charcoal, it reabsorbed phlogiston from it. Charcoal itself was nearly pure phlogiston, which explained why it left so little residue when burnt. Air fully saturated with phlogiston could absorb no more and therefore could not support combustion. An object also stopped burning when all its phlogiston had been released. Phlogiston also circulated living things, being absorbed by plants and then eaten by animals. Stahl also explained other chemical phenomena, like acidity, in terms of phlogiston.

The problem with phlogiston was that calxes actually weighed more than the metal from which they were derived, suggesting that they had not lost but gained something. Stahl treated this as a minor problem, and weighing techniques were not so precise that minor fluctuations in weight were a central chemical issue yet. Such was the power of the phlogiston theory, and Stahl's prestige in German science, that the phlogiston theory was quickly adopted in Germany and Britain. Its adoption in France was slower, but it was effectively promoted by Guillame-François Rouelle (1703–1770) in his lectures at the Royal Botanical Garden in the middle of the century. Phlogiston was incorporated into the new pneumatic chemistry originating mostly in Britain during the late eighteenth century. Joseph Priestley described the new air he had discovered, now called oxygen, as dephlogisticated air, explaining that it promoted combustion by readily absorbing phlogiston. Henry Cavendish suggested that hydrogen—"inflammable air"—was water saturated with phlogiston, and oxygen, water deprived of phlogiston. The concept of phlogiston, however, lost some consistency. Stahl had originally conceived of it as a kind of earth, but it was sometimes treated as an immaterial principle or subtle fluid.

It was a Frenchman, Louis-Bernard Guyton de Morveau (1737–1816), who in a precise series of experiments published as *Academic Digressions* (1772) demonstrated that metals all gained weight when strongly heated in air. Guyton de Morveau was a phlogistonist, however, and suggested that phlogiston was actually of negative weight—an assumption difficult to reconcile with contemporary physics. Reading Guyton de Morveau's essay was one factor driving Antoine-Laurent Lavoisier to question the phlogiston theory of combustion entirely. Another problem was that the calx of mercury could be decomposed back into the metal without the use of charcoal, thus making it difficult to see where the phlogiston was supposed to be coming from—a fact experimentally demonstrated by the Parisian apothecary Pierre Bayen (1725–1798). Lavoisier's problem with phlogiston was not merely its irreconcilability with specific experimental results, but the looseness of the term itself. This was a long-standing grievance of Lavoisier, who went out of his way to dissociate himself from terms like *dephlogisticated air.* Phlogiston, he claimed, was defined differently and ascribed different properties, such as weight and weightlessness, by different chemists, often on an ad hoc basis to explain a particular process. Lavoisier's antiphlogistonic chemistry, first openly set forth in 1785, quickly won over leading French chemists such as Guyton de Morveau and Claude-Louis Berthollet. In Britain Priestley remained a holdout, and progress was slower in Germany, where phlogiston chemistry was identified as German and Lavoisier's as French. By the publication of the ten-volume textbook *A System of Chemical Knowledge* (1800) by Lavoisier's disciple Antoine-François de Fourcroy (1755–1809), phlogiston had clearly been defeated, although mop-up operations such as the revision of pharmacopoeias to reflect the new chemistry continued for decades.

See also Chemistry; Lavoisier, Antoine-Laurent; Priestley, Joseph; Stahl, Georg Ernst.

Reference

Crosland, Maurice P. *Historical Studies in the Language of Chemistry.* Rev. ed. Mineola, NY: Dover, 1978.

Physics

As the eighteenth century began, the word *physics* had the broad meaning that could be traced back to Aristotle (384–322 B.C.) of a science of all natural occurrences. Its meaning narrowed over the course of the century, until by the early nineteenth century the modern meaning was current, although the older, broader conception continued to be used. The disciplines now known as physics can be roughly divided into two categories during the Enlightenment: mathematical and experimental physics. This was an unstable barrier, as experimental disciplines became increasingly mathematized. The mathematized disciplines at the beginning of the eighteenth century were mechanics, terrestrial and celestial, and geometric optics. In the course of the century these disciplines advanced principally through the application of more and more sophisticated mathematical tools, principally by French and Swiss scientists. The experimental disciplines, which became increasingly prominent in the eighteenth century, had to do with phenomena with an obvious mechanical explanation such as heat, light, and electricity. Although scientists from all over Europe and even America contributed to experimental physics, its leading practitioners were the British and, in the early eighteenth century, the Dutch, supported by the strength of their experimental tradition and the quality of their scientific instruments. Experimental physics did not just advance due to its intellectually interesting problems. It was also the bread and butter of the professional lecturer or demonstrator who carried out much physical research, particularly in the early part of the century. Some, indeed, thought physics rather a frivolous discipline.

The two dominant schemes of interpreting experimental phenomena both derived from the "Queries" appended to the later editions of Isaac Newton's *Opticks* (1704). One was based on the short-range attractions and repulsions between particles, which could work like gravity. This school traced itself to the "Queries" appended to the second edition of the *Opticks,* the 1706 Latin translation. The second drew on the further "Queries" appended to the 1717 English edition (which also included the 1706 "Queries"). It explained experimental phenomena through "ethers." Ethers were an ancient concept. As Newton and subsequent physicists defined them, they were extremely subtle media, not directly perceivable, that mediated physical phenomena, including gravity. They were usually thought of as being composed of extremely minute particles that repelled each other, rather than being continuous. Generally, heat, light, electricity, gravitation, and so on were thought to be mediated by different ethers that did not interact with each other. These ethers were sometimes referred to as "subtle fluids" and fulfilled many of the same purposes for Newtonians that "subtle matters" had for Cartesian physicists. This similarity was accentuated as Cartesian physics, which continued well into the second half of the eighteenth century, drew closer to Newtonianism until a Cartesian like the electrician Jean-Antoine Nollet could accept much of Newtonian theory.

The 1717 "Queries" had little immediate impact, as Newtonians continued to concentrate on particle interactions, particularly attractions. Interest in ethers began to increase after around 1740, sparked particularly by developments in electricity and the widespread influence of Hermann Boerhaave's doctrine of a universal fire. In 1744 two early letters by Newton that spoke of an ether were published, lending further Newtonian endorsement to ethers. *Sir Isaac Newton's Account of the Ether* (1745) by Bryan Robinson (1680–1754) was one of the most influential among many expositions of Newtonian ether theory. Ethers and "subtle fluids," like that which was the basis of Benjamin Franklin's electrical theory, had the advantage of providing causal explanations for physical phenomena. Heat was also explained through a subtle fluid, caloric, and in chemistry, phlogiston was sometimes ascribed the properties of a subtle fluid. These fluids were often seen as "imponderable," or weightless, as heated, electrified, or magnetized objects did not seem to gain or lose weight.

Although the ethers and subtle fluids provided neat physical explanations and seemed to lend themselves to quantification, they were open to doubt on empirical grounds. They could not be directly perceived or isolated. They also seemed somewhat arbitrary, and by the end of the century the structure of fluid theory seemed excessively elaborate, with one or two electrical fluids, magnetic fluids, caloric, and light particles. By the late eighteenth century there was a revival of physical theories based on particle interaction. One inspiration for this revival was Ruggiero Giuseppe Boscovich's atomism, which defined the atom as a locus of forces more than as a "Newtonian" physical body with extension. Boscovich's atomism was taken up by British scientists, notably Henry Cavendish, Joseph Priestley, and William Herschel. Opposition to fluid theory continued at the very end of the eighteenth century with a group associated with the newly founded Royal Institution. Benjamin Thompson, Count Rumford, denied the fluid of heat (although he did retain an ether whose vibrations sent particles of matter in motion, producing heat), Sir Humphry Davy defined electricity in terms of chemical interaction rather than electrical fluid (temporarily removing electricity from physics to chemistry), and Thomas Young (1773–1829) revived the seventeenth-century theory of light as a wave.

The other late-eighteenth-century development was the rise of mathematization of experimental physics. Mathematization had been a great success in mechanics, and its application to other branches of physics was a long-held goal. The elusiveness of experimental phenomena and the fundamental indifference to precision and measurement of experimental physicists like Nollet had frustrated it. Improvements in instruments seemed to offer the hope of better quantitative physics—Georg Christoph Lichtenberg suggested in 1784 that the whole of physics be redone with modern equipment. Joseph Black and Johann Carl Wilcke (1732–1796) used improved thermometers and the discrepancies they revealed to establish the concept of a quantitatively defined latent and specific heat, the most important developments in heat theory of the century and not dependent on the acceptance of the caloric theory. The most dramatic examples of mathematization occurred in the fields of electricity and magnetism, specifically in the work of Cavendish and Charles-Augustin de Coulomb.

The most ambitious attempt to remake physics in light of the original Newtonian program of particle interactions was that of the early-nineteenth-century "Laplacian" school, a French group headed by Pierre-Simon de Laplace. Members of the school, including Siméon-Denis Poisson (1781–1849) and Jean-Baptiste Biot (1774–1862), hoped to create a unified physics, subsuming even analytical mechanics, by an unprecedentedly mathematically rigorous treatment and precise measurement of short-range attractions and repulsions between particles (although they did not abandon the caloric theory of heat). This program was put forth in Biot's *Treatise on Experimental and Mathematical Physics* (1816) and led to some important individual achievements, like Poisson's mathematical treatment of electricity. However, it was also challenged, particularly by the revival of the wave theory of light by Young and Augustin-Jean Fresnel (1788–1827). Although the Laplacian program was a failure, it was important in providing a unified idea of the nature and scope of physics.

See also Alembert, Jean Le Rond d'; Ampère, André-Marie; Atomism; Avogadro, Amedeo; Bassi, Laure Maria Catarina; Bernoulli Family; Boscovich, Ruggiero Giuseppe; Cartesianism; Cavendish, Henry; Châtelet, Gabrielle-Émilie du; Coulomb, Charles-Augustin de; Desaguliers, John Theophilus; Electricity; Euler, Leonhard; Franklin, Benjamin; Gauss, Carl Friedrich; Germain, Sophie; Hauksbee, Francis; Heat; Hutchinsonianism; Lagrange, Joseph-Louis; Laplace, Pierre-Simon de; Lichtenberg, Georg Christoph; Lomonosov, Mikhail Vasilyevich; Maupertuis, Pierre-Louis Moreau de; Mechanics; Newtonianism; Nollet, Jean-Antoine; Optics; Ørsted, Hans Christian; Poliniere, Pierre; Ritter, Johann Wilhelm; Technology and Engineering; Thompson, Benjamin (Count Rumsford); Volta, Alessandro Giuseppe Antonio Anastasio; Wolff, Christian.

References

Carter, G. N., and M. J. S. Hodge, eds. *Conceptions of Ether: Studies in the History of Ether Theories, 1740–1900.* Cambridge: Cambridge University Press, 1981.

Hankins, Thomas L. *Science and the Enlightenment.* Cambridge: Cambridge University Press, 1985.

Harman, P. M. *Energy, Force, and Matter: The Conceptual Development of Nineteenth-Century Physics.* Cambridge: Cambridge University Press, 1982.

Heilbron, J. L. *Electricity in the 17th and 18th Centuries: A Study in Early Modern Physics.* 2d ed. Mineola, NY: Dover, 1999.

Schofield, Robert E. *Mechanism and Materialism: British Natural Philosophy in an Age of Reason.* Princeton: Princeton University Press, 1970.

Physiology

The eighteenth century was the scene of a long conflict between mechanistic and vitalistic physiology. The champions of the mechanistic school, the Cartesian Friedrich Hoffman (1660–1742) and the Newtonian Hermann Boerhaave, interpreted the working of the human and other animal bodies as the circulation of fluids governed by mechanical principles. The nerves, for example, were fine hollow tubes through which circulated a subtle liquid secreted by the brain, which Boerhaave believed to be a gland. Hoffman claimed that nervous fluid was forced from the blood into the nerves in the brain. The proper movements of these fluids determined the body's life, death, and state of health. The rival vitalist school, led by Hoffman's colleague at the University of Halle, Georg Ernst Stahl, held that the functioning of living beings was not explainable in solely mechanical or chemical terms. The body was "animated" by a soul or anima. All of these leaders of human physiology were physicians who taught at medical schools. The later eighteenth century would see a growing interest in physiology among scientists who were not medical professionals. Plant and animal physiology was considered part of the physical sciences.

Beginning around 1740 physiology became a more experimental discipline, and physiologists grew more interested in chemical and electrical phenomena than in the purely mechanical. The most important physiologist of the midcentury was Boerhaave's student Albrecht von Haller, author of the eight-volume compendium *Physiological Elements of the Human Body* (1757–1766). Haller made extensive use of experiment and made a number of physiological discoveries, of which the most important was what he called "irritability." Haller viewed the body as composed of fibers, the three types of which were connective, muscular, and nervous. Only the nervous fibers had the ability to sense pain, so they were "sensitive." However, other parts of the body, such as most of the muscular fibers and particularly the heart, reacted when exposed to chemical, electrical, or mechanical stimuli. Haller's extremely influential physiology blended mechanical and vitalist elements, but with the emphasis on mechanism. The tendency in the mid-eighteenth-century vitalism was for matter itself to take on vital qualities—Stahl's use of the "soul" was regarded as suspiciously theological. Another of Boerhaave's students, Julien Offroy de La Mettrie, took the opposite course to Haller, presenting a physiology that was strictly mechanical, but not experimental. His *L'Homme-machine* (1747) had more influence on philosophical debates on materialism than on physiology itself.

The experimental trend in physiology was advanced by René-Antoine Ferchault de Réaumur (1683–1757). His experiments on the digestive process in birds, which involved feeding live birds different substances in tubes or containers, established the relative digestibility of different kinds of food—for example, that meat was more easily digested than starches. By getting a bird to swallow a sponge on the end of a string, which Réaumur then used to pull the sponge out of the bird's stomach, he obtained a nearly pure sample of the gastric juices. Similar digestive experiments were carried out by John Hunter in England and Lazzaro Spallanzani in Italy. The British physician William Hewson (1739–1774) established the presence of a "coagulable lymph" in the blood, necessary for it to clot. (Previous thinking on clotting had viewed it as a process involving the cooling of the blood itself.)

Late-eighteenth-century changes in chemistry and experimental physics directly affected physiology. Pneumatic chemistry, building on the discovery that the air was composed of a number of different gases, raised new questions about the role of air. Antoine-Laurent Lavoisier established that respiration and combustion were processes involving a single substance, the newly discovered gas Lavoisier called "oxygen." Electrical science was also applied to physiology, as in the work of a group of mostly Italian experimenters. The Padua medical professor Leopoldo Caldani (1725–1813) used a Leiden jar to discover the ability of electricity to cause contractions in the skeletal muscles. Luigi Galvani's famous frog experiments seem to demonstrate the existence of an "animal electricity." This was disproved by another Italian, Alessandro Volta, but not to the satisfaction of all.

The year 1795 saw the foundation of the

first specialist journal devoted to physiology, the German *Archive for Physiology,* founded by the Halle medical professor Johann Christian Reil. Reil and the *Archive* presented physiology as a master system for medical knowledge and emphasized the importance of understanding the chemical processes that went on in the body. The journal, along with much of German physiology, was transformed in the early eighteenth century by *Naturphilosophie.* The *Naturphilosophs'* concept of polarity became an organizing concept for physiologists, reaching an extreme in the works of the poet-physician Joseph Gorres (1776–1848), whose *Exposition of Physiology* (1803) and other works set forth an elaborate system of parallels between physiological and cosmic forces. Active argument about the function of organs gave way to descriptions of the harmonies and oppositions of the body, and German physiology in large part lost Reil's interest in chemical processes and turned into anatomical description. It also drifted further from medical practice.

French developments were strikingly different. The vitalist physiology of Marie-François-Xavier Bichat and his successors remained tied to medicine. His *Treatise on Membranes* (1800) broke with the traditional emphasis on organs to discuss the body in terms of tissues. The creation of the organs of the body from a relatively small number of tissue types permitted Bichat and subsequent French physiologists to take an analytic and experimental rather than descriptive approach to the organs and their functioning, focusing on investigations of specific processes rather than constructing theoretical systems. Subsequent French physiologists included François Magendie (1783–1855), a pioneer in the quantitative study of the effects of drugs on the body and the discoverer of the cerebrospinal fluid. Paris became the leading center of experimental physiology. The Academy of Sciences established the Montyon Prize for Experimental Physiology in 1818, and Magendie founded the *Journal of Experimental Physiology and Pathology* in 1821.

See also Bichat, Marie-François-Xavier; Diderot, Denis; Embryology; Galvani, Luigi; Haller, Albrecht von; Hunter Family; Materialism; Medicine; *Naturphilosophie;* Plant Physiology; Polyps; Reil, Johann Christian; Sexual Difference; Spallanzani, Lazzaro; Vitalism.

References

Broman, Thomas. "J. C. Reil and the 'Journalization' of Physiology." In *The Literary Structure of Scientific Argument,* edited by Peter Dear, 13–42. Philadelphia: University of Pennsylvania Press, 1991.

Hankins, Thomas L. *Science and the Enlightenment.* Cambridge: Cambridge University Press, 1985.

Lesch, John E. *Science and Medicine in France: The Emergence of Experimental Physiology, 1790–1855.* Cambridge: Harvard University Press, 1984.

Rothschuh, Karl E. *History of Physiology.* Translated and edited by Guenter B. Risse. Huntington, NY: Robert E. Krieger, 1973.

Plant Physiology

Although seventeenth-century scientists had made some attempts at mechanical or alchemical explanations of the living processes of plants, studies of plant physiology emerged during the Enlightenment with the English clergyman Stephen Hales and his *Vegetable Staticks* (1727). Hales's intellectual background was in Newtonian physics and animal physiology, and he did not concern himself with the central questions of the botanical tradition: the identification and classification of plants and the ascertainment of their medical uses. His experiments on the movement of sap, the absorption of water, and the expiration of gas from leaves disproved the idea of sap circulating like blood, and established the consumption of air by plants. Some of his ideas were less successful; Hales's claim that leaves absorbed water from the air was refuted experimentally by Jean-Étienne Guettard (1715–1786), who also improved on Hales's method for measuring the liquid "perspired" by leaves.

Hales's work was further refined with the emergence of pneumatic chemistry, which enabled more precise identification of the gases absorbed and given off by plants. Joseph

Priestley discovered that plants functioned as the complete opposite of animals by making "bad" air "good," absorbing "fixed air" (carbon dioxide) and giving off "dephlogisticated air" (oxygen). This fitted in well with natural theology, as an example of the harmony of God's creation. The Dutch botanist and physician Jan Ingenhousz (1730–1799) built on Priestley's work in his *Experiments on Vegetables* (1779), demonstrating that only the green parts of plants give off dephlogisticated air, and that they do so only in the presence of light. (In the presence of darkness fixed air is given off instead, although in far lesser quantities.) His widely circulated books inspired further experimentation by the Genevan clergyman and amateur chemist Jean Senebier (1742–1809), who established that the production of dephlogisticated air was dependent on the presence of fixed air, although he believed that it was taken into plants primarily as dissolved in water rather than in the atmosphere. Senebier also established the relation between the volume of oxygen given off by a plant and the intensity of light to which it is exposed. Ingenhousz's and Senebier's work was initially expressed in terms of the old phlogiston chemistry, but was adaptable into the new chemistry of Antoine-Laurent Lavoisier. Ingenhousz himself expressed his analysis in terms of the new chemistry in *On the Nutrition of Plants* (1796), and Senebier in *Vegetable Physiology* (1800).

Another Genevan and an acquaintance of Senebier, Nicolas-Théodore de Saussure (1767–1845), carried out the most extensive and precise experiments on plant physiology yet. He also established the fact that plants respire, give off carbon dioxide, in the light as well as in the dark, and that this is an essential process for their existence. Even more important, he demonstrated experimentally that the absorption of carbon dioxide was followed by the assimilation of the carbon and the release of the oxygen. He established that the nitrogen found in plants is not atmospheric, but, dissolved in water, enters the plant through the roots. Chemical analysis of ash enabled de Saussure to establish the chemical composition of plants, mostly carbon, hydrogen, oxygen, and nitrogen, and the relation of the chemical composition of individual plants to the soil in which they had grown. His work was published in *Chemical Researches on Vegetation* (1804).

See also Botany; Hales, Stephen; Priestley, Joseph.

References

Delaporte, François. *Nature's Second Kingdom: Explorations of Vegetality in the Eighteenth Century.* Translated by Arthur Goldhammer. Cambridge: Massachusetts Insitute of Technology Press, 1982.

Morton, A. G. *History of Botanical Science: An Account of the Development of Botany from Ancient Times to the Present Day.* London: Academic Press, 1981.

Poliniere, Pierre (1671–1734)

Pierre Poliniere was Paris's foremost Cartesian demonstrator in the early eighteenth century. From the provinces, Poliniere arrived in Paris in the 1690s and became interested in mathematics and experimental philosophy. In 1695 he was offering a two-month experimental course in conjunction with the philosophy faculty, and later the medical school, of the University of Paris. In 1709 he published his lectures as *Experiments in Physics* and shortly afterward moved his lectures from college halls to private venues. Such was the impact that Poliniere's demonstrations made on Paris's elite that in 1722 he was invited to present them before the young King Louis XV (r. 1715–1774). Poliniere's physics was Cartesian, and in the early days of his lecturing he combated the still-respectable Aristotelian system. However, Poliniere stressed dramatic experimental phenomena over the explanations for them and did not systematically expound his natural philosophy. He was an expert manipulator of barometers and air pumps, whose demonstrations involved spectacular use of the vacuum. Two marble plates stuck together in the vacuum well enough to be lifted by a hook attached to the top, and an apple exploded as air was removed from a chamber. He also engaged in chemical experimenting, which he referred to as "pyrotechnics" in order to distinguish it from alchemy. The most innovative area of Poliniere's work was in electricity, which he pioneered along with his English contemporary Francis Hauksbee. Poliniere's electrical work was considered important enough to be presented to the Royal Academy of Sciences, although Poliniere himself was never a member of that august institution, and his electrical studies attracted little notice in the academy's official publications.

See also Cartesianism; Electricity; Popularization.

Reference

Sutton, Geoffrey V. *Science for a Polite Society: Gender, Culture, and the Demonstration of Enlightenment.* Boulder: Westview, 1995.

Polyps

One of the most exciting events in mid-eighteenth-century science was the 1744 publication of Abraham Trembley's (1700–1784) study of polyps, now known as hydras, aquatic creatures about a quarter of an inch long. Previous investigators, notably the great microscopist Antoni van Leeuwenhoek (1632–1723), had identified the creatures as plants, but Trembley, a naturalist from Geneva residing in the Dutch Republic, closely observed specimens and discovered that they ate and moved under their own power, defining characteristics of animals. Trembley's really stunning discovery had to do with the creatures' regenerative powers. In some of the earliest studies that can be categorized as experimental biology, Trembley discovered that if a polyp was cut in half, both halves regenerated into a complete creature. If the polyp was divided into many pieces, it regenerated into many polyps. Cutting the head lengthwise while leaving the rest of the body intact produced a polyp with two heads. Eventually, Trembley produced a seven-headed polyp like the Hydra of Greek mythology, leading Carolus Linnaeus to later give *hydra* as the creature's name in his classification system. Finally, if the polyp was actually turned inside out by the use of a bristle inserted within it and then pulled, it regenerated a new outside. This went far beyond previous knowledge of regeneration, that of lizards and crustaceans who could regenerate a severed limb or tail, but not a whole new creature. News of Trembley's discovery began to circulate in 1741, attracting interest all over Europe. Dozens of scholars began chopping thousands of polyps.

Polyps presented a number of intriguing intellectual problems. They had some characteristics of animals, but reproduced asexually like primitive plants. Trembley, and subsequent exponents of the great chain of being such as Trembley's cousin Charles Bonnet, identified them as intermediate between animals and plants. The polyp also called into question preformationist embryology, as in if the polyp was preformed prior to being born, how could it potentially be two polyps? Under the influence of the polyp discoveries, Albrecht von Haller converted from preformationism to the rival doctrine of epigenesis, the belief that reproduction actually involved the formation of a new being. The most fundamental problem the polyps' regeneration presented was one of the nature of the organizing principle of life. If the polyp could create two lives where only one was before, what was its organizing principle or soul, or did one even exist? The polyps seemed to lend support to materialism, by suggesting that matter (in this case, the pieces of the polyp's body) had the ability to organize itself. The polyp discoveries contributed to the rise of materialism, particularly in France, during the later eighteenth century.

See also Embryology; Haller, Albrecht von; Materialism; Physiology; Vitalism.

Reference

Hankins, Thomas L. *Science and the Enlightenment.* Cambridge: Cambridge University Press, 1985.

Popularization

The Enlightenment saw a great expansion in the volume of scientific popularization. Science was popularized for many different, and sometimes opposing, motives. Natural theologians popularized science to encourage wonder at the works of God, whereas anticlericals and materialists popularized for precisely the opposite reason. Some popularizers claimed that the science they presented could be of practical use in a range of activities from manufacturing to agriculture, while others emphasized the amusing, entertaining, or improving qualities of the science they presented. Medical writers, whether orthodox physicians, innovative scientists, or quacks, offered theories of the body and its functioning with the purpose of improving the readers' health.

Science was popularized in many forums and literary genres. Many older genres continued to be popular in the eighteenth century. The Boyle Lecturer and natural theologian William Derham (1657–1735) published his lectures, essentially sermons, in two collections called *Physico-Theology* (1713) and *Astro-Theology* (1715), both of which gave considerable scientific information in the course of their theological arguments. A newer medium that served as a vehicle for popularization was the periodical press. The French *Journal des Scavans* presented a variety of news from the scientific world in a form understandable to educated French men and women, as well as the considerable European audience that read French. In Britain the *Ladies Diary* presented a famous series of mathematical problems that vexed some of the country's leading savants. The French philosophes Voltaire and Denis Diderot embodied scientific knowledge in fictional narrative, such as Diderot's *D'Alembert's Dream* (1769) setting forth his "vital materialism" and Voltaire's science fictional *Micromégas* (1752).

The eighteenth and early nineteenth centuries were the golden age of the professional "lecturer" or "demonstrator." These men offered courses or demonstrations on the open market, sometimes completely independently and sometimes under the aegis of an educational or scientific institution. They prospered according to their ability to combine scientific expertise with dramatic presentation. Early-eighteenth-century lecturers were divided into English and Dutch Newtonians, such as John Theophilus Desaguliers and Pieter van Musschenbroek (1692–1761), and French Cartesians, such as Pierre Poliniere. Despite their philosophical differences, all made use of similar equipment. Lecturers provided an important market for scientific equipment such as air pumps, Leiden jars, and orreries—an eighteenth-century innovation that represented the planets and their orbits by metal balls moved by clockwork on circular tracks around a large ball representing the Sun. (The circularity of the tracks, as opposed to the actual elliptical orbits of the planets, is testimony to the importance of ease of use over strict scientific accuracy.) Equipment tended to grow more elaborate as the century went on—the late-eighteenth-century English lecturer Adam Walker (1731–1821) possessed an orrery twenty feet in diameter, with luminous globes. Many lecturers also published their lectures as textbooks.

This type of scientific lecturing reached its apogee in early-nineteenth-century London, with the founding of the Royal Institution in 1799 as a venue for lectures and presentations, at first to workers and artisans but quickly reorienting its mission to middle-class audiences. Sir Humphry Davy's great reputation in England was founded nearly as much on his spectacular and entertaining demonstrations at the Royal Institution as on his scientific discoveries. Davy was part of a movement whereby chemists imitated the experimental physicists who had dominated the world of demonstrators in the eighteenth century.

Scientific popularization was not merely an affair of reading texts or viewing experiments and demonstrations. Popularizers encouraged the performance of experiments at home as "rational amusement." Electrical experiments were particularly popular, especially after the invention of the Leiden jar opened the door to all sorts of spectacular effects. Leading scientists such as Joseph Priestley and the abbé Jean-Antoine Nollet published directions for home experiments. Another form of popularized scientific activity was the making of botanical or natural-historical collections. The success of the Linnean system of botanical classification was greatly aided by its accessibility to amateurs seeking a relatively easy way to classify their collections or identify species in the wild.

Some scientists distrusted popularization because it gave people a superficial knowledge, which they then confused with real scientific knowledge. This was particularly true in physics and mechanics, where the hard but essential mathematical aspects of the subject were often skipped in favor of analogies and demonstrations. However, many scientists also wrote popularizations, such as Leonhard Euler's three-volume *Letters to a German Princess* (1768–1772). Priestley's works were directed at a broad audience of persons interested in science rather than an audience of scientists only. He viewed the wide dissemination of scientific knowledge as a means of advancing society generally by dispelling superstition and liberating humanity from illegitimate authority, intellectual and institutional. This project, connected with Priestley's theological millenarianism, would arouse great suspicion in Britain after the French Revolution.

Women occupied a particularly important role as an audience for popularizers. While barred from membership or even spectatorship at the meetings of scientific academies such as the Royal Society, women were usually admitted to lectures and demonstrations. Many works, notably Euler's *Letters* and Francesco Algarotti's (1712–1764) *Newtonianism for Ladies* (1737), first published in Italian and translated into many European languages, were specifically addressed to them. Both followed a formula originated in *Conversations on the Plurality of Worlds* (1686) by Bernard Le Bovier de Fontenelle (1657–1757), which despite its Cartesianism was frequently reprinted in the eighteenth century. This formula, also followed by Diderot, was that of an authoritative male interlocutor addressing an intelligent but unlearned female inquirer. By the early nineteenth century women were playing an increasing role as popularizers themselves. Jane Marcet (1769–1858), one of the many popularizers of chemistry in Davy's wake, wrote the very successful *Conversations on Chemistry* (1805) following the same classic form of the dialogue by which male-generated knowledge was presented to a female student, although in Marcet's case both interlocutors were female.

See also Buffon, Georges-Louis Leclerc de; Darwin, Erasmus; Desaguliers, John Theophilus; Diderot, Denis; Education; Encyclopedias; Freemasonry; Hauksbee, Francis; Illustration; Instrument Making; Leiden Jars; Mesmerism and Animal Magnetism; Natural Theology; Periodicals; Poliniere, Pierre; Priestley, Joseph; Whiston, William; Women.

References

Darnton, Robert. *Mesmerism and the End of the Enlightenment in France.* Cambridge: Harvard University Press, 1968.

Golinski, Jan. *Science As Public Culture: Chemistry and Enlightenment in Britain, 1760–1820.* Cambridge: Cambridge University Press, 1992.

Jacob, Margaret. *Scientific Culture and the Making of the Industrial West.* New York: Oxford University Press, 1997.

Porter, Roy. *The Creation of the Modern World: The Untold Story of the British Enlightenment.* New York: W. W. Norton, 2000.

Sutton, Geoffrey V. *Science for a Polite Society: Gender, Culture, and the Demonstration of Enlightenment.* Boulder: Westview, 1995.

Priestley, Joseph (1733–1804)

Joseph Priestley combined great achievements in chemistry with passionate and unorthodox religion. He came from a family of

Protestant Dissenters in the north of England and was educated for the ministry at a Dissenting Academy in Daventry. He held a series of positions as minister to Nonconformist congregations and as schoolmaster, including a stint at the famous Warrington Academy from 1761 to 1767.

Priestley's interest in experimental philosophy led him to research a book on the history of electricity, then a topic of great public interest. He joined the Royal Society in 1766, and the next year he published *History and Present State of Electricity,* which endorsed the theories of his friend Benjamin Franklin. His electrical research led him to one important new discovery: that charcoal was an electrical conductor. The book was a success, which supported Priestley's belief that the history of experimental science was a subject with an unrealized potential. His next major work, *The History and Present State of Discoveries Relating to Vision, Light, and Colours* (two volumes, 1772), was conceived as the first volume in a series, *The History of All the Branches of Experimental Philosophy.* However, the book was a commercial failure, causing Priestley to abandon the project.

A brilliant chemist, fiery minister, radical politician, and diligent historian, Joseph Priestley (shown here in a painting by Gilbert Stuart) represented the apotheosis of the eighteenth-century enlightened man. (Dictionary of American Portraits/Dover Publications)

Priestley's main contribution to eighteenth-century science was in the area of pneumatic chemistry, a rapidly expanding field of knowledge in midcentury Britain. His serious interest in the subject began in the late 1760s, and his first publication in the field was *Directions for Impregnating Water with Fixed Air* (1772), which showed how the "fixed air" (carbon dioxide) recently discovered by Joseph Black could be forced into water to make, Priestley suggested, an artificial form of mineral water. This proved extraordinarily popular. The same year he published an article in *Philosophical Transactions* on how "nitrous air" (nitric oxide), which he had named after producing it by adding spirit of niter to metal, could be used as a test to determine the "goodness" of a sample of air—its ability to support respiration or combustion. This article won him the Royal Society's Copley Medal in 1773. (Priestley later developed an instrument, the eudiometer, for measuring the goodness of air, which had a short-lived vogue.) Priestley also discovered that noxious air could be made "good" again by putting plants in it.

In 1773 Priestley took up the not very demanding duties of a librarian and tutor in the household of William Petty, earl of Shelburne (1737–1805), a leading British statesman who wished to support Priestley's scientific activities. He made his greatest discovery there in 1774, although it took some time for him to realize its significance. This was "dephlogisticated air" (oxygen). (Priestley was actually preceded in this discovery by Carl Wilhelm Scheele.) Priestley obtained this new air from the calx (oxide) of mercury. Experiments demonstrated that this air supported combustion and respiration, not merely as well as common air, but, to Priestley's amazement, even better than common air. He explained this remarkable fact by the phlogiston theory, claiming that this new air was completely devoid of phlogiston. Thus, it absorbed phlogiston with extraordinary ease. Priestley's growing scientific reputation was further enhanced by a trip he took with Shelburne to the European continent. In Paris he encountered some of the leading French chemists, including his future arch-rival Antoine-Laurent Lavoisier. He was shocked to discover that not merely were the leading French scientists not Christian believers, but they had great difficulty in accepting that he was. During his period with Shelburne, he also obtained "vitriolic acid air" (sulfur dioxide) by the application of "oil of vitriol" (sulfuric acid) to charcoal and worked on obtaining more pure acids than were available. Priestley's major discoveries were set forth in two series of three volumes, *Experiments and Observations on Different Kinds of Air* (1774, 1775, 1777) and *Experiments and Observations Relation to Various Branches of Natural Philosophy: With a Continuation of the Observations on Air* (1779, 1781, 1786).

In 1780 Priestley left Shelburne's household to take up a position as minister at Birmingham's New Meeting. The retirement of the president of the Royal Society, Sir John Pringle (1707–1782), a supporter of Priestley and pneumatic chemistry, in 1778, and his replacement by Sir Joseph Banks, whose interests lay more in natural history, further encouraged Priestley to base himself outside London. In Birmingham he became the scientific star of the Lunar Society, which shifted its meeting day from Sunday to Monday to accommodate Priestley's schedule. Wealthy members of the Lunar Society, such as Josiah Wedgwood (1730–1795) and Erasmus Darwin, donated money and equipment to support Priestley's researches.

Priestley was an enormously prolific if not very polished writer. In addition to a myriad of scientific publications, he published books on education and many theological works. His books on experimental philosophy were addressed to a broad audience rather than specialists and were among the most popular works of the genre. Priestley's philosophy of science was Baconian in emphasizing experiment and data gathering over theory, and in its suspicion of claims to exclusive knowledge. He saw his works in the context of a general rational enlightenment, with applications in politics and religion. His writings emphasized the cheapness of his equipment, opening to his readers the possibility of duplicating his experiments or extending natural knowledge even further. He encouraged public lecturers to disseminate his results, offering them training in his laboratory. This emphasis on the openness of science to all led him to oppose the claims of Lavoisier's antiphlogistic "new chemistry." Lavoisier's experiments depended on very expensive and finicky equipment, like the calorimeter, and could not be widely duplicated like Priestley's. Priestley also believed that Lavoisier had simply not performed enough experiments before starting his theorizing.

Priestley led a long and increasingly isolated battle against the new chemistry. He argued against the new chemists that oxygen and "inflammable air" could form nitrous oxide as well as water, and that "inflammable air" could be produced from the heating of iron calx with charcoal. Contemporary chemists judged that the production of nitrous oxide was a result of experimental contamination, and in 1801 William Cruikshank identified the "inflammable air" produced by Priestley's calx experiment as a new gas, an oxide of carbon (carbon monoxide). However, Priestley remained undaunted, the last major chemist to support the phlogiston theory.

Priestley's religious views grew more radical over the course of his career. He became one of Britain's best-known Unitarians, denying the divinity of Jesus Christ (although continuing to believe that he was the Messiah). This view was technically illegal. He was also a fervent millenarian, interpreting the events of the French Revolution and the rise of Napoléon as signs of the forthcoming Apocalypse. Priestley was unusual among Christians in being a philosophical materialist, but claimed that materialism provided a better support for Christianity than supernaturalism. For example, emphasizing humans' completely material nature made their resurrection, the Last Judgment, a true miracle, rather than simply the survival of a soul already by its nature immortal. Because matter accounted for all the things that spirit had been traditionally invoked for, such as thought, Priestley required that matter be active and forceful, rather than inert. He endorsed the "point-atom" matter theory of Ruggiero Giuseppe Boscovich.

The reaction to the French Revolution made Priestley's intellectual and political radicalism increasingly dangerous. In 1791 a politically conservative "Church and King" mob attacked his dwelling, destroying much of his scientific equipment. The mob acted with the tacit support of local magistrates and Church of England clergy. Priestley left Birmingham, and with his family and what possessions were left to him took up residence in the London suburb of Hackney, emigrating to the United States in 1794. There he turned down an offer to be professor of chemistry at the University of Pennsylvania and established what he hoped would become a colony for exiled English radicals in Northumberland, Pennsylvania. Priestley's last years were spent in an exile from the centers of science both intellectual, as a phlogistonist, and physical. The new chemistry had preceded him to America, and he engaged in the last scientific controversy of his life defending phlogiston chemistry from the American chemists John McLean of Princeton and James Woodhouse (1770–1809), a chemical lecturer in Philadelphia. Priestley's scientific legacy to his adopted country was his disciple and executor Thomas Cooper (1759–1839), who became a prominent chemistry teacher.

See also Chemistry; French Revolution; Lavoisier, Antoine-Laurent; Lunar Society of Birmingham; Materialism; Natural Theology; Phlogiston; Popularization; Religion.

References

Brock, William H. *The Norton History of Chemistry.* New York: W. W. Norton, 1993.

Gibbs, F. W. *Joseph Priestley: Revolutions of the Eighteenth Century.* Garden City, NY: Doubleday, 1967.

Golinski, Jan. *Science As Public Culture: Chemistry and Enlightenment in Britain, 1760–1820.* Cambridge: Cambridge University Press, 1992.

Tapper, Alan. "The Beginnings of Priestley's Materialism." *Enlightenment and Dissent* 1 (1982): 73–82.

Probability

Mathematical probability in the eighteenth century developed as a science of practical application. It was conceived to be of use in such disparate areas as gambling, insurance (not so separate from gambling as it is today), medicine, and even politics. Enlightenment probabilists generally related their subject to the expectations of a hypothetical "reasonable man," rather than viewing it as something purely "out there," independent of human subjectivity.

Several significant mathematicians worked on probability questions in the early eighteenth century. The most important was Jakob Bernoulli, whose posthumously published *Art of Conjecturing* (1713) put forth the earliest version of what came to be known as the law of large numbers. Bernoulli demonstrated that the larger the number of trials, the closer the results will be to the probability of the events. Thus, the greater number of throws of an honest die, the more likely the six numbers will appear an equal number of times. Bernoulli also originated the "urn model," in which probability problems were conceptualized as drawing different colored balls from an imaginary urn, a metaphor still commonly used today, and the first step in abstracting probability from the everyday reality of cards and dice. Abraham de Moivre (1667–1754), a French Protestant living in

England, published *The Doctrine of Chances* (1718). He applied probability to questions of annuities, a popular subject for probabilists in the following decades.

The foremost intellectual antagonist to developing probabilistic mathematics was Jean Le Rond d'Alembert. Probability mathematics was particularly vulnerable to d'Alembert's criticisms due to a failure to come up with a convincing solution to the Petersburg paradox, so-called because Daniel Bernoulli published an analysis in the *Commentaries* of the Imperial Academy of Sciences of St. Petersburg. The Petersburg paradox was an analysis of a game based on coin flips, where one player paid another two coins if heads came up on the first flip, four if it came up on the second, eight on the third, and so on. Probabilistic analysis showed that the expectation of the second player was infinite, and therefore he should pay the first player any amount to play the game. In practice, no one would pay so much. Bernoulli claimed that the paradox was resolved by pointing out that in reality, no one could pay an infinite stake, and therefore the second player's expectation was not really infinite. Georg Christoph Lichtenberg and Georges-Louis Leclerc de Buffon tackled the problem empirically, with repeated trials of coin tossing. After a child in his employ tossed a coin more than 2,000 times, Buffon computed that the second player's expectation was about five coins. Buffon pointed out that in practice, rational people disregarded extremely small probabilities—he set the limit of concern at one-one-hundredth of 1 percent.

D'Alembert argued that probabilists had no firm foundation for their concept of expectation, and in fact were often wrong. His debate with Daniel Bernoulli over smallpox inoculation and his other statements on the inadequacy of probability theory was marked by an insistence on the distinction between mathematical expectation and "real-life" expectation. D'Alembert claimed that between a 99 percent chance of 1,000 coins and a 1 percent chance of 99,000 coins, any "reasonable man" would choose the first, even though they are exactly the same in terms of mathematical probability. The confusion between probabilistic rationality, which would indicate paying any amount to be the second player in the Petersburg game, and real-life rationality, which would indicate paying only a small sum, was responsible for the paradox. The criticisms of d'Alembert and others were one force pushing probabilists to a more "objective" view of probability later in the century, partly through his influence on his protégés, Marie-Jean-Antoine-Nicolas de Caritat, Marquis de Condorcet, and Pierre-Simon de Laplace.

"Inverse probabilities," the derivation of the probabilities from the outcomes rather than the projection of the outcomes from the probabilities, was developed independently in the late eighteenth century by the Englishman Thomas Bayes (1702–1761) and Pierre-Simon de Laplace. Bayes's "Essay towards Solving a Problem in the Doctrine of Chances" appeared in *Philosophical Transactions* in 1764. Bayes's theorem is a way of determining from the frequency of an event the chance that it will fall between two given levels of probability. Subsequent development of inverse-probability reasoning is sometimes called "Bayesianism." Despite Bayes's priority, Laplace's similar theorem announced in 1774 was more influential at the time, however, given Laplace's reputation and position at the center of French mathematics.

The range of problems open to probabilistic analysis and probabilistic mathematics broadened in the Enlightenment. David Hume (1711–1776) used a probabilistic argument to destroy the credibility of religious miracles. The Marquis de Condorcet had an ambitious plan to reform the French judicial and legal system on a probabilistic basis. Many Enlightenment champions believed that probability offered one way to reform the hopelessly corrupt and unjust criminal justice system of their time. The hope of directly applying mathematical probability theory to questions of the reliability of witnesses and the verdicts of juries persisted until Siméon-Denis Poisson (1781–1840) published *Researches on Probability of Civil and Criminal Judgements* (1837).

Laplace was the leading probabilist of the early nineteenth century. His *Analytical Theory of Probabilities* (1812) applied the most sophisticated mathematics yet to probabilistic issues, and his *Philosophical Essay on Probabilities* (1814) explained probability for the lay reader.

See also Alembert, Jean Le Rond d'; Bernoulli Family; Buffon, Georges-Louis Leclerc de; Laplace, Pierre-Simon de; Mathematics; Smallpox Inoculation.

References

Daston, Lorraine. *Classical Probability in the Enlightenment.* Princeton: Princeton University Press, 1988.

Hankins, Thomas L. *Science and the Enlightenment.* Cambridge: Cambridge University Press, 1985.

Maistrov, L. E. *Probability Theory: A Historical Sketch.* Translated and edited by Samuel Kotz. London: Academic Press, 1974.

Psychology

There were several attempts in the Enlightenment to raise psychology to the status of a science. Although the word *psychology,* derived from the Greek *psyche,* meaning soul, had been in use since the sixteenth century, it was established as the canonical term for the study of the mind in the eighteenth century, beginning in Germany, and arriving in the English-speaking world in the early nineteenth century. This process began with Christian Wolff's two Latin texts, *Psychologia Empirica* (1732) and *Psychologia Rationalis* (1734). Like many German writers, Wolff drew on a rationalist tradition stemming from Gottfried Wilhelm Leibniz (1646–1716). Writers in this tradition were "faculty psychologists," dividing the mind into specific functions such as memory, imagination, reason, and so on. Wolff's work contributed to a great deal of interest in psychology in Germany, although Immanuel Kant claimed that because it could not be quantified and was dependent on introspection, psychology could never be a true science. The influential pedagogue Johann Hebart (1776–1841), author of *Psychology As Science* (1824–1825), attempted to refute Kant by creating a quantitative science treating the force and interaction of ideas. Hebart's quantitative psychology was unsuccessful, but important in enhancing the status of psychology and moving away from the faculty approach.

The most dynamic psychological doctrine in the English-speaking world was "associationism," building on the work of the English philosopher and physician John Locke (1632–1704). Associationists sought to identify the laws by which ideas, originally based on perceptions, combined in the mind. Associationism was systematized by another English physician, David Hartley (1705–1757), in his *Observations on Man, His Frame, His Duty, and His Expectations* (1749). Hartley attempted to analyze mental processes in Newtonian terms. Physiologically, he analyzed thought in terms of vibrations in the nerves and spongy matter of the brain. Like other early writers, he linked his position on the mind to religious issues about the soul.

Hartley's work had its major impact on late-eighteenth- and early-nineteenth-century British culture not in its original form, but through an abridgement of *Observations on Man, Hartley's Theory of the Human Mind* (1775), edited and with annotations by Joseph Priestley. Priestley, a great admirer of Hartley, eliminated the material on vibrations as well as Hartley's theological speculations to concentrate on associationism. In this form Hartley's psychology greatly influenced subsequent British Utilitarians, especially James Mill (1773–1836), author of *Analysis of the Phenomena of the Human Mind* (1829). There was a similar but more radically materialistic movement in Enlightenment France, also deriving from Locke and emphasizing mechanism and the primacy of the senses. Its members included Étienne Bonnot de Condillac and Pierre-Jean-Georges Cabanis (1757–1808), a physician who famously equated the brain's production of thought with the liver's production of bile.

The opposition to Hartley and other associationists in Britain came from the Scottish school of "commonsense" philosophers whose leading member was Thomas Reid (1710–1796), author of *An Inquiry into the Human Mind, on the Principles of Common Sense* (1764), *Essays on the Intellectual Powers of Man* (1785), and *Essays on the Active Powers of Man* (1788). The commonsense school held that certain mental presuppositions were innate and universal, including the existence of an external world and other minds. Reid mistrusted the conceptualization of mental activity in terms of "ideas," which he held led to skepticism. He also treated mental and physical phenomena as entirely distinct. Reid was predominantly a philosopher rather than an investigator, but he performed and recorded a few simple psychological experiments to determine questions such as the minimum span of time of which human beings could be conscious. Outside Scotland the commonsense school was particularly influential in America.

The early nineteenth century saw significant advances in the study of the anatomy and functions of the brain and nerves. Unsurprisingly, there were also attempts to integrate this new knowledge into the study of the mind. One of the most influential in the nineteenth century, although it ultimately proved a failure, was phrenology, as propagated by the German brain anatomist Franz Joseph Gall (1758–1828). Gall attempted to localize psychological functions, whose definition was in some ways a throwback to faculty psychology, by assigning them to areas in the brain. Later phrenologists reduced his theory to a matter of locating "bumps" on the skull, correlated with different psychological qualities.

See also Anatomy; Madness; Medicine; Mesmerism and Animal Magnetism; Wolff, Christian.

References

Dematteis, Philip B., and Peter S. Fosl, eds. *British Philosophers, 1500–1799.* Dictionary of Literary Biography, no. 252. Detroit: Gale, 2002.

Hearnshaw, L. S. *The Shaping of Modern Psychology.* London: Routledge and Kegan Paul, 1987.

R

Race

The eighteenth century saw the development of the scientific concept of human races, and of its application to explain and justify social inequalities. The classification of human beings into separate entities called "races" was in part a result of the drive to classify that was a prominent feature of the eighteenth-century life sciences. The earliest scientific classification of human beings to have much influence was that of the Frenchman François Bernier (1620–1688), whose *New Division of the World among the Different Species or Races of Men That Inhabit It* (1684) divided humanity into a small number of groups based primarily on physical characteristics, such as skin color and presence or absence of facial hair in males. (The relative profusion of facial hair in the European male, particularly as compared to the African, the Native American, and all females, brought forth profuse tributes to the dignity and majesty of the "philosopher's beard.") The greatest of the classifiers of plants, Carolus Linnaeus, also set forth a system for the classification of human beings based on skin color and habitat: white Europeans, red Americans, black Africans, and yellow Asians. Linnaeus's fourfold classification was influenced by archaic theories of the four humors and the four elements. The Göttingen professor Johann Friedrich Blumenbach produced the most detailed and influential system of racial classification, drawing on extensive evidence to divide humans into five categories: Caucasian, Mongolian, Ethiopian, American, and Malay.

A key component of the emerging concept of race was the emphasis on heredity rather than environment as the most important factor in shaping human beings. The idea that the different physical qualities of different peoples was a result of the different climates and environments they were exposed to was an old one. For example, African people were believed to be black because of their exposure to the heat of the tropical sun. Although climactic considerations were not forgotten, European thinkers increasingly ascribed differences between human groups to inherited differences, independent of environment.

Early discussions of human physical differences in Europe had begun with the most obvious: skin color. During the eighteenth century the conformation of the skull became more prominent as a criterion of race. The pioneer was the Dutch anatomist and painter Pieter Camper (1722–1789). His concern with the accurate artistic representation of different human groups led him to devise the concept of the facial angle. This was the angle formed by the horizontal line joining the ear to the base of the nose and the vertical line joining the upper jaw to the most protruding point of the forehead. The facial angle provided a way to quantify differences in the shape of skulls. Craniometricians, in an intellectual project extending well into the nineteenth century, ranged human skulls by the facial angle, finding it most acute in Africans and least acute in Europeans. Whereas Camper had considered the least-acute facial angles to be simply the most beautiful, subsequent craniometricians elevated the facial angle to a general measurement of human worth. Collections of skulls became an important part of the equipment of racial classifiers, such as Blumenbach and Georges Cuvier. Both Blumenbach, a supporter of the fundamental equality of human beings, and Cuvier, a believer in racial hierarchy, appealed to the evidence of the skulls.

Emerging systems of racial hierarchy invariably placed Europeans on top, and usually placed black Africans at the bottom. The African presence in Europe was growing during the eighteenth century, partially as a result of the expansion of the slave trade. (Afro-Europeans such as Ottobah Cugoano and Olaudah Equiano [c. 1750–1797] intervened in the racial debate, upholding the essential sameness of Africans and Europeans.) Traditional European antiblack attitudes had emphasized the ugliness of blacks in European eyes and the biblical story of the curse of Ham, the son of Noah who was supposedly the ancestor of Africans. Although these arguments continued to be made, there was a growing emphasis in European antiblack literature on blacks' alleged physical, intellectual, and moral deficiencies. The German anatomist Samuel Thomas von Soemmerring (1755–1830) dissected a number of bodies of Africans to publish *On the Physical Differences between the Moor and the European* (1784), which asserted that Africans were physically closer to apes than to Europeans. Black inferiority was often used to justify slavery, particularly by people coming from European colonies where black slavery was the basis of the economy. The Jamaican Edward Long (1734–1813), in *The History of Jamaica* (1774), described blacks as a separate species similar to apes. However, opponents of slavery did not necessarily view blacks as equals. Soemmerring himself opposed slavery, and another opponent, the philosopher David Hume (1711–1776), held extreme views on black intellectual inferiority, comparing blacks who had acquired proficiency in European intellectual disciplines to trained parrots. (Alternatively, intellectually able blacks like the American surveyor and astronomer Benjamin Banneker [1731–1806] were described as persons of mixed race, whose abilities sprang from their white inheritance.)

Most eighteenth-century racial theorists did not view races as fixed and entirely separate categories, but emphasized the gradations and intermediate conditions. Because all varieties of humans known appeared to be interfertile, racial theorists such as Blumenbach acknowledged the "unity of mankind," scorning those who identified blacks with apes, for example. However, even Blumenbach identified the "Caucasian" (a term for whites that he invented) as possessing the most beautiful and perfect of human bodies, from which other types were variations or degenerations. The most extreme view of race, which became known as "polygenism," was associated with radical materialists and religious skeptics like Hume, who denied the common descent of humanity from Adam and Eve and saw racial differences as entirely innate and fixed. The existence of persons of mixed race was particularly annoying for such theorists. Although polygenism was supported by only a minority in Europe, it became highly influential among nineteenth-century defenders of slavery in the southern United States.

See also Anatomy; Blumenbach, Johann Friedrich; Colonial Science; Exploration, Discovery, and Colonization; Sexual Difference.

References
Hannaford, Ivan. *Race: The History of an Idea in the West.* Washington, DC: Woodrow Wilson Center Press, 1996.
Schiebinger, Londa. *Nature's Body: Gender in the Making of Modern Science.* Boston: Beacon, 1993.

Rain Gauges

Although rain gauges have appeared in many civilizations, scientifically designed instruments for measuring the amount of rain to fall in a given area in a given time first appeared in Europe during the seventeenth century. They were initially put forth as a means of answering the question of whether it was possible for streams and lakes to be replenished entirely by rain, but were revealed to have meteorological uses as well. Although they varied greatly in size, early rain gauges usually worked on the principle of a funnel, which channeled the rain into an enclosed, narrow-necked glass container. This arrangement worked to minimize evaporation. The rain was usually, in early rain gauges, measured by weight, not volume. In 1725 Johann Georg Leutmann (1667–1736) described a heated gauge for winter use. The proximity of a furnace to the gauge prevented it from being blocked by ice. Observers noted that gauges placed higher—many early gauges were placed on roofs—generally showed smaller amounts of rain than those placed near the ground, an effect generally ascribed to wind. Strategies to minimize the effect of wind and splashing included putting raised rims on the funnels, square funnels, and placing the gauges closer to the ground. The great English meteorologist Luke Howard (1772–1864) placed the body of his rain gauges underground.

See also Instrument Making; Meteorology.

Reference
Middleton, W. E. Knowles. *Invention of the Meteorological Instruments.* Baltimore: Johns Hopkins University Press, 1969.

Reil, Johann Christian (1759–1813)

Johann Christian Reil bridged the Enlightenment and romantic eras in Germany as one of its leading medical thinkers. The son of a Lutheran pastor, Reil studied to become a physician at the Universities of Göttingen and Halle. After some years in practice, he returned to Halle as a medical professor in 1788. In 1795 Reil founded the medical journal *Archive for Physiology,* the first medical journal primarily aimed at advancing medical research and theory rather than case reports or popular medical advice. Like many German scientists, Reil was influenced by Immanuel Kant's philosophy. He was a "vital materialist," believing in the powers of life inhering in the matter of life itself. He set forth this philosophy in a long article, "On the Lifeforce," appearing in the first volume of the *Archive for Physiology.* The early volumes of the *Archive for Physiology* transmitted the work of French chemists on bodily substances to a German audience, whereas the later volumes were more anatomical and influenced by *Naturphilosophie,* reflecting a shift in Reil's own thinking. Like most medical professors, Reil continued to practice medicine and served for a time as the personal physician of Johann Wolfgang von Goethe.

Reil's growing interest in *Naturphilosophie* was accompanied by interest in psychological medicine—he coined the word *psychiatry.* His *Rhapsodies on the Use of Psychological Treatment Methods in Mental Breakdown* (1803) emphasized moral over physical means of treatment. Reil gave a number of suggestions for dispelling a patient's obsessions, including theater and therapeutic terror. He founded another journal, *Journal of Psychological Therapy,* in 1805. Reil was also interested in brain anatomy and gave a full description of what became known as the "Islands of Reil" in the cerebral cortex. Such was Reil's popularity as a medical professor that he accepted an invitation to join the medical faculty of the new Humboldt University of Berlin in 1810.

He died of typhus contracted while tending the wounded of the Battle of Leipzig.

See also Anatomy; Goethe, Johann Wolfgang von; Materialism; Medicine; *Naturphilosophie;* Periodicals; Physiology; Romanticism; University of Halle; Vitalism.

References

Broman, Thomas. "J. C. Reil and the 'Journalization' of Physiology." In *The Literary Structure of Scientific Argument,* edited by Peter Dear, 13–42. Philadelphia: University of Pennsylvania Press, 1991.

Porter, Roy. *The Greatest Benefit to Mankind: A Medical History of Humanity.* New York: W. W. Norton, 1998.

Religion

The eighteenth-century Enlightenment was marked by contrasting developments in the relation of science and religion. The alliance between the two forces established in the scientific revolution continued, particularly in England. Even on the Continent, the clergy continued to provide many scientific workers and thinkers. The general waning of religious conflict in the eighteenth century fostered the growth of science by making it easier for scientists to collaborate across confessional lines. But conflicts also emerged between science and religion. For those Enlightenment thinkers, particularly common in Catholic Europe, who were dissatisfied with the repressive churches of eighteenth-century European states or with Christianity itself, science provided an alternative system of intellectual and moral validation to religion. Religious explanations, in terms of the relationship of particular natural phenomena to God's providential plan, were gradually being eliminated from scientific arguments, although this process was far from complete. In some limited areas specific intellectual conflicts between science and Christian orthodoxy did emerge.

The continuation of the alliance of science and religion was part of the conservative Enlightenment in England. After the defeat of the Deists in the early part of the century, the English Enlightenment was dominated by relatively conservative intellectuals, many of them members of the Church of England or Dissenting clergy. Those religious movements, such as Methodism or Hutchinsonianism, that opposed the established authorities in the Church of England tended to be anti-Enlightenment and skeptical about the worth of science as well. The Dissenting denominations produced several significant scientific workers and thinkers, such as the Unitarian minister Joseph Priestley. The conservative reaction in Great Britain after the French Revolution led a mob to sack Priestley's house and destroy much of his laboratory, but both Priestley and the mob were motivated by religious rather than antireligious ideas.

One way in which the role of religion changed in eighteenth-century science was the gradual marginalization in natural theology and science of the idea that the universe needed regular divine intervention. The irregularities in planetary motions whose cumulative effect, according to Isaac Newton (1642–1727), would require God to intervene to correct them were explained in terms of Newtonian mechanics to be caused by the attraction of the planets to each other. The solar system as a whole was demonstrated, most important by Pierre-Simon de Laplace, to be stable. The cumulation of this trend was the famous story of Laplace's remark to Napoléon—who had asked why his celestial mechanics contained no mention of God—that he had no need of that hypothesis. Laplace's nebular hypothesis about the creation of the solar system, which dispensed with any necessity for intelligence or design, seemed particularly challenging to the religious account of Creation based on Genesis, and in the early nineteenth century cultural conservatives pointed to Laplace with horror as the archetypal atheist scientist. However, with a little ingenuity even Laplace's system could be interpreted as testifying to the power and providence of God.

The reputation and intellectual dominance of religion was declining in eighteenth-century European society for many reasons unconnected with science, such as the growing knowledge of sophisticated non-Christian civilizations like China and the rejection of religious persecution. For both scientific and nonscientific reasons, antireligious sentiments did spread in some scientific and philosophical circles. This happened to some extent in England, where Deists and skeptics became influential in the Royal Society after the death of Isaac Newton but did not publicly promote their beliefs. Due to the lack of religious freedom and the persistence of (only intermittently effective) censorship in France, antireligious polarization was much greater there than in Britain. Some French science developed in the direction of materialism, which in the eighteenth century was always an implicit challenge to Christianity. There was a growing interest, with both medical and Cartesian roots, in treating human beings as material rather than spiritual creatures. The understanding of human beings as natural, biological phenomena in Enlightenment psychology undermined religion in several ways. Enlightenment psychologists were not exactly optimistic about human nature, but emphasized malleability and the effects of upbringing and cultural environments on the way humans think. The emphasis on malleability contradicted a fundamental psychological truth for all varieties of Christianity: the doctrine of original sin. Enlightenment psychology, like Newtonian-influenced science in general, was also more deterministic and tended to attack ideas about free will. Scientific materialism and determinism (not the only ideological options for a scientist) were clearly incompatible with traditional religion, and by the late eighteenth century some French philosophers and scientists were openly calling themselves atheists. Priestley, by contrast, tried to ally materialist science and Christianity by claiming that the dualism of matter and spirit was not truly Christian, but a result of early Christianity's corruption by Greek philosophy.

Some specific conflicts between religion and science did emerge in the eighteenth century. As science developed the idea of a rationally ordered world, less and less credibility could be given to miracles. The Scotsman David Hume (1711–1776), the greatest philosopher of the Enlightenment and a strong, although discreet, opponent of Christianity, made an important argument against the belief in miracles, which depends on the notion of a scientific law. Hume claimed that reports of miracles must be evaluated by asking whether it is more likely that a law of nature be violated or that someone would lie. This argument, unlike arguments made against contemporary Catholic reports of miracles by Protestants (many Protestants accepted the doctrine of the "cessation of miracles," meaning that miracles had ceased to be performed shortly after the deaths of the Apostles), would exclude the miracles of Jesus and other biblical accounts as well as more modern ones. It should be noted that Hume's argument, which philosophers still debate, was not pressed with such vehemence as to prevent him from being friends with leaders of the established Presbyterian Church of Scotland. On a more modest level scientific causation tended in the course of the eighteenth century to drive out providential causation—people were less likely to explain bad weather or disease by the wrath of God and more likely to point to scientific reasons.

Another area where conflicts between religion and science emerged was the challenge to biblical chronology on the part of natural historians and geologists. This was a sensitive issue that could be addressed only discreetly for most of this period. Georges-Louis Leclerc de Buffon argued publicly that the world is many thousands of years old, while believing privately that it is millions of years old. The emergence of the doctrine that the kinds of creatures living on the Earth have changed in the past—that a kind of animal living at one time became extinct, or another suddenly came into existence, or that a species "transmuted" to another—opened up vistas of vast stretches of time, and also denied a literal interpretation of the Genesis account. With the opening up of the fossil record, and its reconstruction by naturalists such as Georges Cuvier, changes in the animals on Earth became harder and harder to deny (although Cuvier himself denied transformation while accepting extinction). One thing that moderated conflicts between religion and science in debates over the age of the Earth was internal to religion: a decline in biblical literalism. The most prominent example of the abandonment of the Bible as a guide to science is the silent renunciation of anti-Copernicanism by the Catholic Church. By the time the works of Copernicus were removed from the Index of Forbidden Books in 1822 the seventeenth-century prohibition of heliocentric astronomy had long been a dead issue for Catholic scientists and teachers. The development of biblical criticism made it possible to interpret the Genesis account as an allegory or folk legend to be examined for its religious truth rather than as a literal account of the Creation. Biblical literalism was strongest in the British Isles and the United States, where a movement called "Mosaic geology" tried to incorporate a literal reading of the Bible, particularly Genesis, with recent scientific discoveries.

The idea of an overarching conflict between two monolithic bodies called "religion" and "science" was first fully articulated in the late Enlightenment in the posthumously published work of the Marquis de Condorcet, Marie-Jean-Antoine-Nicolas de Caritat, *Sketch for a History of the Progress of the Human Mind* (1795). In the reaction to the French Revolution and industrialization, conservative romantics who opposed both the Revolution and Enlightenment would also posit the conflict, but unlike Condorcet they would support religion against presumptuous human reason. Joseph-Marie de Maistre (1753–1821), the reactionary intellectual, held that reason, exceeding its proper, limited sphere in the natural sciences, had challenged legitimate political and religious authority. Some radicals, such as the poet William Blake (1757–1827), denounced the scientific heroes Francis Bacon (1561–1626) and Isaac Newton for different reasons. Blake believed that science and natural theology, which he detested, reduced God into a divine clock maker, destroying the love and attempt at union with God, which were at the heart of religion.

See also Enlightenment; Freemasonry; Jewish Culture; Materialism; Natural Theology; Newtonianism; Universities.

References

Brooke, John Hedley. *Science and Religion: Some Historical Perspectives.* Cambridge: Cambridge University Press, 1991.

Lindberg, David C., and Ronald L. Numbers, eds. *God and Nature: Historical Essays on the Encounter between Christianity and Science.* Berkeley: University of California Press, 1986.

Ritter, Johann Wilhelm (1776–1810)

Johann Wilhelm Ritter was the foremost physicist in German romanticism and *Naturphilosophie.* The son of a Silesian minister, Ritter began experimenting with Luigi Galvani's recent electrical discoveries in 1796, while attending the University of Jena. In 1798 Ritter published a book on galvanism, *Proof That a Continuous Galvanism Accompanies the Process of Life in the Animal Kingdom.* This work combined inductive reasoning, mostly repeating the experiments of Alessandro Volta, with a speculative romantic theory of the universe as an essentially living organism. Ritter continued his experimental program by extending his analysis of galvanism from organic to inorganic matter, demonstrating that the production of electricity was the result of a chemical process. His subsequent experimental work combined effective experiments, including the creation of an electrical accumulator, with grandiose speculations inspired by *Naturphilosophie.* Ritter saw the universe as a cosmic organism whose functions at all levels were interrelated. For example, planetary orbital periods were mirrored by the fluctuations of galvanic actions. Periods of the maximum inclination of the ecliptic coincided with great discoveries in electricity, such as the Leiden jar in 1745 and the voltaic pile in 1800. This led him to predict another great electrical discovery in 1820, a prediction fulfilled by Ritter's friend Hans Christian Ørsted, who discovered electromagnetism that year.

Ritter's *Naturphilosophie* led him to another important physical discovery in 1801, when he applied the fundamental *Naturphilosoph* principle of complementary duality to William Herschel's recent discovery of infrared light and demonstrated the existence of ultraviolet light by showing how it darkened silver chloride. Ritter continued to publish in the most highly regarded German physics journal, the *Annals of Physics,* until 1805, but his philosophical speculations, combined with his penchant for borrowing money without returning it, were pushing him further from the community of European physicists. In 1805 he moved to Munich as a member of the Bavarian Academy of Arts and Sciences and involved himself in a series of experiments on the electrical nature of dowsing. These experiments led to nothing but the destruction of Ritter's remaining reputation. A romantic to the end, Ritter died of the quintessentially romantic disease tuberculosis, after years of abusing his body in electrical experiments.

See also *Naturphilosophie;* Ørsted, Hans Christian; Romanticism.

Reference

Wetzels, Walter D. "Johann Wilhelm Ritter: Romantic Physics in Germany." In *Romanticism and the Sciences,* edited by Andrew Cunningham and Nicholas Jardine, 199–212. Cambridge: Cambridge University Press, 1990.

Romanticism

The romantic movement of the late eighteenth and early nineteenth centuries related to science in two major ways. Romantics evolved a new critique of the scientific project, one that continues to have powerful resonances to modern times. On the other hand, romantics also actively practiced science, and many aspects of their romanticism influenced the development of science.

Romanticism is notoriously difficult to define, and there was much dissimilarity between various romantics or groups of romantics. The most useful definition in considering romanticism's relation to science is as a loosely connected group of individuals and movements opposed to what they viewed as the excessive rationalism of the Enlightenment, and they emphasized sincerity and truth to the heart. Romanticism emerged in the late eighteenth century and gathered strength as a reaction to the French and Industrial Revolutions. The intellectual roots of the romantic critique of science were in the writings of the Enlightenment philosopher Jean-Jacques Rousseau (1712–1778), whose famous *Discourse Concerning the Arts and Sciences* (1750) argued that the arts and sciences have not ministered to the benefit of mankind, and that as people have grown more knowledgeable and able to exert more power over nature they have grown more corrupt, unequal, and further from the natural order.

Romanticism was compatible with a broad range of political affiliations. Conservative romanticism took shape as opposition to the French Revolution. Conservative romantics such as the reactionary Catholic Joseph-Marie de Maistre (1753–1821) argued that revolutionary violence and terror were the inevitable result of the abandonment of religious and traditional values in favor of innovation and an arrogant, scientistic Enlightenment approach to politics and society. De Maistre particularly disliked Francis Bacon (1561–1626), a great hero of the Enlightenment narrative of the rise of science. Conservative romantics shared with other romantics a tendency to exalt earlier, ostensibly prescientific times such as the Middle Ages, which the Enlightenment had never had much time for, and to view the modern era as one that had lost fundamental truths.

The left-wing romantic critique of science and technology emerged in Britain as a response to the early Industrial Revolution. The intellectual and cultural links of science and technology were increasingly visible and accepted throughout the early Industrial Revolution, particularly in Britain. Romantic mistrust of technology, science, and scientists was memorably expounded in the classic *Frankenstein* (1818) by Mary Wollstonecraft Shelley (1797–1851). William Blake (1757–1827), the English working-class poet and opponent of industrialism, was more explicit: "For Bacon and Newton, sheath'd in dismal steel, their terrors hang / Like Iron scourges over Albion." Blake opposed both what he believed to be the oppression of the English workers by industrialism and the fetters placed on the mind by scientism. Blake's romantic critique of science, like many others, was also religious, as Blake, who had connections with various underground London heretical groups, believed that science reduced God into a divine clock maker, destroying the love and attempt at union with God, which were at the heart of religion. Blake hated and despised natural theology.

The belief that the poetic mind and scientific mind were innately opposed was common in British romanticism, although much rarer on the Continent. (Even in Britain it was far from universal—Sir Humphry Davy was a good friend of leading romantic poets such as Samuel Taylor Coleridge [1772–1834] who admired and took an informed interest in his science, and Davy wrote rather mediocre but thoroughly romantic verse himself.) By dispelling the mystery of nature science was believed to work against poetry. The romantic poet John Keats (1795–1821) in his poem *The Lamia,* about a wise philosopher who dispels the illusions created by a sorceress, asked:

> Do not all charms fly
> At the mere touch of cold philosophy?
> There was an awful rainbow once in heaven
> We know her woof, her texture; she is given
> In the dull catalogue of common things
> Philosophy will clip an Angel's wings
> Conquer all mysteries by rule and line,
> Empty the haunted air, and gnomed mine
> Unweave a rainbow.

But not all romantics opposed science, and many vociferously supported it or were even scientists themselves. Romanticism most affected actual scientific practice in Germany and areas under German cultural influence like Scandinavia. Although romanticism was an international phenomenon, Germany was where it was strongest, and German writers and intellectuals seized romanticism as a way of vindicating what they believed to be a more spiritual Germanic approach to reality against French rationalism and English materialism. German romantic science tended to work against what Germans considered the materialism and mechanism of the Enlightenment. Romantics did not usually argue that mechanical science was wrong, but that it gave a partial and incomplete view of the universe. In physics, for example, romantic physicists like Johann Wilhelm Ritter and Hans Christian Ørsted did not usually challenge the Newtonian system head-on, but concentrated their efforts on phenomena outside Newtonian explanation, like heat and electricity. Germans did not define science in opposition to poetry, and industrialization was not a strong cultural presence in Germany yet. Major German romantics, including Achim von Arnim (1781–1831) and Friedrich Leopold von Hardenberg (1772–1801), better known by his pen name, Novalis, had scientific educations and were interested in natural science, and only a few manifested the distrust of it expressed by some French and British romantics. Johann Wolfgang von Goethe, the recognized genius of Germany in the romantic era (although his own relationship to romanticism was complex), was both the leading German writer and a leading scientist. Romantic scientists also made a greater use of literary devices such as metaphor in presenting their science than had Enlightenment scientists.

Romantics preferred to view the universe as a living organism rather than a mechanical clock. Romantics supported a vitalistic rather than mechanical approach to life, asserting that living things are fundamentally different from nonliving. Romantic biologists rejected the Newtonian system as a means of understanding living things. The science of "biology" was founded in *Biologie,* an 1802 textbook by the German natural historian Gottfried Reinhold Treviranus (1776–1837), as a replacement for "natural history." Life, as a scientific concept, was a product of the romantic era, and biology, unlike natural history, firmly separated the study of animals and plants from mineralogy. By defining life as an irreducible quality, romantic vitalists also asserted that the methodology of the physical sciences was not appropriate for its study. Atomistic physics was based on the idea that the entities scientists studied could be divided into very small parts, but applying that method to a living thing killed it.

Romantics viewed living things not in a purely functional or taxonomic fashion, but as embodiments of an underlying idea, which was frequently categorized as a spiritual pattern that transcended the material, a concept with affinities to Platonic philosophy. For example, romantic botanists viewed all plant structures as modifications of the basic archetypal form of the leaf. The "philosophical anatomy" of the French zoologist Étienne Geoffroy Saint-Hilaire stressed the homologies of organs of different animals, seeing all animals as variations on a basic plan, and was opposed to the Enlightenment functionalism of Georges Cuvier, which analyzed the organs of animals solely in terms of their use. Romanticism also tended to have a strongly historical dimension, emphasizing the development of the universe over time. As opposed to the universe of the eighteenth-century Newtonians, in which change over time played little role, the romantic universe was one in a state of change that was not random but oriented in a specific direction. The universe was seen as in a state of development—one frequently used metaphor was that of a pregnant woman. Geology, a science that appealed to many romantics, began to be conceptualized not in terms of simply describing the state of the Earth now, but as discovering its past history. (This was not exclusive to romantics, and was also practiced by the heirs of the Enlightenment such as Cuvier. It is interesting to note, though, that Cuvier's geological history of the Earth and its past species was one aspect of his works that romantics admired.) Romantics were fascinated with caves as offering a route to the past of the Earth. They also found them intriguing because they were part of the wild Earth, less shaped by human activities, along with mountains and remote wildernesses. This contrasts strongly with the Enlightenment, which preferred well-ordered landscapes. The most extreme example of romantic developmentalism, as well as other trends in romantic science, was German *Naturphilosophie.*

See also Brunonianism; Davy, Sir Humphry; Enlightenment; Geoffroy Saint-Hilaire, Étienne; Geology; Goethe, Johann Wolfgang von; Humboldt, Alexander von; Literature; *Naturphilosophie;* Ørsted, Hans Christian; Ritter, Johann Wilhelm.

References

Cunningham, Andrew, and Nicholas Jardine, eds. *Romanticism and the Sciences.* Cambridge: Cambridge University Press, 1990.

Lenoir, Timothy. *The Strategy of Life: Teleology and Mechanics in Nineteenth-Century German Biology.* Dordrecht, Netherlands: D. Reidel, 1982.

Royal Academy of Sciences

The Royal Academy of Sciences, founded in Paris in 1666, was the leading scientific institution of Enlightenment Europe. Virtually every important scientific innovation of the eighteenth century was introduced or debated at a meeting of the academy. Admission to the academy was the goal of every socially ambitious male French scientist, and inclusion in

This late-seventeenth-century imaginary scene combines the activities of the Royal Academy of Sciences and the Academy of Fine Arts. (Edgar Fahs Smith Collection, University of Pennsylvania Library)

its small group of foreign associates was one of the highest honors a European scientist could aspire to. The academy also served as a model for the subsequent scientific academies founded in Berlin, St. Petersburg, and many other places.

The influence of the academy in the eighteenth century was built on reforms made in 1699 under the direction of the abbé Jean-Pierre Bignon (1662–1743), and confirmed by royal letters patent in 1713. The academy served the French state, carrying out government functions, such as the examination of new inventions for which inventors sought royal patents. Members of the academy were treated as technical resources, available for their expertise on the many technological and industrial questions facing the French government. The academy also supervised the Paris Observatory and bore an ultimate responsibility for its work in navigation and cartography. As an institution the academy also advanced the cause of science by sponsoring prize competitions and scientific expeditions. It took a leading role in the coordination of observations of the transits of Venus in 1761 and 1769 as well as other astronomical phenomena. The academy met twice a week, on Wednesdays and Saturdays. Some of the meetings were taken up with the reading of scientific papers, but more with committee reports and the reading of correspondence.

In the hierarchical and corporatist society of old regime France, the academy sought to defend and expand its and its members' privileges. Long campaigns were fought on such questions as whether the academicians had the right enjoyed by other academies of genuflecting in the king's presence. One highly important privilege was academicians' exemption from censorship of their works, which on approval by the academy could be printed without further ado. The academy's publications included its annual *Histories and Memoirs,* an almanac, the monthly *Journal des Savants,* and a massive collection of works on technologies employed in various industries, *Description of Arts and Occupations,* produced in seventy-four treatises between 1761 and 1782. Academy-sponsored publication in the course of the century was increasingly extended to nonacademicians who submitted their work to it. A journal, *Memoirs Mathematical and Physical,* was founded in 1750 specifically to publish the work of nonacademicians.

The Royal Academy of Sciences defended its monopoly over scientific affairs by opposing, with varying degrees of success, the emergence of other societies, whether state-sponsored or voluntary. The only scientific area free from its control was medicine. It did have a formal relationship with the Royal Society of Sciences, a group founded in 1706 in Montpellier, the home of the most important French medical school. The Royal Society of Sciences was the only provincial French scientific society the academy recognized.

Unlike the rival institution devoted to literature and language, the French Academy, which admitted no distinctions among its "forty immortals," the Royal Academy of Sciences was complexly divided, both among disciplines and hierarchically. The disciplines were divided into two broad categories: the mathematical sciences, divided into the three categories of mathematics, astronomy, and mechanics, and the physical sciences, divided into anatomy, chemistry, and botany. A reorganization led by Antoine-Laurent Lavoisier in 1785 added the categories of experimental physics as a mathematical science and natural history, including mineralogy, as a physical one. The hierarchical division placed a class of twelve honorary members, including wealthy and powerful leaders of French society, at the top, ahead of the working scientists. It was from this group that the academy's president and vice president were selected every year by the French government.

The highest class among the scientists themselves were the pensionaries, so-called because they were paid for their participation. The pensions averaged 2,000 livres, a substantial sum of money but not enough to live a bourgeois life. Pensionaries had to supplement their income. Fortunately, membership in the academy also opened the doors to other honorable and lucrative positions in France. In theory there were supposed to be three pensionaries for each of the six disciplines, but in practice "supernumerary" pensionaries were common. The next group were the associates, divided into two groups. Eight positions were reserved for foreign scientists, and a dozen for Frenchmen. Several clergymen-scientists became associates, as they were ineligible to receive pensions. The third group was known as adjuncts. They received no pensions and were ineligible to vote, but could participate in meetings. (Lavoisier abolished this rank, effectively raising the adjuncts to the status of associates.) These classes were supposed to be elected by the pensionary and honorary members of the academy, but in practice the government often interfered. The academy sometimes vindicated its independence—twice in the 1770s the academy's protests led to government-chosen academicians retiring and being replaced by the academy's candidates. Finally, there were persons having an official status as correspondents, living outside Paris and sending scientific information to one or another member of the academy. Like all scientific academies outside Italy, the academy excluded women from its ranks.

A leading figure in the academy was the perpetual secretary, who oversaw the publication of the academy's annual *Memoirs.* The first was Bernard Le Bovier de Fontenelle (1657–1757), secretary from 1697 to 1740, and the last was the Marquis de Condorcet, Marie-Jean-Antoine-Nicolas de Caritat. An important genre developed by the secretaries of the academy, particularly Fontenelle and Condorcet, was the *eloge,* a speech given in praise of a deceased member. In printed form the *eloges* remain an important source for the history of eighteenth-century science, but their main purpose was not biographical but hagiographic, in giving a picture of the selfless, modest scientist, wholly devoted to seeking out the truth.

By the late eighteenth century the academy's strengths—its high standards, its social standing, its alliance with the state—had also become liabilities. To young French intellectuals, many of whom took an interest in natural science, the academy seemed a close-minded and corrupt body, more concerned with protecting its narrow interests than in promoting the growth of natural knowledge. The academy had always attracted the resentment of circle-squarers and inventors of perpetual-motion machines, but the real problem was structural and demographic. The academy had simply failed to keep pace with the expansion of the French scientific community. More and more good scientists spent more and more of their careers outside the academy. The average age of admission to the academy went up about a decade in the course of the century. Even the creation of new scientific institutions in the last decades of the old regime, such as the Royal Society of Medicine founded in 1778, failed to alleviate the problem, as academicians simply added positions in the new societies to their positions in the academy.

The Royal Academy of Sciences would pay for the hostility it had aroused during the radical phase of the French Revolution, which brought to power or influence many old enemies of the academy, such as Jacques-Pierre Brissot de Warville (1754–1793) and the frustrated scientist Jean-Paul Marat (1743–1793). At first the Revolution aroused hope of the academy's reform. Condorcet and others put forth plans for reducing or eliminating honorary members and placing working members on a more equal footing. In addition to having to deal with the rapid flux of revolutionary politics, the academy started to lose its control over French scientific life with the general loosening of central control. A number of independent scientific journals and voluntary scientific societies were formed, continuing a trend begun in the last decade of the old regime. The academy's biggest problem, shared by the other academies of France, was that by its very name, which proclaimed its royal status, it was thoroughly bound up in the aristocratic and monarchical French culture that was now being destroyed. After a long struggle the academy was closed on 8 August 1793. A similar institution, the First Class of the Institute of France, was established in 1795, and renamed the Royal Academy of Sciences after the Restoration in 1815, but these new institutions, ill-fitted to an age of increasing specialization, never matched the glory of the eighteenth-century academy.

See also Academies and Scientific Societies; Alembert, Jean Le Rond d'; Ballooning; Berthollet, Claude-Louis; Buffon, Georges-Louis Leclerc de; Condorcet, Marie-Jean-Antoine-Nicolas Caritat, Marquis de; Cartesianism; Coulomb, Charles-Augustin de; French Revolution; Laplace, Pierre-Simon de; Lavoisier, Antoine-Laurent; Macquer, Pierre Joseph; Maupertuis, Pierre-Louis Moreau de; Mesmerism and Animal Magnetism; Nollet, Jean-Antoine; Transits.

References

Gillispie, Charles Coulston. *Science and Polity in France At the End of the Old Regime.* Princeton: Princeton University Press, 1980.

Hahn, Roger. *The Anatomy of a Scientific Institution: The Paris Academy of Sciences, 1666–1803.* Berkeley: University of California Press, 1971.

Paul, Charles B. *Science and Immortality: The "Eloges" of the Paris Academy of Sciences (1699–1791).* Berkeley: University of California Press, 1980.

Royal Botanical Expedition (Spain)

Spanish authorities in the late eighteenth century took a renewed interest in the scientific exploration and mapping of their large empire. This was partly inspired by the desire to emulate the French and British, but more by the belief that the Spanish Empire contained valuable resources that could succor the state. The leading spirit in the organization of the expedition that left Spain in 1787 was a physician, Martin de Sesse y Lacasta (1751–1808). He was supported by the powerful director of the Royal Botanical Garden in Madrid, Casimiro Gomez Ortega (1740–1818). The expedition was originally assigned a six-year term, during which it would be headquartered in Mexico City, then the capital of the viceroyalty of New Spain.

One of the missions of the expedition was to apply the Linnaean system of botanical classification to the flora of Mexico. This led to conflict with the Mexican botanical community, suspicious of metropolitan systematizing. The expedition also did a great deal of natural-historical collecting, sending many of the items back to the Royal Botanical Garden in Madrid.

The expedition led to an expansion of the Mexican botanical infrastructure, with the founding of the Royal Botanical Garden in Mexico shortly after the expedition's arrival (although it did not find a permanent home until 1791) and a special course of botanical lectures by the pharmacist Vicente Cervantes (1755–1829). (Cervantes, who stayed on as director of the Royal Botanical Garden after the expedition left to return to Spain, and through the Mexican War of Independence, also translated Antoine-Laurent Lavoisier's *Treatise on Chemistry* into Spanish for the benefit of Mexican chemists.) One student of the lectures was José Mariano Mocino (1757–1820), a Mexican of Spanish descent who became a leading botanical explorer in his own right. He wrote up the results of an exploration into northern and western Mexico in 1790 and 1791 as *Plants of New Spain,* although the manuscript was never published.

The Royal Botanical Expedition cooperated with another Spanish mission, the expedition led by the naval officer Alejandro Malaspina (1754–1809). This expedition focused on the west coast of the Americas, where Spanish predominance was being challenged by the British and the Russians. Mocino accompanied the Malaspina group in their major survey of Nootka Sound in the Far North in 1792. The Royal Botanical Expedition's mandate was renewed in 1794, and it extended its explorations to Guatemala and the West Indies. Despite the importunities of a Spanish government increasingly impatient for the expedition's return, Sesse and his associates, including Mocino, did not go back to Spain until 1803. Although the Royal Botanical Expedition and the Malaspina expedition both collected many specimens and much information, relatively little of it was published until the end of the nineteenth century, due to political turmoil in Spain during the Napoleonic Wars.

See also Botany; Colonial Science; Exploration, Discovery, and Colonization; Nationalism.

References

Engstrad, Iris H. W. *Spanish Scientists in the New World: The Eighteenth-Century Expeditions.* Seattle: University of Washington Press, 1981.

Lafuente, Antonio. "Enlightenment in an Imperial Context: Local Science in the Late-Eighteenth-Century Hispanic World." *Osiris,* 2d ser., 15 (2000): 155–173.

Royal Society

Britain and Europe's oldest scientific society, the Royal Society, continued as a leading scientific body in the eighteenth century. The organization of the London-based society differed from what became the dominant model, the hierarchically arranged academy with salaried academicians on the pattern of the Parisian Royal Academy of Sciences. The Royal Society was a voluntary organization that supported its day-to-day operations by collecting dues rather than a state subvention, and its membership was open to men other than professional scientists. A majority of the membership, the governing council, and usually its officers were nonscientists during the period. In theory all members were equal and could vote in the society's elections. This practice was reinforced by the custom of referring to all fellows as "gentlemen." Although aristocrats always found entrance into the society easier than commoners, the social base of membership broadened in the eighteenth century to include middle-class men such as surgeons, apothecaries, schoolmasters, and instrument makers.

The personality of the president made an enormous difference in the society's operation. During the presidency of Isaac Newton (1646–1727) from 1703 to 1727, the society emphasized experimental science. Newton, assisted by his successor as president Sir Hans Sloane and others, also reorganized the society's finances and administration, and in 1710 procured its first permanent home in Crane Court. Newton also brought the society into his battle with Gottfried Wilhelm Leibniz (1646–1716) over priority in the invention of the calculus. Sloane, president from 1727 to 1741, placed more emphasis on natural history, and under his rule the experimental tradition of the society declined sharply.

Major areas of activity by the fellows of the Royal Society in the eighteenth century included natural history, medicine, and applied mathematics in astronomy, cartography, and geography. Royal Society science tended to eschew theory and pure mathematics and proclaim its loyalty to facts and exactitude. The society was also fiercely loyal to Newton and Newtonianism. Its cooperative efforts with other societies also grew throughout the period. It responded to outreach from the Imperial Academy of Sciences of St. Petersburg with a standing exchange of publications beginning in 1726, and started another with the Bologna Academy of Sciences in 1732. A publication exchange with the Royal Academy of Sciences was set up in the period 1750–1753. An attempt to coordinate meteorological observations with other societies and individuals in 1723 proved a failure, but the society participated in joint observations of the two transits of Venus, playing a secondary role in the 1761 transit but the lead one in the 1769 transit. This transit led to perhaps the most important society-sponsored scientific venture: the voyages of Captain James Cook.

The society continued to grow as an institution throughout the eighteenth century. Its membership increased from 131 at the beginning of the century to 531 at its end. It turned a legacy from Sir Godfrey Copley (1653–1709) into the annual Copley Medal beginning in 1736. The Copley Medal was awarded for an outstanding achievement in any field of science, and was second only to the prizes of the Royal Academy of Sciences in prestige. In 1796 Benjamin Thompson, Count Rumford, endowed another prize, the biannual Rumford Medal for discoveries in light and heat first awarded to Rumford himself. A dining-club meeting after the society's weekly meetings, first known as the Society of Royal Philosophers and then as the Royal Society Club, was formed in 1743. In 1752 the society took over *Philosophical Transactions,* and made it their official, as it had long been their unofficial, organ.

Despite the society's financial independence from the British government, it was intimately bound up with the institutions of the British state and ruling class. In 1710 the society's oversight of the Royal Observatory at Greenwich was formalized. The president of the Royal Society sat ex officio on the Longitude Board and the Board of the British Museum, and the society worked closely with these institutions. It also collaborated with the Admiralty on funding and equipping expeditions. The society was responsible for specific tasks for the government, including resolving the dispute between pointed and rounded lightning rods.

By the mid-eighteenth century the society was growing increasingly unwieldy. Many of its officers were antiquarians rather than scientists, and treated their positions as part-time occupations. The standards for membership, particularly for foreign members, had been substantially lowered. The number of foreign fellows was capped at 100 in 1776. Further efforts for administrative reform continued under Sir Joseph Banks, president from 1778 to his death in 1820, the longest presidency in the Royal Society's entire history. Banks was responsible for moving the society out of its increasingly cramped quarters in Crane Court into Somerset House in 1780, a move not without its difficulties.

Much of the society collections had to be given to the British Museum as there was no place to put them in the new building. In 1783, with the support of the leading physicist Henry Cavendish, Banks defeated a revolt of some mathematical scientists afraid that the society under Banks would tilt too far toward natural history.

The Royal Society's dominant role in English science was challenged in the early nineteenth century first by societies in the north of England and Scotland such as the Manchester Literary and Philosophical Society and second by specialized societies such as the Royal Geological Society, founded in 1807 over Banks's protests. (The Society for Animal Chemistry, founded in 1809, was not a threat, as it was an "assistant" society to the Royal Society rather than an independent group.) The northern societies had a more professional and less social ethos, one that was increasingly common among fellows of the Royal Society itself. The election of a scientist, Sir Humphry Davy, as the society's president after Banks's death was part of the effort to remake the society as a more professional body dominated by working scientists. This effort would take many years.

See also Academies and Scientific Societies; Banks, Sir Joseph; Bradley, James; Brown, Robert; Cavendish, Henry; Cook, James; Davy, Sir Humphry; Desaguliers, John Theophilus; Hales, Stephen; Hauksbee, Francis; Herschel Family; Jewish Culture; Lightning Rods; Nationalism; Periodicals; Sloane, Sir Hans; Transits.

References

Lyons, Henry. *The Royal Society, 1660–1940: A History of Its Administration under Its Charters.* Cambridge: Cambridge University Press, 1944.

McClellan, James E., III. *Science Reorganized: Scientific Societies in the Eighteenth Century.* New York: Columbia University Press, 1985.

O'Brian, Patrick. *Joseph Banks: A Life.* Boston: David R. Godine, 1993.

Sorrenson, Richard. "Towards a History of the Royal Society in the Eighteenth Century." *Notes and Records of the Royal Society* 50 (1996): 29–46.

Royal Society of Edinburgh

Although a shadowy philosophical society was founded in Edinburgh in 1705, the first scientific society there to leave a record was the Society for the Improvement of Medical Knowledge, founded in 1731 by the Edinburgh professor of anatomy and surgery Alexander Monro I (1697–1767). Like all subsequent Edinburgh scientific societies, it was closely attached, although not formally affiliated with, the University of Edinburgh. In 1737 it was reorganized as the Society for Improving Arts and Sciences, commonly referred to as the Edinburgh Philosophical Society. This signified a broadening of interest beyond medicine and was promoted by another Edinburgh professor, the Newtonian mathematician Colin Maclaurin (1698–1746). The society shut down after the taking of Edinburgh by the Jacobite army in 1745, and was not restored until 1750. Its most illustrious member was the philosopher David Hume (1711–1776), who served as its secretary. It published three volumes of transactions, the first volume of which in 1754 Hume edited.

The Philosophical Society was reborn in 1783 as the Royal Society of Edinburgh. This society was modeled on the London Royal Society, with a charter and a similar voluntary structure. Unlike the Royal Society, it was not a pure scientific society but was divided into a scientific and a literary section (although the scientific side increasingly predominated) and met monthly rather than weekly. The society invited scientists from all over Scotland, usually professors, to join, although the Edinburgh element was dominant and the society met in the library of the University of Edinburgh. The society offered a better venue than the university for new ideas. Sir James Hall (1761–1832) introduced the new chemistry of Antoine-Laurent Lavoisier to Edinburgh in a paper read before the society early in 1788. The society's principal impact on the scientific world at large came through its *Transactions,* first published

Sociability was central to the Scottish Enlightenment, as can be seen from this affectionate caricature of two members of the Royal Society of Edinburgh, James Hutton (left) and Joseph Black (right), in conversation. (National Library of Medicine)

in 1788. This became an important journal, particularly in geology, the main object of the society's interest. James Hutton, a member, read a paper before it in 1785, which propounded his geological theory, and the conflict between Huttonians and Wernerians was fierce. Other Huttonians besides Hutton himself who were members of the society were Hall and John Playfair (1748–1819), general secretary of the society from 1798 to 1819. Huttonian dominance in the early nineteenth century provoked a secession and the founding of the Wernerian Society by Robert Jameson (1774–1854) in 1808.

See also Academies and Scientific Societies; Black, Joseph; Geology; Hutton, James; University of Edinburgh.

References

McClellan, James E., III. *Science Reorganized: Scientific Societies in the Eighteenth Century.* New York: Columbia University Press, 1985.

Porter, Roy. *The Making of Geology: Earth Science in Britain, 1660–1815.* Cambridge: Cambridge University Press, 1977.

Royal Society of Medicine (France)

Founded in 1778, the French Royal Society of Medicine was conceived as a way of introducing modern, more scientific practice to the French medical profession. Its leading spirit and secretary, Félix Vicq d'Azyr (1748–1794), claimed that it would follow the spirit of the Royal Academy of Sciences. As a scientist himself, Vicq d'Azyr made his principal mark on brain anatomy. He had also acquired fame, if not much popularity, among French peasants, with his prompt and vigorous action as head of a government mission to deal with a cattle plague in January 1775. Vicq d'Azyr had supported drastic measures, such as the slaughter and burial of entire herds when one individual showed signs of infection, and carried on experiments establishing the mechanisms of transmission. A group of Paris physicians gathered by royal commission in 1776, in the wake of the cattle plague, to deal with medical emergencies and public health was the nucleus of the Royal Society of Medicine.

Like the Royal Academy of Sciences, the society occupied a quasi-official role. The French government consulted it on medicine and public health. It judged medical books and remedies submitted to it and carried on an enormous correspondence with provincial and colonial physicians. The full members were forty-two, thirty physicians and twelve associates. The society also had sixty provincial associates and more than a thousand correspondents, but lacked the complex internal hierarchy of the Academy of Sciences. Its connections with so many provincial doctors enabled it both to gather information on weather, disease conditions, and the effects of various treatments and to route medical information and government decrees throughout France. The society also concerned itself with the French colonial empire, gathering information on tropical diseases and their treatments. It met twice a week, on Tuesdays and Fridays, held prize competitions, and published annual memoirs.

The principal problem the society faced was the difficulty of fitting a new institution in the highly organized world of French medicine. Its claim to jurisdiction over drugs and mineral waters was opposed by the apothecaries, and the Faculty of Medicine, the body associated with the University of Paris, which traditionally sat at the top of the hierarchy of French physicians, was highly suspicious of the new body. (Vicq d'Azyr probably chose the name "Society" rather than "Academy" to lull the suspicions of the faculty.) The faculty tried to take over the new society, protested to the king about it, and eventually threatened to go on strike if the society were not suppressed. A key weapon of the new society was the scientific expertise of many of its members. Its chemists included Pierre Joseph Macquer and Antoine-François de Fourcroy (1755–1809), and its botanists Antoine-Laurent de Jussieu (1748–1836). It also formed alliances with medical institutions outside of Paris. The society, strengthened by its relationship with the government, survived the assaults of its bureaucratic foes, but as an elitist body it was suppressed along with the Royal Academy of Sciences after the French Revolution.

See also Academies and Scientific Societies; Agriculture; French Revolution; Medicine; Nationalism; Royal Academy of Sciences.

Reference

Gillispie, Charles Coulston. *Science and Polity in France At the End of the Old Regime.* Princeton: Princeton University Press, 1980.

Royal Swedish Academy of Science

Sweden acquired its national scientific organization in 1739. The "Hat" Party, then in control of the Swedish Diet, was interested in applying science to economic development, and some Hat leaders were influenced by the Newtonian engineer Martin Triewald, a keen promoter of Swedish science. Sweden's greatest scientist, Carolus Linnaeus, was the president of the new society, and a young nobleman, Anders von Hopken, its first secretary. The society's statutes were confirmed by the king in 1741, and it took the name Royal Academy of Science.

The new organization blended the two dominant models of a national scientific society: those of the British Royal Society and the French Royal Academy of Sciences. Like the Royal Society, the Swedish Royal Academy did not have an internal hierarchy of grades among its members, and supported itself by dues rather than government subvention. Like an academy, its membership was limited, initially to 100 members. The Swedish Royal Academy was also marked by a strong practical focus on agriculture and economic development. Its quarterly journal, *Handligar,* was published in Swedish, addressing a domestic rather than an international scientific audience. Such was its interest, however, that it was translated in its entirety into German, and selected articles were translated into several other European languages. The academy was also connected with the gathering of statistics, a practice in which the Swedish government led Europe.

Changes in the late 1740s moved the society more toward the French model of theoretical science and participation in the international scientific community. In 1747 it received a state monopoly on almanacs and calendars, similar to that which the Berlin Academy possessed in Prussia. (Demand for almanacs and calendars was particularly high given Sweden's impending change to the Gregorian calendar.) The same year the academy began work on an observatory, completed in 1753, and admitted its first foreign members. The following year it hired a new secretary, the astronomer Pehr Wilhelm Wargentin (1717–1783), one of the great scientific organizers of the eighteenth century. Wargentin, whose astronomical work had already convinced him of the value of international scientific communication, increased the volume of the society's correspondence, raising its profile in the European community. The Swedish Academy of Science participated vigorously in the transit observations of 1761, providing a contingent second in number only to the French, and of 1769, where Swedish access to the Far North proved particularly important. The academy also benefited from the reputation of its leading members, including Linnaeus, the mineralogist and chemist Torbern Olaf Bergman (1735–1784), and Carl Wilhelm Scheele.

The golden age of the academy proved short-lived. In 1772 the Hats lost power as the "royal coup" of King Gustav III (r. 1771–1792) made Sweden monarchical rather than parliamentary. The new regime was less interested in supporting science, and Wargentin's death in 1783 led to the marginalization of the academy in the international community. It would revive in 1818, when Jöns Jakob Berzelius became its president.

See also Academies and Scientific Societies; Berzelius, Jöns Jakob; Linnaeus, Carolus; Nationalism; Scheele, Carl Wilhelm.

References

McClellan, James E., III. *Science Reorganized: Scientific Societies in the Eighteenth Century.* New York: Columbia University Press, 1985.

Widmalm, Sven. "Instituting Science in Sweden." In *The Scientific Revolution in National Context,* edited by Roy Porter and Mikulas Teich, 240–262. Cambridge: Cambridge University Press, 1992

S

Scheele, Carl Wilhelm (1742–1786)

The pharmacist Carl Wilhelm Scheele, from his apothecary's shop in the small Swedish town of Koping, discovered a plethora of organic and inorganic chemical substances. However, his remoteness from the major centers of European science meant that he was often not credited with priority in discovery. The most famous example is oxygen, which Scheele called "fire air." Scheele had prepared the gas and identified it as a new and different substance about two years before Joseph Priestley's independent discovery, but since Priestley's achievement was publicly announced, he was acclaimed as the discoverer. Scheele did not make his discovery public until he published *Chemical Treatise on Air and Fire* (1777).

Scheele's remoteness was his own choice. A German born in Swedish Pomerania, a part of Germany under Swedish rule, Scheele entered pharmaceutics in hopes of having time and equipment for chemical experiments. Although he was admitted as a member of the Royal Swedish Academy of Science in 1775 and published in its journal, he never went to Stockholm after settling in Koping, and turned down several offers of university positions. He did correspond with chemists outside Sweden, and his works were published in French and English.

Although Scheele held to a modified version of the phlogiston theory, his interest in chemistry was not theoretical, but practical and analytic. In his Koping laboratory, Scheele discovered, isolated, and identified the properties of a long list of chemicals. These included hydrofluoric acid, referred to at the time as "Swedish acid"; silicon fluoride; molybdic acid; tungstic acid; arsenic acid; and chlorine, whose discovery he announced in 1773. Scheele's studies in organic chemistry included many types of food, such as the fruits and berries that he examined for their acids. Scheele isolated citric acid from lemons in crystallized form, lactic acid from milk, oxalic acid from sorrel, malic acid from apples, and mucic acid from milk sugar. Scheele was also interested in the chemistry of dyeing. An eighteen-year study of the color "Prussian blue" led to the isolation of prussic acid, and the devising of a new green coloring, copper arsenite, known as "Scheele's green." This turned out to be a disastrous discovery, as Scheele's green, widely used to color confections, was actually poisonous.

In animal chemistry Scheele devised a way to extract phosphorus from animal bones, rather than urine as was the common practice, and separated "acid of calculus," later known as uric acid, from kidney stones and

This somewhat simplified image of a chemical laboratory was used on the title page of Carl Wilhelm Scheele's Chemical Treatise on Air and Fire *(1777). (Edgar Fahs Smith Collection, University of Pennsylvania Library)*

urine. He used such classical chemical concepts as crystalline form, solubility, and points of boiling or melting to distinguish his substances, but he also made use of taste, a common chemical habit. Considering the highly toxic nature of some of the substances he worked with, this may account for his early death.

See also Chemistry; Phlogiston; Royal Swedish Academy of Science.

Reference

Dictionary of Scientific Biography, s.v. "Scheele, Carl Wilhelm."

Sexual Difference

The Enlightenment period saw a sharpening of the scientific concept of sexual difference, with the sexes increasingly defined as opposite. Male physicians and scientists continued to view the male as the norm, but moved from the traditional picture of the female as an inferior male to one of the female as a radically different kind of being. Science was only one of many cultural spheres where this distinction was increasingly important—others were politics, particularly after the French Revolution introduced the political equality of men, and the moral philosophy of Jean-Jacques Rousseau (1712–1778).

The ancient model of sexual difference that had viewed female reproductive anatomy as fundamentally a variation on male anatomy, with the genitals inverted rather than forced out (the vagina as the equivalent of the penis), already challenged during the scientific revolution, disappeared from scientific discourse by the early nineteenth century (although it persisted in popular culture and medical handbooks aimed at a popular audience). The identification of the male spermatozoon and what was thought to be the female egg (the graafian follicle was wrongly thought to be the egg until Karl Ernst von Baer [1792–1876] discovered the true female egg) made it clear that female and male contributions to conception were naturally different, rather than being simply a mingling of fluids. Menstruation was conceptually separated from male nosebleeds or hemorrhoidal bleeding, a connection that appears as late as the work of Albrecht von Haller. The ovaries gained that designation, rather than being referred to as the "female testes." The idea that women, like men, needed to climax sexually for a child to be conceived was also abandoned by many learned physicians and scientists, and women were increasingly regarded as the less sexually passionate of the two genders. This was a startling inversion of traditional Western thinking on the genders, which had identified women's sexual lusts as much greater than those of men.

This tendency to conceive of genders opposed worked for the exclusion of women from science, as rationality was gendered as a masculine quality. Women were characterized not as beings whose reasoning power was inferior to men's, but as completely nonreasoning beings. This allowed fewer openings for the "exceptional" intellectual and scientific woman, who now was labeled unnatural. Although not all male philosophers and scientists held such beliefs, they were held by such leading male thinkers as Rousseau and Immanuel Kant.

Not all scientific innovations in the period contributed to the new idea of opposite sexes. Advances in embryology led to more awareness of the homology of the penis and clitoris in developing fetuses. However, evidence of similarities between the sexes had less influence on discourse generally than evidence of difference. The growing emphasis on the difference between male and female sexual organs contributed to conceptions of overall female nature. The role of reproduction as the key differential between the sexes can be seen in the criticism of the female skeletal illustration produced by the German anatomist Samuel Thomas von Soemmerring (1755–1830) as possessing an insufficiently broad pelvis. Representations of female skeletons were more common in eighteenth-century anatomy books, but they were there to illustrate specifically female characteristics—generically human ones were illustrated by males. Women's functions in reproduction, identified with the uterus and ovaries, were considered determinants of the proper social role of woman as a wife and mother. The breast was also increasingly conceptualized primarily in the role of nursing children rather than as an adornment.

The concept of sexual difference extended far outside the human realm. No one did more to promote sexual difference as a key organizing characteristic for all kinds of life than Carolus Linnaeus. Linnaeus's "sexual system" of plant classification, building on the idea of plant sexual reproduction, relied on close analogies between plant and human sexes. Plant and human sexual organs were considered to be closely homologous, with, for example, the filaments of the stamens in plants homologous with the vas deferens in animal males. Linnaeus, whose domestic arrangements were thoroughly conventional, and his disciples proclaimed the social as well as the functional homology of plant and human sexes, referring to plants as "husbands" and "wives." This tendency reached its peak in Erasmus Darwin's epic, *The Loves of the Plants* (1789). The radical atheist Darwin had a different view of plant sexual relations than Linnaeus, replacing botanical marriage and domesticity with plants with multiple and incestuous lovers. His work was sometimes claimed to be improper for female readers, although so was the less obviously scandalous work of Linnaeus and his popularizers.

Linnaeus's botanical system enshrined the dominance of the masculine category. Plants were first divided into classes based on the number and arrangement of the "male" parts, the stamens. Only then were the classes divided into orders on the basis of the female parts, the pistils. Hermaphroditic plants and organisms such as ferns lacking obvious sexual parts proved a problem for Linnaean botanists.

Sexual difference also played a central role in Linnaean animal classification. Linnaeus made a sex-linked characteristic, the female breast, the defining characteristic of the class Mammalia, the class to which humans belong. (Linnaeus strongly advocated breast-feeding.) The identification of the breast as the mark of the class, rather than hair, the four-chambered heart, or other uniquely "mammalian" characteristics, led to a prolonged controversy in the early nineteenth century as to whether the recently discovered platypus, the females of which had mammary glands but oozed milk through pores rather than through nipples, was a mammal. Like plants, animals, particularly apes, were increasingly anthropomorphized in their sexual relations.

Naturphilosophs and other romantic scientists, who saw a universe structured by cosmic dualities, identified these dualities as male and female and correlated them with other dualities. *Naturphilosophs* conceived of the animate universe as a pregnant female. This strongly gendered cosmology also tended to exclude women from active participation in science.

See also Anatomy; Botany; Education; Embryology; Masturbation; *Naturphilosophie;* Physiology; Popularization; Women.

References

Laqueur, Thomas. *Making Sex: Body and Gender from the Greeks to Freud.* Cambridge: Harvard University Press, 1990.

Schiebinger, Londa. *Nature's Body: Gender in the Making of Modern Science.* Boston: Beacon, 1993.

Sloane, Sir Hans (1660–1753)

The physician Sir Hans Sloane was the greatest collector in early modern science and antiquarianism. He was born in Ireland, making his way to London in 1679, where he made friends with the scientists Robert Boyle (1627–1691) and John Ray (1627–1705). In 1683 he went to Paris, where he studied at the Royal Botanical Garden and made friends with the botanist Joseph Pitton de Tournefort (1656–1708). On his return to London the sociable Sloane joined Boyle and Ray as a fellow of the Royal Society in 1685 and became a fellow of the Royal College of Physicians in 1687. The same year he departed as physician to the newly appointed governor of Jamaica.

In 1689 Sloane returned with a huge plant and animal collection (none of his living animals survived the voyage) and the body of the governor, which he had pickled for preservation. Sloane set up what became a hugely successful medical practice in London and began to publish his observations of Jamaica's natural history in *Philosophical Transactions.* The first book-length fruit of Sloane's Jamaican sojourn was a botanical reference in Latin, *Catalog of Plants of the Island of Jamaica* (1696). This book's value was enhanced by its cross-references to other books of Caribbean botany, a rarity in the world of natural history. Sloane also promoted new world products as medically valuable, notably quinine and chocolate, of which he was the principal early promoter in England.

Sloane became a leader in the London medical and scientific world. He was prominent among the informal group of London botanists who sometimes met at the Temple Coffee House. He became secretary of the Royal Society in 1693, keeping the position for twenty years and reorienting *Philosophical Transactions* toward natural history and away from mathematics and physics. Sloane was an avid correspondent throughout his career, and restored some of the links with continental science that had been allowed to wither since the death of *Philosophical Transactions*'s first editor, Henry Oldenburg (c. 1619–1677). Not everyone was pleased with the direction Sloane was taking *Philosophical Transactions.* His interest in wonders and marvels made him a target of satire, and Sloane's rival physician and collector the prickly John Woodward (c. 1665–1728) attacked him as one who reduced science to trivialities. Sloane also published another book drawing on his Caribbean experiences, this one in English, *Voyage to the Islands Madera, Barbados, Nieves, S. Christophers, and Jamaica* (two volumes, 1707 and 1725).

In medicine Sloane received many offices, becoming physician to Christ's Hospital in 1694, a royal physician in 1712, and president of the Royal College of Physicians in 1719. He was the first physician to receive the hereditary title of baronet in 1716, and one of the first in England to endorse and promote smallpox inoculation. Combined with his marriage to a Jamaica planter's widow, Sloane's medical practice made him very wealthy. His wealth was applied to the great passion of his life: collecting. He had already made a great collection in Jamaica, and he added to it by buying or otherwise acquiring entire collections. He bought the collection of physician and natural philosopher Walter Charleton (1620–1707) on Charleton's death, and that of apothecary James Petiver (c. 1663–1718) on Petiver's. Sloane also acquired 8,000 specimens from the herbarium of Leonard Plukenet (1642–1706). The Reverend Adam Buddle (c. 1660–1715), the leading expert on England's grasses and mosses, bequeathed Sloane his well-preserved and cataloged collection of dried plants and insects. Many of the collections Sloane acquired were not as well organized as Buddle's, and much of his and his assistants' labors were directed toward organizing and cataloging the collections, which by 1729 took up eleven rooms in Sloane's house. The collection and Sloane's extensive library, including the superb catalogs of his collections, were accessible to interested persons.

Sloane became president of the Royal Society after Isaac Newton's death in 1727. His elevation took place after some debate, as he unquestionably represented a very different scientific tradition than did Newton. Sloane proved a good president, restoring the society's finances and hiring his personal assistant, Cromwell Mortimer (1698–1752), to reorganize the society's neglected collections. Deteriorating health forced Sloane to step down as president in 1741. In his remaining years the question of the disposition of his collection, in which several groups were interested, became an increasing preoccupation. His final solution was to give the British Parliament first refusal, on the condition that if they accepted the collection they pay his heirs 20,000 pounds and take responsibility for housing and maintaining it. This was done, and Sloane's collection became the nucleus of the collection of the newly founded British Museum.

See also Botany; Colonial Science; Medicine; Museums; Periodicals; Royal Society; Smallpox Inoculation.

References

Jessop, L. "The Club at the Temple Coffee House: Facts and Supposition." *Archives of Natural History* 16 (1989): 263–274.

MacGregor, Arthur, ed. *Sir Hans Sloane: Collector, Scientist, Antiquary, Founding Father of the British Museum.* London: British Museum Press, 1994.

Smallpox Inoculation

Smallpox was one of the most feared killers in early modern Europe, and even survivors frequently bore ugly scars or "pocks." The eighteenth century saw the first real advances in its treatment with the introduction of inoculation, first by variolation and then by vaccination. Variolation, the infecting of a healthy person with a mild smallpox to create immunity, had had a long history in Asia and Africa and was mentioned in scattered European publications beginning in the seventeenth century; accounts appeared in *Philosophical Transactions* in 1714 and 1716. One of the main sources for the introduction of the actual practice to Europe, however, was Lady Mary Wortley Montagu (1689– 1762). Montagu was the wife of the British ambassador to Turkey and observed the practice among Turkish peasants. On her return to England in 1721 she had her five-year-old daughter immunized by the surgeon Charles Maitland (1677–1748). Sir Hans Sloane was also working to promote inoculation in England. The practice was the subject of statistical

Britain's finest caricaturist, James Gillray, produced this memorably grotesque image mocking those who predicted dire results from vaccination. (National Library of Medicine)

demographic studies by the secretary of the Royal Society, James Jurin (1684–1750), and the physician Thomas Nettleton (1683–1742). Inoculation spread quickly in Britain and the Continent, particularly after George, prince of Wales, the future George II (1683–1760) had two of his daughters inoculated. (Royalty setting an example would be a pattern repeated many times in the spread of inoculation. Catherine the Great of Russia [r. 1762–1796] used matter from her own pustules to inoculate her courtiers, assimilating the new practice to the traditional role of the monarch as healer.) For many Enlightenment thinkers, inoculation became a symbol of the progress of modern science, but not all of its champions were Enlightened thinkers by any standard—in New England the Puritan minister and fellow of the Royal Society, Cotton Mather (1663–1728), who had lost wives and children to smallpox, was a leading champion of inoculation. He worked primarily from the reports in *Philosophical Transactions* and reports on African practices from his slave Onesimus. He began to openly promote inoculation during a smallpox epidemic in Boston in 1721. Mather suffered considerable personal attack from those who thought the procedure too risky. A bomb was actually tossed into his house, but fortunately it failed to explode. (Mather believed that the devil had taken possession of those who opposed inoculation and attacked him personally.)

Like many medical procedures, inoculation could be done as an elaborate ritual with purges and bloodletting by a physician, or quickly and cheaply by a surgeon. One dynasty of British surgeons, Robert Sutton (1707–1788) and his son Daniel (1735–1819), claimed to have inoculated 400,000 people in thirty years, with a very low death rate. Inoculation was never risk-free, and the fact that inoculated people did sometimes die of smallpox did hinder its spread. It also led to a controversy between Daniel Bernoulli and Jean Le Rond d'Alembert in the 1760s over Bernoulli's assertion that probability theory endorsed taking the small risk of inoculation over the large risk of dying of smallpox. D'Alembert, who endorsed inoculation, claimed that Bernoulli's argument was weak and based on inadequate data. He claimed that it was unrealistic to expect a person considering inoculation to treat the immediate risk of death by inoculation as strictly comparable to the risk of dying of smallpox spread out over the person's remaining lifetime.

The further refinement of smallpox immunization to eliminate that small risk of death was the work of the surgeon Edward Jenner (1749–1823), a product of the London experimental medical circle around John and William Hunter. Jenner learned that it was general knowledge among the country people where he practiced that dairymaids, subject to a mild disease called cowpox, never got smallpox. (This is why the word *vaccination* derives from the Latin word for *cow*.) In 1796 Jenner inoculated a boy named James Phipps with matter taken from a cowpox pustule on the dairymaid Sarah Nelmes. After a slight fever the boy was inoculated six weeks later with smallpox, which failed to have any effect. Jenner announced his procedure in 1798, in *Inquiry into the Cause and Effects of the Variolae Vaccinae*. This book had an immediate effect, running into a third English edition by 1801 as well as an American edition and translations into Latin, German, Dutch, Italian, French, Spanish, and Portuguese by 1803, the same year the Royal Jennerian Society was founded to promote vaccination. Sweden made vaccination compulsory, in one of the first public health campaigns by a government, and Napoléon had his army vaccinated. The excitement the procedure aroused led some to speak of the possibility of eradicating smallpox entirely. Although it was generally accepted that the procedure worked, how and why it worked remained a mystery.

See also Hunter Family; Medicine; Sloane, Sir Hans; Surgeons and Surgery.

References

Daston, Lorraine. *Classical Probability in the Enlightenment*. Princeton: Princeton University Press, 1988.

Middlekauf, Robert. *The Mathers: Three Generations of Puritan Intellectuals*. Oxford: Oxford University Press, 1971.

Porter, Roy. *The Greatest Benefit to Mankind: A Medical History of Humanity*. New York: W. W. Norton, 1998.

Spallanzani, Lazzaro (1729–1799)

The cleric and university professor Lazzaro Spallanzani was the most meticulous and creative biological experimenter of the Enlightenment. Spallanzani was initially intended by his father to enter the law, and his interest in natural science emerged late, fanned by his cousin Laure Bassi at the University of Bologna. Spallanzani was ordained a Catholic priest in 1757 and was appointed a professor at the newly founded University of Reggio the same year. He worked his way up the university ladder, moving to Modena in 1760, and finally accepting the chair of natural history at the University of Pavia at the invitation of the Hapsburg empress Maria Theresa (1717–1780) in 1768. He built up a great natural-history collection, under the care of his sister, Marianna Spallanzani, a fine naturalist herself, and became a highly respected scientist in Italy and elsewhere, admitted to many scientific academies and societies.

Spallanzani's concern for experimental technique came into play in his controversy with the English Catholic priest and experimenter John Turberville Needham (1713–1781) and Needham's ally Georges-Louis Leclerc de Buffon over spontaneous generation. Needham had boiled some beef broth for a few minutes, then put it in a flask sealed with cork. He claimed that the boiling would kill any germs in the broth and the sealing prevent other germs from entering. Thus, when the flasks were opened later, and microscopic examination found living organisms, they could have been produced only by spontaneous generation. In experiments carried out in 1765 Spallanzani established that boiling needed to go on for far longer to kill the originally existing germs, and that cork sealing was inadequate to keep out atmospheric germs. Instead, he used narrow-necked glass flasks that could be hermetically sealed by melting the glass. In *Accounts of Microscopic Observations Concerning the System of Generation of Needham and Buffon* (1765) Spallanzani claimed that by following these procedures, the broth or any other liquid remained free of organisms, thus disproving spontaneous generation. (Like Spallanzani's other scientific writings, this book was originally written in Italian, and widely translated.) The fact that breaking the glass, thus exposing the liquid to air, led to many organisms being found in it demonstrated that the air was full of organisms. However, Needham and his supporters claimed that prolonged boiling destroyed the capacity of the liquid to generate life, and the controversy over spontaneous generation remained unsettled.

Spallanzani also studied the regeneration of animals, a very popular topic in the midcentury. His careful experiments included the transplantation of the head of a snail onto another snail. He established that a younger animal was more regenerative than an older one, a principle sometimes known as "Spallanzani's law," and that limbs and even heads, but not internal organs, could be regenerated.

Another of Spallanzani's great experimental labors was establishing the necessity of sexual generation in the formation of the young of most animals. As an embryologist, Spallanzani was an ovist preformationist who believed that the spermatozoa had no role in conception. He claimed that they were parasites of the blood that were drawn to the testes before mating, possibly explaining the agitated behavior of males during the mating season. However, he did believe that contact with the semen, if not the sperm, was essential for an egg to develop into a living organism. He established this through one of the most famous experiments in the history of biology—the trousered frogs. Spallanzani had already observed that fertilization of frog eggs takes place outside rather than within the female's body. To establish the fact that the male's semen was necessary, Spallanzani fashioned and fitted trousers on the frogs, then allowed them to mate. The eggs did not develop into tadpoles. His subsequent experiments, involving the filtration of sperm, demonstrated that the fertilizing power dwelled within the viscous mass, rather than either a vapor or an aura, or the clear liquid that remained after the semen had been filtered. To further establish this point Spallanzani carried out the first successful experiments in artificial fertilization and insemination, including a famous one on a spaniel. Spallanzani continued to be an ovist, holding that the young were already formed in the egg. He claimed that the semen was necessary not in forming the animal, but in stimulating its heart to beating.

Although Spallanzani's experiments were often harsh on the animals, he could be nearly as harsh on himself. The series of experiments by which he established the power of the gastric juices to dissolve food involved swallowing cheesecloth bags with food in them and cords tied to them, then using the cord, which hung outside of his mouth, to pull the bag back and examine its contents. Spallanzani also established that the saliva plays a role in digestion, rather than merely lubricating and softening the food for swallowing.

In addition to his laboratory work, Spallanzani also enjoyed scientific traveling, and went on a number of journeys in Europe and Turkey, acquiring specimens and observing collections. Under the guiding hands of Lazzaro and Marianna, the museum at Pavia became one of the finest in Italy. Spallanzani remained intellectually active to the end of his life, supporting his Pavia colleague Alessandro Volta in his dispute with Luigi Galvani over animal electricity.

See also Bassi, Laure Maria Catarina; Embryology; Museums; Universities.

References

Hankins, Thomas L. *Science and the Enlightenment.* Cambridge: Cambridge University Press, 1985.

Pinto-Correia, Clara. *The Ovary of Eve: Egg and Sperm and Preformation.* Chicago: University of Chicago Press, 1997.

Stahl, Georg Ernst (1660–1734)

The German physician Georg Ernst Stahl was the most influential chemist of the early eighteenth century, as well as the foremost champion of medical vitalism, in opposition to the medical mechanism associated with Hermann Boerhaave. From a family of officials, Stahl received his M.D. from the University of Jena in 1684. He was appointed physician to the duke of Weimar in 1687, holding this position until 1694, when Frederick III, elector of Brandenburg (1657–1713), invited him to be second professor of medicine at the newly founded University of Halle. The religious atmosphere of Halle, the center of Lutheran Pietism, would have been congenial to Stahl. Like many medical professors, he carried on a medical practice while lecturing, and he briefly edited a journal, *Curious Observations Chemical-Physical-Medical,* which ran from July 1697 to May 1698 and covered practical topics like fermentation. In 1715 he left Halle to become physician to Frederick III's successor, Frederick I of Prussia (r. 1713–1740).

Stahl's chemistry is marked by its continuity with the older alchemical tradition, particularly as passed down by Johann Joachim Becher (1635–1682), and his avoidance of the new mechanical philosophies of René Descartes (1596–1650) and Isaac Newton (1642–1727). He found mechanical explanations of chemical phenomena, whether the oddly shaped corpuscles with hooks and eyes of the Cartesians or the short-range forces of the Newtonians, incomplete and often fanciful, and preferred the older chemistry's explanations in terms of principles. Stahl's chemistry was based on four fundamental elements. One was water, and the others were three varieties of earth that he took over from Becher. They were vitrifiable or fusible earth; inflammable earth, which Stahl named phlogiston; and mercurious or liquid earth. Combustion was caused by the emission of phlogiston, and the heating of a metallic calx (now called an oxide) with charcoal restored the metal by its reabsorption of the phlogiston in the charcoal. The four elements come together to form stable substances that cannot be decomposed by the chemist, such as the metals. Stahl derived a number of ingenious explanations of chemical processes based on his four elements and their combinations and transfers. Stahl also followed Becher and the German chemical tradition in his concern for practical applications. His first publication, *Zymotechnia Fundamentalis* (1697), was concerned with fermentation of beer and wine. Stahl's chemistry was passed on both through his writings and through his pupils and their pupils, who dominated German and Scandinavian chemistry and mineralogy. Stahl's chemistry was introduced into France by Guillame-François Rouelle (1703–1770) in midcentury. The French Enlightenment philosopher Baron Paul-Henri-Dietrich d'Holbach (1723–1789) also promoted Stahl's chemistry in France, translating Stahl's German exposition of phlogiston theory, *Occasional Thoughts on the Debate over the So-Called Sulfur,* into French in 1766, and another work, *Treatise on Salt,* in 1771. Stahl's systematic textbook *Fundamenta Chymia Dogmaticae et Experimentalis* was translated into English by Dr. Peter Shaw (1694–1763) in 1730 as *Philosophical Principles of Universal Chemistry.*

Stahl's medicine, like his chemistry, rejected mechanical theories, which Stahl found reductionistic and incapable of explaining the operation of a living body. He was even unwilling to apply his own chemistry to organic materials. Although inorganic chemical processes could be reversed, once a living thing had been altered or destroyed chemically, it could not be restored. Stahl was a nonmaterial vitalist, viewing living beings as animated by a soul, anima, which was nonmaterial. All living beings possessed souls, although only humans had immortal souls in the religious sense. The soul was responsible for regulating the body's functions and for carrying on purposive activities, and disease was a defense on the part of the soul against foreign matter invading the body. The decay of dead bodies shows that they cannot subsist once the anima is taken away. Stahl's writings had a great deal of influence on the development of eighteenth-century theories of irritability, beginning in the 1740s. In the medical world his theory competed with the iatromechanism of Boerhaave, but it was accepted at many medical schools in Germany, and in France at Montpellier's famous medical school.

See also Boerhaave, Hermann; Chemistry; Medicine; Phlogiston; University of Halle; Vitalism.

References

Brock, William H. *The Norton History of Chemistry.* New York: W. W. Norton, 1993.

Oldroyd, David. "An Examination of G. E. Stahl's *Philosophical Principles of Universal Chemistry.*" *Ambix* 20 (1973): 36–52.

Porter, Roy. *The Greatest Benefit to Mankind: A Medical History of Humanity.* New York: W. W. Norton, 1998.

Surgeons and Surgery

During the Enlightenment surgery became an increasingly specialized, prestigious, and scientific discipline. Surgeons traditionally extracted teeth and dealt with tumors, chancres, and other features on the outside of the body, and also performed operations such as amputation and the removal of stones from the bladder, lithotomy. These operations, extremely painful and dangerous under the best conditions, were performed only as a last resort and were not the basis of surgical practice. The treatment of sexually transmitted diseases was a long-term surgical mainstay, which they justified as the treatment of the sores and other disfigurements on the outside of the sufferer, although physicians occasionally attempted to take over the field. Surgeons were also branching out into new fields, including man-midwifery and hospital practice. Surgery was becoming more specialized—dental surgeons, the ancestors of the modern dentist, first appeared as a separate body of practitioners in eighteenth-century France. In Britain the expansion of the British fleet made the naval surgeon a common figure.

The surgeons had been rising in prestige during the seventeenth century, particularly in France, the intellectual center of European surgery. One milestone was achieved by the French surgeon C. F. Felix (1650–1703) in his removal of an anal fistula from Louis XIV (r. 1643–1715) in 1687, for which he was lavishly rewarded. The French surgeons benefited from their long-standing alliance with the monarchy, which contrasted with the domination of the physicians by the university faculty. The king's surgeon was considered the head of his profession, as the king's physician never was. In the eighteenth century the surgeons made a rapid series of gains. In 1724 the organization of Paris surgeons, the College of Saint-Come, received from the king the right to give public courses in anatomy. Among its teachers was Jean Louis Petit (1674–1750), inventor of the screw tourniquet. In 1731 François de la Peyronie, surgeon to King Louis XV (r. 1715–1774), secured the founding of a surgical academy, which subsequently asserted its identity as a learned society rather than a craft guild by publishing volumes of papers. The separation of the surgeons from the barbers, long a reality, was officially recognized by royal decree in 1743. The same decree recognized surgery as a liberal art, requiring its practitioners to have an M.A. The academy of the surgeons received the designation "royal" in 1748, and the College of Saint-Come was transformed into the College of Surgery in 1750. Surgical education increasingly moved to a classroom orientation with lectures and public demon-

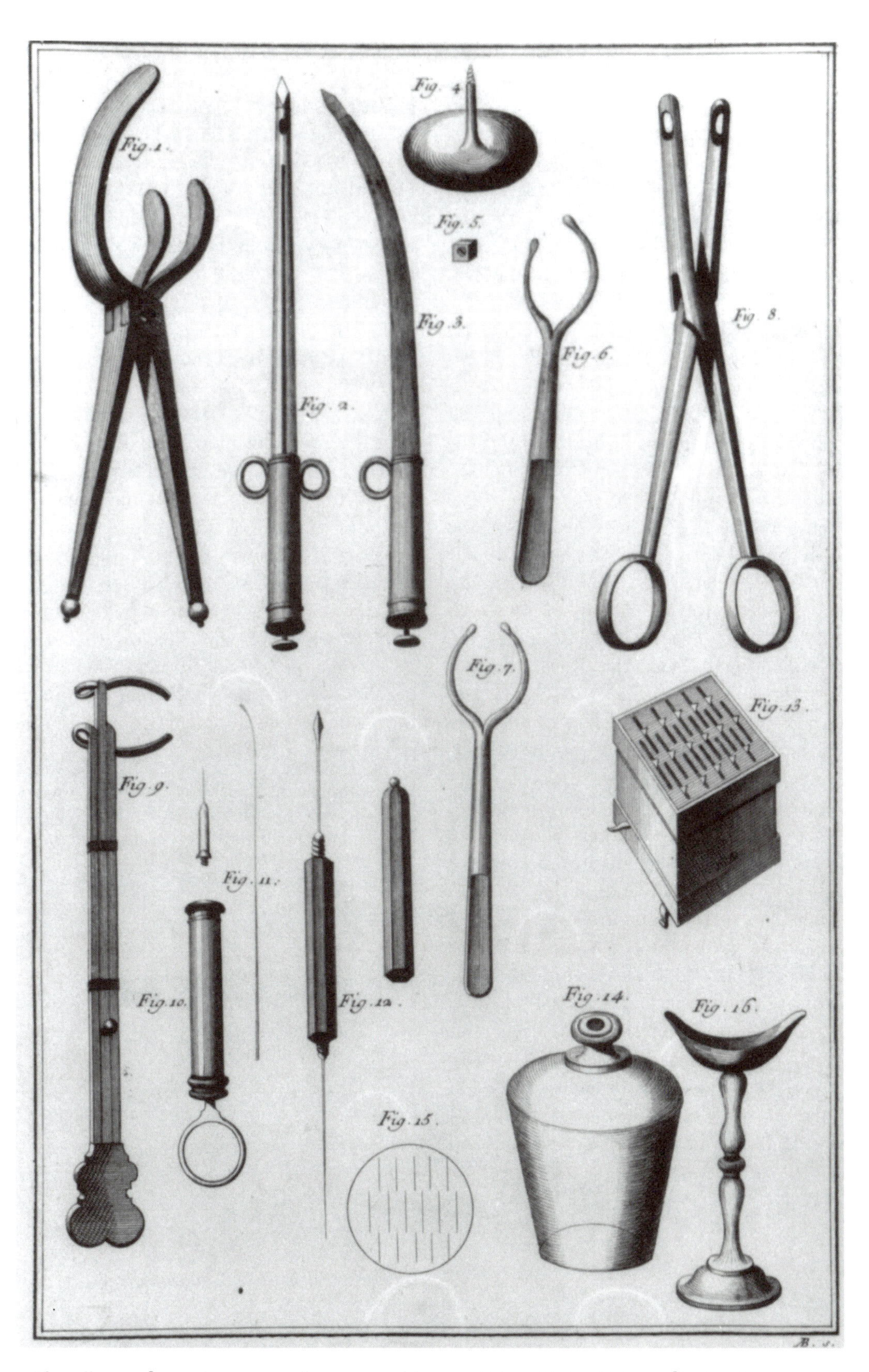

This collection of surgical instruments illustrates a work by Denis Diderot. (National Library of Medicine)

strations and away from the old apprenticeship model with secrets passed down from master to disciple.

The creation of the Saint-Come lectures and the subsequent gains for the surgeons aroused the ire of the Paris doctors. A pamphlet controversy raged between 1724 and 1750, involving mainly the elite of both medicine and surgery. In this struggle the surgeons cast themselves as the champions of a pragmatic medical reform based on empirical and clinical study. They allied with Enlightenment champions, notably Julien Offroy de La Mettrie, who although a physician himself supported the surgeons in his medical writings. The surgeons' victory was symbolized by both the Royal Academy and the College of Surgery moving into new quarters at the newly built Palais des Écoles in 1775. Although the French revolutionary government suppressed the Royal Academy of Surgeons, along with many other privileged bodies, the surgeons gained from the Revolution. In 1794 the government merged surgery and learned medicine, essentially on the surgeons' terms. The surgical tradition contributed greatly to the formation of early-nineteenth-century French clinical medicine.

There were fewer conflicts between the two professions in Britain, whose surgeons came to rival those of France during the eighteenth century. The appointment of the great surgeon Alexander Monro (1697–1767) as professor of anatomy and surgery at the University of Edinburgh in 1726 established an alliance between physic and surgery at Britain's premier medical school. The new surgical knowledge was brought to London by the Scottish Hunter brothers, William and John. William ran a school of anatomy and surgery, while John was the premier experimenter among surgeons. His and his brother's training also produced a generation of star surgeons, such as Edward Jenner (1749–1823), the pioneer of vaccination, and Sir Astley Paston Cooper (1768–1841), who repeated Felix's feat of social climbing by removing a cyst from the head of King George IV (r. 1820–1830) and being rewarded with a baronetcy.

Actual surgical practice changed little during the period. Speed and strength remained important qualifications—the surgeon Robert Liston (1794–1847) claimed that a lithotomy should never take more than two or three minutes. Surgeons developed a greater willingness to perform operations, and some new operations were created, such as the lateral cyostomy performed by the iterant French surgeon Jacques de Beaulieu (1651–1719) or the ovariotomy performed by the American surgeon Ephraim McDowell (1771–1830) in 1809. Real advances in surgical practice would occur in the nineteenth century, with the development of anesthesia and sterilization.

See also Bichat, Marie-François-Xavier; Brown, Robert; Hospitals; La Mettrie, Julien Offroy de; Medicine; Midwives; Smallpox Inoculation; War.

References

Gillispie, Charles Coulston. *Science and Polity in France At the End of the Old Regime.* Princeton: Princeton University Press, 1980.

Porter, Roy. *The Greatest Benefit to Mankind: A Medical History of Humanity.* New York: W. W. Norton, 1998.

Wellman, Kathleen. *La Mettrie: Medicine, Philosophy, and Enlightenment.* Durham: Duke University Press, 1992.

T

Technology and Engineering

The eighteenth century was a time of many important developments in different fields of technology, and the rise of the engineering profession. The relation of science and technology was complex and was not recognized in the eighteenth century in the way it is today. Some technological innovations, such as the use of chlorine for bleaching, originated in scientific discoveries, whereas others, such as the water-frame spinning machine of Sir Richard Arkwright (1732–1792), patented in 1769, had little to do with science. Science also benefited from technology, both in the creation of new and more precise scientific instruments and in the new problems posed by new machinery, as the steam engine influenced the theory of heat. Other scientific innovations would only begin to exert their technological effect in this period but would have an immense one later—the greatest example being Alessandro Volta's electric battery, invented in 1800.

Much of Europe's technological heritage by the eighteenth century had developed in practice, by trial and error and countless variations. This kind of technological knowledge was the province of skilled craftspeople rather than educated engineers, and much of it was passed down orally rather than written down. The imposition of the "scientific" or "rational" way of doing things on Europe's traditional manufacturing cultures was fraught with difficulty. Since the seventeenth century scientists had sought to capture artisans' knowledge in textual form, what the English called "Histories of Trades." In the eighteenth century several projects, including the *Encyclopédie,* for compiling this knowledge were undertaken. The collaborative *Description of Arts and Trades* (1761–1782) was supervised by the academician Henri-Louis Duhamel du Monceau (1700–1782), under the overall supervision of the Royal Academy of Sciences. Duhamel du Monceau's own *Treatise on Shipbuilding* (1747) was a handsome and beautifully illustrated volume that sought to reform and standardize shipbuilding practices as well as record them. It contains experiments on such subjects as how to arrange cords into ropes of the maximum tensile strength, and contributed to the superiority of French shipbuilding in the eighteenth century. He was a natural choice to head the description project, to which he contributed descriptions of twenty trades, ranging from commercial fishing to making tobacco pipes. Seventy-four treatises appeared in the series, some produced by academicians, others by educated artisans or amateurs.

The two most technologically innovative societies of the time, Britain and France, saw the profession of engineer develop in startlingly different ways. In France the engineer, whether military or civilian, was usually a state employee, and, increasingly, one trained in a state institution. British engineers were more likely to be associated with entrepreneurs than with state organizations and were often self-taught or trained by apprenticeship rather than institutionally educated. Although British engineers often moved in the same social and intellectual circles as scientists, they made fewer contributions to science than did French engineers. Despite British superiority in steam-engine building and operation, the most important theoretical description of the steam engine came from a French engineer trained at the Polytechnic School, Sadi Carnot (1796–1832), son of the great engineer Lazare-Nicolas-Marguerite Carnot (1753–1823). Sadi Carnot's *Reflections on the Motive Power of Fire* (1824) gave the first mathematical theory of steam power and the workings of heat engines.

The most important of the French civilian technological institutions for most of the eighteenth century was the famous School of Bridges and Roads, founded around 1747 and headed until his death by the great civil engineer Jean-Rodolphe Perronet (1708–1794). Students learned geometry, trigonometry, cartography, mechanics, and a number of other scientific and mathematical subjects. Thanks to the engineers the school trained, the Corps of Bridges and Roads, France became the first country in Europe since the Roman Empire to have an adequate system of highways. (Indeed, in many parts of Europe the best roads available in the eighteenth century were still those built by the Romans.) However, the graduates of the School of Bridges and Roads included no one of the scientific excellence found among France's military engineers, notably Charles-Augustin de Coulomb and Lazare Carnot. The tradition was further expanded after the French Revolution with the founding of the Polytechnic School in 1794. Inspired by the French example, the Austrian government founded Polytechnic Schools in Prague in 1805 and Vienna in 1815.

German principalities that were large enough to encourage technological development mostly followed the French model, with the founding of technical institutions such as the famous Freiberg School of Mining in the electorate of Saxony in 1767. Freiberg's head for many years, Abraham Gottlob Werner, transcended the role of technical instructor to be the leader of European geology. Another mining institution was founded in Hungary, a kingdom belonging to the Hapsburgs, at Schemnitz. Although teaching of chemistry, mechanics, and other relevant disciplines had been going on there for many years, the Royal Hungarian Mining Academy was founded in 1770. (France also had a mining school, founded in 1783.)

Although the British did not develop the technical training institutions that the French did, they did create effective voluntary associations dedicated to the advance of technology. Of these the most important was the Society for the Encouragement of Arts, Commerce, and Manufactures, founded in 1754. More informal groups, such as the Lunar Society of Birmingham, brought together manufacturers, engineers, and scientific researchers. British engineering education was on a much more informal basis than French, but nonetheless effective. Iterant Newtonian lecturers and demonstrators, such as John Theophilus Desaguliers or Benjamin Martin (1704–1782), used illustrations from contemporary machinery, including the steam engine, to demonstrate the principles of mechanics. The Royal Society recognized what would now be considered purely technical as well as scientific achievements, awarding its highest medal, the Copley Medal, in 1759 to the civil engineer John Smeaton (1724–1792), a fellow of the society. Smeaton also published in the society's journal, *Philosophical Transactions,* and encouraged engineers to think of themselves as a profession. He is credited with coining the phrase *civil engineer,* as the primary meaning of the term *engineer* was still a military one. (The Institution of Civil Engineers was founded in 1818.) The British civil engineer was not a state functionary, as his counterpart in France, but a contractor who worked for different clients, the state being one among them. The excellent educational system of Scotland and the poverty of the country made it particularly a nursery of engineers—Britain at the end of the eighteenth century began to catch up with the French road builders thanks largely to the canals, bridges, and roads built by the great Scottish engineers Thomas Telford (1757–1834), John Rennie (1761–1821), and John Loudon McAdam (1756–1836), who denounced French-style road building as too cumbersome and expensive and invented the "macadam" road, built using compacted broken stone.

The driving force in British industrial technology was not the state, as in France or Saxony, but private capitalist enterprise. One aspect of this was the dominating position in international science of the London instrument manufacturers. The instrument-making trade was a nursery of scientist-engineers who went on to other things, including Smeaton and James Watt (1736–1819). British manufacturers also looked to science and technological improvement to provide them with ways of making more profitable products. Watt's development of the separate condenser for the steam engine was financed by two entrepreneurs, first the ironmonger John Roebuck (1718–1794) and then, after Roebuck's business failure, the manufacturer and fellow member of the Lunar Society with Watt, Matthew Boulton (1728–1809).

A good example of the interplay of science and technology across national lines is the development of chlorine bleaching. Chlorine, originally known as "oxymuriatic acid," became known as part of the burst of scientific interest in isolating specific gases in the mid-eighteenth century. It was first isolated by the Swedish chemist Carl Wilhelm Scheele in 1773. The potentialities for oxymuriatic acid to replace the traditional method of bleaching, which relied on exposing cloth to the Sun for long periods of time, were realized at the French government dye works, the Gobelins. The director of dyeing at the Gobelins was Claude-Louis Berthollet, who spent years in the 1780s developing the chlorine bleaching process and rendering the highly poisonous gas to a state where it was safe to handle yet retained its bleaching effectiveness. However, the process was most effectively exploited not in France, but in the booming textile industry of Great Britain. Watt and Boulton, who learned of the process through Watt's correspondence with Berthollet, introduced it, and they were followed by other British textile entrepreneurs. A Scottish chemical manufacturer, Charles Macintosh (1766–1843), best known for his invention of waterproof fabric, further refined the process in 1799 by mixing chlorine with dry slaked lime, inventing bleaching powder. The most famous example of the direct application of science to technology, however, came not from a trained engineer or scientist, but from the amateur Benjamin Franklin: the lightning rod.

See also Berthollet, Claude-Louis; Böttger, Johann Friedrich; Education; Industrialization; Instrument Making; Lightning Rods; Longitude Problem; Lunar Society of Birmingham.

References

Alder, Ken. *Engineering the Revolution: Arms and Enlightenment in France, 1763–1815.* Princeton: Princeton University Press, 1997.

Cardwell, Donald. *The Norton History of Technology.* New York: W. W. Norton, 1994.

Gillispie, Charles Coulston. *Science and Polity in France At the End of the Old Regime.* Princeton: Princeton University Press, 1980.

Jacob, Margaret. *Scientific Culture and the Making of the Industrial West.* New York: Oxford University Press, 1997.

Telescopes

Both of the major types of telescope devised in the seventeenth century, the refracting and the reflecting models, underwent continuous improvement and refinement in the eighteenth century as London instrument makers displaced the Italians as leaders in telescopes and astronomical equipment. Improvements in the older refracting model at first seemed blocked by the authority of Isaac Newton (1642–1727), who had stated that the elimination of the phenomenon whereby light coming from the object observed was split into different colors—chromatic aberration—was impossible. This problem was solved on a practical level by an English landowner and amateur scientist, Chester Moor Hall (1703–1771), who had the idea of combining a convex lens of crown glass with a concave lens of flint glass. Workmen he contracted and supervised created the first achromatic refracting telescopes in 1733. Hall was not interested in fame or profit from his innovation from his discovery, however, and these first "achromats" had no immediate successors. Word of his innovation circulated in the community of London instrument makers, however, and a leading optician, John Dollond (1706–1761), began producing achromatic reflectors around midcentury, winning the Copley Medal from the Royal Society for his work in 1761. On a theoretical level, eliminating chromatic aberration had been shown to be possible by Leonhard Euler in a paper delivered to the Berlin Academy in 1747. Theoretical analysis of the achromat was further advanced by the Swedish physicist Samuel Klingenstierna (1698–1765) in 1754 and 1760. Dollond's son-in-law, Jesse Ramsden (1735–1800), continued the tradition of making fine refracting telescopes, fitted with micrometer microscopes.

Despite the success of the achromatic telescope, refracting telescopes were losing their usefulness for astronomical discovery. Most advanced astronomical work employed reflecting telescopes. The London instrument makers again took the lead. John Hadley (1682–1744) presented a six-inch "Newtonian" reflector to the Royal Society in 1721. James Short (1710–1768), a London Scotsman, refined the design of the reflector and produced more than 1,000 of them, although many were sold to astronomical dilettantes and had little effect on the progress of science. The reflecting telescope was described in a very popular book by Robert Smith (1689–1768), *A Compleat System of Opticks in Four Books* (1738). It was from this book, which was translated into German and French, that William Herschel and scores of others learned how to make and use reflecting telescopes.

Herschel established the reflecting telescope as the superior instrument for astronomical discovery, although the refractor was still widely used for close observation. The largest reflecting telescope, forty feet in length, was built for Herschel, who used it for observation intermittently from 1789 to 1813. However, the instrument was too cumbersome and took too much time to set up, so Herschel did most of his observations from other telescopes.

Supremacy in the making of telescopes and other optical instruments passed from England to Germany in the early nineteenth century. The British government, strapped for cash in the Napoleonic Wars, introduced a new tax on glass, devastating the industry. Technologically, the next important step was taken by a Swiss glassmaker, Pierre-Louis Guinand (1748–1824), who devised ways of producing optical glass of unequaled purity and consistency. He was temporarily lured to Munich in 1805, where his new techniques were absorbed by his German assistant Joseph von Fraunhofer (1787–1826). Fraunhofer applied much more rigorous tests to the glass he produced before employing it for telescopic lenses, whereas previous makers had taken what they were

This map shows the large and small Herschel telescopes amid the constellations. (Österreichische Nationalbibliotek, Vienna)

given. Fraunhofer's experiments on glass led to the identification of spectral lines, although not their explanation.

See also Astronomy; Herschel Family; Instrument Making; Observatories; Optics.

References

King, Henry C. *The History of the Telescope.* London: Charles Griffin, 1955.

North, John. *The Norton History of Astronomy and Cosmology.* New York: W. W. Norton, 1995.

Thermometers

The thermometer had been invented in the seventeenth century. In the eighteenth century it was greatly improved, and standardized temperature measurement was created with the modern Fahrenheit and Celsius temperature scales. Daniel Gabriel Fahrenheit (1686–1736), a German, worked on thermometers in the second decade of the century. He published descriptions of mercury and alcohol thermometers in *Philosophical Transactions* in 1724. Fahrenheit's original temperature scale was based on the fixed points of a mixture of ice, salt, and water at 0 degrees and human blood heat at 96 degrees. The use of 212, the boiling point of water, as a fixed point on the Fahrenheit scale began after his death. (Fahrenheit himself was aware that the boiling point varied according to atmospheric pressure.) Another early temperature scale was the work of the French scientist René-Antoine Ferchault de Réaumur (1683–1757). Réaumur's thermometers used alcohol, which he preferred for its greater volatility, rather than mercury. His temperature scale relied on two fixed points: the freezing point of water at 0 degrees and the cooling point of boiling water at 80 degrees. Réaumur's scale, whose use persisted for decades in France, was adapted to mercury thermometers by the Genevan meteorologist Jean André Deluc (1727–1817).

Building on French work, the Swedish astronomer Anders Celsius (1701–1744) introduced another scale based on the difference between the freezing and boiling points of water in 1742. Celsius's scale differed in several ways from Réaumur's. He used a mercury thermometer, divided the interval into 100 rather than 80 degrees, and inverted it so that 0 was the boiling point and 100 the freezing point. His most important innovation, however, was specifying a specific atmospheric pressure, measurable on the barometer, for the boiling of the water. Celsius's scale was reinverted to form the modern Celsius or centigrade scale shortly after he introduced it. Henry Cavendish further improved the thermometer by using the steam given off by boiling water rather than the water itself to set the high point. Interest in thermometers ran in Cavendish's family—his father, Lord Charles Cavendish, had been an early deviser of a thermometer capable of registering the maximum and minimum temperatures attained in the course of a period of time. These "registering thermometers" were further developed in the eighteenth century.

See also Cavendish, Henry; Heat; Instrument Making.

Reference

Middleton, W. E. Knowles. *Invention of the Meteorological Instruments.* Baltimore: Johns Hopkins University Press, 1969.

Thompson, Benjamin (Count Rumford) (1753–1814)

The American Benjamin Thompson lived one of the most adventurous and cosmopolitan lives of any Enlightenment physicist. From a New England farm family, Thompson was interested in science from an early age, and his apprenticeships with various tradesmen were cut short by his scientific activities. The turning point of his early career was the American Revolution, in which Thompson, a political conservative, joined the Loyalist Army. His patron, the royal governor of New Hampshire, Sir John Wentworth (1737–1820), secured his young protégé a commission as major in the New Hampshire militia. Early in the war Thompson spied for the British around Boston—an activity that suited both his devious nature and his knowledge of invisible ink. In 1776 he went to England, where he became an adviser on American affairs to the British secretary of state for the colonies, Lord George Germain (1716–1785). Relations between Thompson and his new patron were close enough for there to be rumors of their being lovers.

Thompson published his first paper, in *Philosophical Transactions,* while in London. It discussed measuring the strength of gunpowder, and gave a method for measuring it, which Thompson claimed as original but which he actually derived from an earlier source. Thompson's experiments demonstrated the superiority of dry powder to wet. In 1781 he returned to America as lieutenant colonel of a Loyalist regiment he was raising, the King's American Dragoons. After some successful encounters with the Swamp Fox, Francis Marion (c. 1732–1795), around Charleston, Thompson was transferred to New York, where he made himself thoroughly unpopular on Long Island by razing a church to use its timbers for fortifications, confiscating headstones to use in ovens, and generally ravaging the countryside. Thompson, an inveterate tinkerer, also designed a new portable gun carriage and a cork life preserver enabling a horse to swim across a river carrying a cannon.

On the war's end Thompson went back to England, but finding few opportunities he entered the service of the elector of Bavaria in Munich. He launched an ambitious program to reform the Bavarian military and attempted to end beggary with organized workhouses. In Munich Thompson began what were to be his chief claim to scientific fame: his long series of experiments on heat. His experiments to determine the best material for soldiers' uniforms led him to invent a "passage thermometer" to measure the conductivity of various fabrics. He claimed that the insulation properties of a cloth were the function of the air it contained. This was a significant discovery, and on its publication in 1792 Thompson received the Copley Medal from the Royal Society. Thompson, an avid collector of honors, may have been more pleased by the elector of Bavaria's decree the same year elevating him to the rank of count with the designation of Rumford, after a piece of land near his ancestral home in New England. In Bavaria he also met Countess Nogarola, one of his many mistresses, but a particularly useful one as she was interested in science and translated many of his publications into Italian.

A portrait of Benjamin Thompson, Count Rumford (Dictionary of American Portraits/Dover Publications)

Rumford's studies of physics always had a practical end and paid off richly with a series of innovations in ovens, cookers, and fireplaces. His innovations in fireplaces while on a visit to England in 1795–1796 made him extremely wealthy, which he commemorated by donating huge sums of money to establish prize funds at the Royal Society and the Boston-based American Academy of Arts and Sciences. (As he had prearranged, he himself was the first recipient of the Royal Society's Rumford Medal.) Back in Bavaria, he carried out a series of experiments on heat generation and conduction in solids and liquids. Rumford discovered conduction currents in heated liquids, and a famous experiment with the heat generated from boring a cannon led him to become a leading opponent of the dominant theory of heat: that it was a substance called caloric. Rumford claimed that the heat produced by friction in boring the cannon was "inexhaustible," and therefore not produced by a finite substance like caloric but by an ether-propagated vibration of the individual particles of the heated substance. The politics of the French revolutionary wars made Rumford's continuance in Bavaria impossible, and he returned to London in 1798. In London he turned down an offer to head the United States Military Academy, but helped found the Royal Institution, which he envisaged as a place for mechanics and laborers to be taught the latest technologies, which they would then disseminate throughout the country. The count was the leading spirit in the building of the institution, which was heated by an innovative and efficient system of steam heat of his own design, but the project developed away from his original vision as a venue for the delivery of lectures and demonstrations to an educated audience. One of Rumford's most important contributions to the future of the institution, which opened its doors in 1800, was the introduction of the young Humphry Davy. Rumford was less impressed by Davy's scientific work than by his potential use as a tool in the institution's power struggles.

A visit to Paris in 1801 introduced Rumford to the French scientific community and also to Napoléon Bonaparte, whom he greatly admired for crushing the pernicious revolutionary principles of liberty and equality. Rumford became friends with Pierre-Simon de Laplace, but his most important connection was with Marie-Anne Pierrette Lavoisier (1758–1836), the widow of Antoine-Laurent Lavoisier. So impressed was Rumford with the hospitality and charming ladies of Paris that after a brief return to London, where the Royal Institution was suffering severe growing pains, he permanently relocated to the Continent. By 1803 he had secured the necessary permission of the French government to reside, as an enemy national, in Paris. He was also admitted to the First Class of the Institute of France and gave several papers there. Rumford's chief scientific project now was the overthrow of caloric and its replacement by his own theory of heat as a vibration in an immaterial ether. He was planning to marry Madame Lavoisier, and was taken by the symmetry of her first husband overthrowing phlogiston and her second, caloric. After some difficulties obtaining the death certificate of Rumford's long-abandoned American first wife, the two were married in October 1805.

Their relationship did not long survive their marriage, and Rumford sequestered himself in his laboratory to escape his wife. (This added to the friction, as Madame Rumford had hoped to be an intellectual partner to Rumford as she had been to Lavoisier.) French xenophobia and Rumford's intellectual distance from the highly mathematical culture of Parisian physics added to his difficulties. He partially alleviated them by separating from his wife and moving to Auteuil outside Paris. There he continued to give papers at the Institute and work on improved designs for such mundane objects as lamps and coffeemakers as well as scientific instruments such as the photometer. Any grudge he may have borne against the United States was long forgotten, as can be seen from his will leaving his books to the still-unfounded American military academy and endowing the still-existing Rumford Professorship "of the Physical and Mathematical Sciences As Applied to the Useful Arts" at Harvard.

See also Colonial Science; Davy, Sir Humphry; Heat; Nationalism; Physics; Technology and Engineering; War.

Reference

Brown, Sanford C. *Benjamin Thompson, Count Rumford.* Cambridge: Massachusetts Institute of Technology Press, 1979.

Toft Case

The 1726 case of Mary Toft, the Surrey woman who claimed to have given birth to rabbits, involved much of London's medical elite and posed important scientific questions. Toft, the wife of a poor Surrey cloth worker, seems to have been driven by economic desperation and the hope that she could make money by being exhibited in London. She claimed her rabbit deliveries were caused by having been startled by rabbits when pregnant. This fitted in with the popular theory, among both ordinary people and scientists, that a fetus could be affected by the strong emotions of the mother—the "power of the imagination." Toft's claim was extreme, however, in that the normal result of the incident according to the theory would have been the birth of a human child with rabbitlike features. Toft was revealed as a fraud, thanks to an investigation by a Surrey magistrate who found out that her husband had been buying young rabbits for some time before her first delivery, and the arrest of a porter who was smuggling another rabbit to her. The leading London physician Sir Richard Manningham, F.R.S. (1690–1759), a skeptic from the beginning, extorted a confession from Toft by threatening to perform a painful operation on her.

The affair was a disastrous embarrassment for many London doctors, surgeons, and man-midwives. Both those who had supported Toft's claims and those that were skeptical, including Manningham and the Scottish anatomist James Douglas, F.R.S. (1675– 1742), were ridiculed in the popular press as credulous impostors. The most lasting scientific result was a book by the physician James Blondel (d. 1734), *The Strength of the Imagination in Pregnant Women Examined* (1727). In the first book-length attack on the theory of the power of the imagination, Blondel refuted it on the basis of a mechanist and preformationist embryology excluding maternal influences on development. This led to a controversy with the physician Daniel Turner (1667–1741).

See also Embryology; Medicine; Popularization; Whiston, William.

Reference

Todd, Dennis. *Imagining Monsters: Miscreations of the Self in Eighteenth-Century England.* Chicago: University of Chicago Press, 1995.

Transits

Observations of the transits of Venus across the face of the Sun in 1761 and 1769 were the greatest feats of international cooperation in eighteenth-century science. The idea of using transits to measure the solar parallax—the angular size of Earth as seen from the Sun—and thus arriving at the distance of Earth and the Sun by observing and timing Venus from different parts of Earth had been put forth by Edmond Halley (1656–1742) in 1716. It was known that the next two transits would be in 1761 and 1769, after which there would be no transits for more than a century. Transits of Mercury were more common, but less suited for drawing conclusions about astronomical distances. The transits of Mercury in 1723 and 1753 did provide practice for the transits of Venus, as well as helping maintain interest in the subject.

The most important astronomer to plan for the transits of Venus by midcentury was the Frenchman Joseph-Nicolas Delisle (1688–1768), a member of the Royal Academy of Sciences who refined the method and coordinated observations. The academy sponsored three expeditions to observe and time the transit of 6 June 1761 from remote areas. Guillame-Joseph-Hyacinthe-Jean-Baptiste Le Gentil de la Galasière was to observe from the French base at Pondicherry in India, Jean-Baptiste Chappe d'Auteroche (1722–1769) from Tobol'sk in Siberia, and Alexandre-Gui Pingue from Isle Rodrigue in the Indian Ocean. The Royal Society of London sent out two groups. Nevil Maskelyne went to St. Helena, and the astronomer Charles Mason (1728–1786) and the surveyor Jeremiah Dixon (d. 1777)—the first joint endeavor of the pair who were to become famous for the Mason-Dixon line—were to go to Bencoolen in Sumatra. In the British colonies of North America, the Harvard professor John Winthrop (1714–1779) led an expedition to observe the transit from St. Johns in Newfoundland.

These expeditions met with varying degrees of success. The whole project was complicated by the fact that Britain and France were on opposite sides in the Seven Years' War. Pondicherry's capture by the British frustrated Le Gentil, whereas Mason and Dixon were forced to make their observations from the Cape of Good Hope. Maskelyne was frustrated by a cloud passing over at the crucial moment, whereas Chappe d'Auteroche, Pingue, and Winthrop were successful. Of course, many other observations were made in more prosaic locales, with the Royal Swedish Academy of Science being particularly active in sponsoring astronomers in the north. Although many amateur astronomers observed the transits, the total number of observations of sufficient reliability for European scientists to incorporate them into their calculations was 120, made at 62 observing stations scattered around the globe. Subsequent calculations established the solar parallax as between 8.5 and 10 seconds of arc, more exact than previous calculations but still frustratingly imprecise.

The observations of the transits of 3 June 1769 benefited from the experience of 1761, as well as the fact that Britain and France were now at peace. Whereas the French had taken the lead in 1761, in 1769 the British were responsible for more observations—69 out of 150—than anyone else. The most famous British expedition was James Cook's in the *Endeavour,* accompanied by astronomer Charles Green (1735–1771). For an observation in Hudson Bay, it was necessary to build the first portable observatory, because of the lack of wood to construct one on-site. Some of the observers of 1761 repeated their involvement. The luckless Le Gentil, who had spent the intervening time in the East so as to be ready, was frustrated by a cloud. Chappe d'Auteroche was even more unfortunate, as his expedition to San Juan del Cabo was nearly wiped out, and he himself killed by an epidemic. Only one member survived, to return with the observations. As in 1761 many observations were made by Jesuits, in what was virtually the order's last great scientific endeavor before its suppression in 1773. Subsequent calculators, benefiting from Leonhard Euler's masterly mathematical treatment of the problem published in 1769, narrowed the range to 8.43 to 8.8 seconds of arc. They were frustrated by the "black drop" effect, which made it difficult to ascertain the exact moment when the disk of the planet crossed that of the Sun.

The scientific contribution of the transit voyages was far greater than merely the transit observations. They encouraged cooperation among scientists and scientific societies. They also provided an opportunity for many other kinds of scientific information gathering. The most famous example is Sir Joseph Banks's natural-historical work on the *Endeavour,* but there are many others. Le Gentil was the first European to study Indian astronomy, and the expeditions made many navigational and natural-historical observations.

See also Academies and Scientific Societies; Astronomy; Cook, James; Exploration, Discovery and Colonization; Maskelyne, Nevil.

Reference

Woolf, Harry. *The Transits of Venus: A Study in Eighteenth-Century Science.* Princeton: Princeton University Press, 1959.

U

Universities

Universities played a smaller role in European science during the eighteenth century than before or since. They did absorb, however slowly, new scientific developments. The transition from Aristotelian to mechanical philosophy was under way in Protestant Europe by the second half of the seventeenth century. It had reached the leading Catholic schools by 1700. Spain, dominated by reactionary Catholics, was the last holdout. Its universities continued to teach Aristotelian natural philosophy until the mid-eighteenth century. The form of mechanical philosophy that most continental European universities had adopted in the early eighteenth century, however, was Cartesianism, and it was necessary for another intellectual revolution to occur in the mid-eighteenth century for them to adopt Newtonianism. The pattern was somewhat different in the universities of Protestant Germany and Scandinavia, where Newtonianism faced a formidable rival in the modified Cartesian-Leibnizian physics of the Halle professor Christian Wolff. Universities also adapted to the newly emerging discipline of experimental physics, although this presented a problem due to the cost of acquiring and maintaining the equipment.

However quickly the universities adapted to changing science, they were not expected to be at its forefront. A contrast often drawn in the eighteenth century was between scientific academies, expected to be scientifically innovative and research-oriented, and universities, whose main function was pedagogical.

Of the four faculties into which early modern universities were divided—the undergraduate faculty of arts, or philosophy, and the graduate faculties of law, theology, and medicine—science was taught in the arts and medical faculties. Medicine was the major exception to the rule that universities were not centers of new science. A university M.D. or at least a university background remained virtually indispensable for anyone who wanted the illustrious title of physician, and thus medical faculties retained a crucial role as gatekeepers. New medical ideas were disseminated through university medical schools. (The great exception was England, where the university medical faculties were moribund, and a network of schools in London took their place.) The most important physician in the early eighteenth century was the Leiden professor Hermann Boerhaave, whose students and followers colonized other institutions such as Edinburgh and Vienna. The breadth of scientific knowledge covered in university medical departments was narrowing, however. Disciplines such as chemistry and botany, nearly always taught in the medical school previously, were acquiring independent status and moving to the arts faculty. In Sweden, for example, the chemistry chairs established in the second half of the eighteenth century were included in the faculty of arts, where they were supposed to teach statecraft and economic development. University medicine did broaden in one area, though, when in 1773 the ancient medical center of Padua became the first university to offer a course in veterinary medicine.

The role of universities in science varied tremendously by country. In France the universities became essentially irrelevant in nonmedical science, although the medical faculties of Paris and Montpellier still held their own. Their place was taken by the scientific academies and a broad range of secondary schools that emerged outside the university system. In England Cambridge early adopted a Newtonian curriculum emphasizing mathematics, but it never became a center of advanced science. In Scotland, by contrast, the universities, particularly Edinburgh, were major scientific powers, mostly in the medical faculties. Universities in Catholic Europe outside France were somewhat hindered in adopting the new physical theory by the Church's anti-Copernicanism. Catholic higher education benefited from the freer atmosphere caused by the suppression and eventual abolition of the Jesuit order, although it also lost many good teachers.

The center for development of universities as research institutions was Germany. German universities had always played a more central role in intellectual life than universities elsewhere, partly because there were so many of them—Germany's political fragmentation had led universities to proliferate, as many local rulers wanted one in their territory. By midcentury Germany's leading institution was Göttingen, founded in 1734 in the territory of Hanover. Göttingen was strongly backed by the electors of Hanover, who were also kings of Great Britain, and benefited from a policy that suppressed the stifling confessionalism that was the bane of German universities. Its prominent science professors in the late eighteenth century included Johann Friedrich Blumenbach, Georg Christoph Lichtenberg, and the chemist Johann Friedrich Gmelin (1748–1804). Göttingen differed from most German universities in that it incorporated small-group learning, where students worked directly with a professor rather than just listening to a lecture, into the formal curriculum. It also placed more emphasis on professorial research and publication than was common, although the modern distinction between scholarly and nonscholarly publication did not exist. Another star university was emerging around the same time in Italy, where Pavia benefited from the patronage of the Hapsburg dynasty. Its leading science professors were Lazzaro Spallanzani and Alessandro Volta.

Germany was the birthplace of a new idea of the university, which challenged the professional orientation that had been characteristic of European universities since the Middle Ages. Philosophers such as Immanuel Kant said that the ideal of education was *Bildung,* or cultivation, and the faculty of arts, where this was taught, should be the most prestigious, rather than the least prestigious, of the faculties.

University life in Europe suffered terribly from the Napoleonic Wars. Napoléon, in his rationalization of higher education, suppressed many universities in Germany and the Netherlands. In France itself the Napoleonic reorganization of higher education left the preeminent scientific and mathematical role of nonuniversity schools, such as the Polytechnic, largely unaffected. However, the period also saw the foundation of a new institution, drawing on the traditions of Halle and Göttingen and the ideas of philosophers such as Kant, that would eventually lead to the resurgence of universities in science. This was the University of Berlin, founded in 1809, whose guiding hand was Wilhelm von Humboldt (1767–1835), brother of Alexander von Humboldt. The new university abandoned the four-faculty arrangement and emphasized professorial research and scholarly publication. Teaching was in German rather than Latin. A particular mark of the new university's intended preeminence in Berlin science was that the Berlin Academy was subordinated to it. Along with Göttingen, Berlin would pioneer the major role of German universities in nineteenth-century science.

See also Botanical Gardens; Education; Kant, Immanuel; University of Edinburgh; University of Halle; University of Leiden; Wolff, Christian.

References

De Ridder-Symoens, Hilde, ed. *Universities in Early Modern Europe.* Cambridge: Cambridge University Press, 1996.

Shaffer, Elinor S. "Romantic Philosophy and the Organization of the Disciplines: The Founding of the Humboldt University of Berlin." In *Romanticism and the Sciences,* edited by Andrew Cunningham and Nicholas Jardine, 38–54. Cambridge: Cambridge University Press, 1990.

Turner, R. Steven. "University Reformers and Professorial Scholarship in Germany, 1760–1806." In *The University in Society II: Europe, Scotland, and the United States,* edited by Lawrence Stone, 495–531. Princeton: Princeton University Press, 1974.

University of Edinburgh

The University of Edinburgh emerged from obscurity in the seventeenth century to leadership among European universities by the mid-eighteenth century. It was already among the very first institutions to teach Newtonian natural philosophy, as Newton's Scottish disciple David Gregory (1659–1708) had taught there from 1683 to his departure for Oxford in 1691. Reform of the Edinburgh medical faculty had begun by 1708, with the appointment of Robert Eliot to a newly created professorship of anatomy. In 1720 he was succeeded by Alexander Monro (1697–1767), launching the most long-lived dynasty in academic history, as Monro's son, Alexander Monro II (1733–1817), and grandson, Alexander Monro III (1773–1859), would hold the chair without a break until 1842. In 1726 Edinburgh became one of the first schools to establish a chair of midwifery. In collaboration with the Edinburgh College of Physicians, the medical school established the Royal Infirmary in 1729, and clinical teaching became an Edinburgh specialty (although fewer than half the medical students took the clinical course). Monro and the other leading medical men of early-eighteenth-century Edinburgh were products of Hermann Boerhaave's medical school of Leiden, and the medical philosophy of the early-eighteenth-century medical school was Boerhaavian.

There was no separate science school at Edinburgh. Science chairs were divided between the medical school and the arts faculty. The first "star" appointment in the sciences Edinburgh made was in the arts faculty. That was the Newtonian mathematician Colin Maclaurin (1698–1746), appointed to the chair of mathematics in 1725. Maclaurin was already employed by the University of Aberdeen, and his appointment launched the town council of Edinburgh, which was responsible for hiring professors, on its long career of poaching top talent from other Scottish universities. Like other British schools, Edinburgh taught calculus by the Newtonian system of "fluxions," isolating its students and professors from many developments on the European continent.

By the second half of the century Edinburgh had replaced Leiden as Europe's leading medical school. It possessed a galaxy of scientific talent, led by the chemistry professors William Cullen (1710–1790) and Joseph Black (both of whom Edinburgh lured from the University of Glasgow), who established chemistry as a discipline in its own right, separate from medical uses. Cullen's theory of medicine, based on the nervous system, displaced Boerhaave's in the medical school, although Edinburgh retained a practical rather

The University of Edinburgh in the early nineteenth century (National Library of Medicine)

than theoretical emphasis. The distinction between physicians and surgeons was largely ignored. Other leading professors included John Hope (1725–1786), professor of botany from 1761 to 1786, who greatly advanced knowledge of Scottish flora and reformed the university museum and botanical garden, and the professor of materia medica Francis Home (1719–1813), an industrial chemist.

The majority of the 17,000 medical students who passed through Edinburgh in its first century did not actually take degrees, but acquired qualifications. Edinburgh attracted English Dissenters, barred from the Church of England–affiliated English universities, as well as students from the British colonies in North America and the Continent. The first medical school in British North America, the medical school of the University of Pennsylvania, was founded in 1765 by Edinburgh alumnus John Morgan (1735–1789) and modeled on Edinburgh. Colonial and early national America's leading physician, Benjamin Rush (1745–1813), a signatory of the Declaration of Independence, was an Edinburgh graduate. Edinburgh's attraction for American medical students was not diminished by the American Revolution.

During the British wars with revolutionary France and Napoléon several new chairs were established by royal authority in disciplines supposed to be of immediate use, such as military surgery, a topic conceived to embrace the health and sanitation of armies rather than simply immediate battlefield surgery. However, the Edinburgh medical faculty in the early nineteenth century was considered to be in decline, as more and more professorial chairs went to the sons of professors, and the wars greatly diminished the flow of foreign students. Among the arts faculty the leading professors included the mathematician and geologist John Playfair (1748–1819), the popularizer (although not the inventor) of "Playfair's Axiom" in geometry. Edinburgh's luster was somewhat dimmed by the revival of the English universities and the rise of Glasgow in the nineteenth century.

See also Black, Joseph; Brunonianism; Medicine; Royal Society of Edinburgh; Universities; University of Leiden.

References
Horne, David Bayne. *A Short History of the University of Edinburgh, 1556–1889*. Edinburgh: University Press, 1967.
Porter, Roy. *The Greatest Benefit to Mankind: A Medical History of Humanity*. New York: W. W. Norton, 1998.

University of Halle

Halle, founded in 1693–1694 as part of a program of Prussian cultural aggrandizement, was Germany's leading institution for science and medicine in the first half of the eighteenth century. It was Lutheran, but its Lutheranism was of the unorthodox variety known as Pietism. It was less theology-dominated and intellectually conservative than were the other older German Lutheran institutions. Oriented to the administrative needs of the Prussian state, Halle became a leader in introducing science and lecturing in German, not Latin, to the German university world. It also became one of Germany's largest, with an enrollment of more than 500 students. Despite the school's troubles with Christian Wolff, who was driven away in 1723 as religiously suspect, it became a center of the dissemination of the Wolffian philosophy after Frederick the Great (r. 1740–1786) forced it to rehire him in 1740. Because the Pietists of Halle were quite concerned with missionary work, Halle became a center for the accumulation of scientific and ethnographic knowledge from many parts of the world. From the early days, Halle's medical school was particularly distinguished, with two of Europe's most prominent theorists on the faculty: the vitalist Georg Ernst Stahl and the mechanist Friedrich Hoffman (1660–1742). In 1754 Halle, for the first time anywhere, granted an M.D. to a woman, the feminist writer Dorothea Erxleben (1715–1762).

By the midcentury intellectual leadership of Protestant Germany had passed from Halle to the recently founded (1734) University of Göttingen in Hanover (although Halle remained the larger of the two). Halle missed one opportunity to build its strength in the sciences through stinginess, when it refused to buy Wolff's large and fine collection of experimental equipment on his death in 1754. Halle did not get an adequate equipment collection until the end of the century. However, it did continue to boast some outstanding professors, such as Johann Christian Reil and the chemist Friedrich Albert Carl Gren (1760–1798), a leading German opponent of Antoine-Laurent Lavoisier's new chemistry. Napoléon closed Halle in 1806, following the Prussian defeat at the Battle of Jena, but it survived and merged with the University of Wittenberg in 1817.

See also Reil, Johann Christian; Stahl, Georg Ernst; Universities; Wolff, Christian.

References
De Ridder-Symoens, Hilde, ed. *Universities in Early Modern Europe*. Cambridge: Cambridge University Press, 1996.
Heilbron, J. L. *Electricity in the 17th and 18th Centuries: A Study in Early Modern Physics*. 2d ed. Mineola, NY: Dover, 1999.
Hufbauer, Karl. *The Formation of the German Chemical Community, 1720–1795*. Berkeley: University of California Press, 1982.

University of Leiden

In the early eighteenth century, the University of Leiden in the Dutch Republic was the center for the dissemination of Newtonianism on the European continent and, under Hermann Boerhaave, Europe's leading medical school. Boerhaave's appointment in 1704 attracted others to both medicine and natural philosophy. The numbers of medical students there increased as they were declining at other Dutch medical schools. In natural philosophy Cartesian domination was waning with the older generation of professors. Two Leiden students in Boerhaave's time, Willem Jakob 's Gravesande (1688–1742) and Pieter van Musschenbroek (1692–1761), would adopt and promulgate Newtonianism as Leiden professors. Gravesande became professor of mathematics and astronomy in 1717, partly on the recommendation of Isaac Newton (1642–1727), whom Gravesande had met on a visit to England. Gravesande was not an innovative scientist, but he was an outstanding teacher (he was Voltaire's instructor in Newtonian philosophy) and textbook writer. His *Mathematical Principles of Physics* (1721) was influential in France, Germany, and England as well as the Dutch Republic itself. Working with Musschenbroek's brother, the instrument maker Jan van Musschenbroek (1687–1748), he also built an outstanding collection of experimental instruments that the university purchased after his death. Gravesande was not a slavish Newtonian and took a Leibnizian position on the *vis viva* (living force) controversy.

The death of Boerhaave in 1738 and Gravesande in 1742 ended the great age of Leiden science. The rise of other medical schools—many, like Edinburgh, staffed by Leiden graduates—cut into Leiden's medical preeminence. However, the university retained its position as the republic's leading scientific institution. Its lobbying helped prevent the establishment of a national scientific academy that could have been a rival, and it retained some significant teachers. Pieter van Musschenbroek, who became professor of mathematics in 1740, carried on the Newtonian tradition established by Gravesande in a more experimental as opposed to mathematical fashion, and was an important electrical experimenter—as witness the name of the Leiden jar. Boerhaave's pupil Bernard Siegfried Albinus (1697–1770), professor of anatomy from 1721, was an anatomist of distinction whose works on the muscles set a new standard for painstaking accuracy of illustration; and professor Hieronymus David Gaubius (1705–1780) was a leader in applying chemistry to medicine.

Although Leiden survived the Napoleonic reorganization of the Dutch university system in 1811, it shared in the general lassitude of early-nineteenth-century Dutch science.

See also Boerhaave, Hermann; Universities.

References

Ruestow, Edward G. *Physics at Seventeenth and Eighteenth-Century Leiden: Philosophy and the New Science in the University.* The Hague: Nijhoff, 1973.

Van Berkel, Klaas, Albert Van Helden, and Lodewijk Palm, eds. *A History of Science in the Netherlands: Survey, Themes, and Reference.* Leiden: Brill, 1999.

V

Vitalism

Although the eighteenth century is often identified as an age of mechanical science, the tradition that life and living things could not be reduced to matter in motion—vitalism—was by no means eclipsed. Vitalism had a long history in European thought, although the term itself was not always used. Although mechanism was dominant in the early eighteenth century, vitalism had an influential champion in Georg Ernst Stahl, who, drawing on Aristotelian and alchemical traditions, identified the vital force as a "soul." Later eighteenth-century vitalists interpreted the vital force in more material terms, as a natural force whose effects could be measured even if its true nature could not be known, like gravity.

Vitalism emerged in the early eighteenth century from the failure of pure mechanism to explain processes that seemed unique to living things, such as digestion and reproduction. Physicians and physiologists by midcentury had put forth a number of principles that seemed innate to living things and resistant to mechanical explanation. The Edinburgh professor Robert Whytt (1714–1766), in *Essay on the Vital and Other Involuntary Motions of Animals* (1751), claimed that a vital principle of irritability—a concept dating back to the seventeenth century—explained some actions of living bodies. The foremost champion of irritability was Albrecht von Haller, who performed many experiments to establish which body parts were irritable, and the functions and purposes of irritability. Haller was reluctant, however, to generalize beyond the evidence into theories of a universal vital force. In France vitalism in the Stahl tradition flourished at the leading medical school, the University of Montpellier. Vitalistic Montpellier graduates included Theophile de Bordeu (1722–1776), Gabriel-François Venel (1723–1775), Henri Fouquet (1727–1806), Paul-Joseph Barthez (1734–1806), and Jean-Joseph Menuret de Chambaud (1739–1815). The influence of these Montpellier vitalists in French and European culture was enhanced by the fact that all, led by Venel, were contributors to the *Encyclopédie* of Denis Diderot. Bordeu's medical philosophy identified the fibers of the body as possessors of "animality" and treated the life of the whole body as the sum of the vitality of the individual organs. Barthez, whose 1773 *The Principle of Human Life* was responsible for popularizing the word *vitalism,* took a more holistic approach, identifying vitality as a property of the whole body.

Some Enlightenment *philosophes,* most notably Bordeu's friend Diderot, were attracted to vitalism as a way to explain the actions of living things without invoking the notion of a "soul," identified with religious superstition. Diderot's "vital materialism" broadened to take the idea of a self-moving, active matter to nonliving as well as living things. The difference between organic and inorganic matter was one of degree, rather than kind. Other kinds of vitalism emphasized the distinction between vital matter and inorganic, "mechanical" matter. Drawing on the ideas of Louis Bourguet (1678–1742), Georges-Louis Leclerc de Buffon claimed that the smallest units of vital matter were "organic molecules" that living organisms took in from their environments, and from which they were composed.

Vitalism continued to be supported by the German *Naturphilosophs* and by some French physiologists, notably Marie-François-Xavier Bichat. Early-nineteenth-century vitalists like Bichat used the concept to establish a firm division between living and nonliving things, in doing so inventing the concept of biology, the science of what lives. Organs' and tissues' ability to sense and to react to stimuli was unique to living things. The German chemist Leopold Gmelin (1788–1853) incorporated vitalism into chemistry. His *Handbook of Theoretical Chemistry* insisted that organic compounds could be formed only by living bodies due to their complexity. This idea did not survive the synthesis of urea by Friedrich Wöhler (1800–1882) in 1828, but vitalism continued to claim the loyalty of many life scientists in the nineteenth century.

See also Bichat, Marie-François-Xavier; Diderot, Denis; Materialism; *Naturphilosophie;* Physiology; Polyps.

References

Hall, Thomas S. *History of General Physiology, 600 B.C. to A.D. 1900, Volume Two: From the Enlightenment to the End of the Nineteenth Century.* Chicago: University of Chicago Press, 1969.

Hankins, Thomas L. *Science and the Enlightenment.* Cambridge: Cambridge University Press, 1985.

Volcanoes

Volcanoes were relevant to several major issues in eighteenth- and early-nineteenth-century science. There were two major theories as to their origin. One was that volcanic eruptions were the escape of the Earth's central heat. The other was that eruptions had local causes, the accumulation of flammables, most commonly thought to be coal, beneath the Earth. Some geologists, known as vulcanists, believed that volcanic action was a major cause of geological transformation, but others thought that the rarity of volcanoes indicated that they were only a minor factor in Earth history. Volcanic action was particularly often associated with the presence of the mineral basalt. This theory received a powerful boost from the work of Nicolas Desmarest (1725–1815) who in 1763 and 1766 visited the Auvergne region of France, identifying a series of cone-shaped mountains as extinct volcanoes and associating them with flows of basalt. (Desmarest was preceded in the identification of the Auvergne volcanoes by another Frenchman, Jean-Étienne Guettard [1715–1786], who did not believe basalt volcanic.) However, the dominant geological school of the late eighteenth century, the "neptunist" followers of Abraham Gottlob Werner, identified basalt as sedimentary, and those few active volcanoes available for close study by Europeans, Etna and Vesuvius, emitted lava flows that did not resemble basalt.

The preeminent observer of active volcanoes in the late eighteenth century was the British diplomat Sir William Hamilton (1730–1803). His work on Vesuvius and his observations of its eruptions partially rehabilitated the vulcanist interpretation of Earth history and established the long period of time over which a volcano could display activity. Although some Wernerians by the end of the century accepted that basalt was volcanic, the theory espoused by both Werner and Desmarest of volcanoes being formed by underground coal beds remained dominant.

In the early nineteenth century Sir Humphry Davy put forth an alternative chemical theory of volcanic action, in which eruption was caused by the heat generated by the reaction of metallic oxides with water. Davy, an inveterate showman, even produced a model volcano to demonstrate his theory, which gained popularity among many Wernerian geologists. In 1818 the Wernerian geologist Leopold von Buch (1774–1853) employed Davy's model and his own observations of European volcanoes, including an eruption of Vesuvius, and evidence of Mexican and South American volcanoes provided by Alexander von Humboldt to distinguish between two types of volcanoes. "Craters of elevation" were formed when heat within the Earth raised land over a wide area, forming a large dome, and the other were true volcanoes formed when the Earth's heat actually punctured the crust.

The idea of volcanoes being outlets for heat emerging from the deep depths of the Earth, rather than a coal fire or chemical reaction near its surface, was held by James Hutton and his Huttonian followers, although their theory of geological transformation emphasized heat generally rather than volcanoes specifically. Hutton viewed volcanoes as providentially designed to emit heat in a local way, thus preventing the elevation of land over a large area that caused destructive earthquakes. Despite the fact that the eighteenth century witnessed several disastrous volcanic eruptions with great loss of life, Hutton and other geologists emphasized that they should not be viewed as destructive, and their eruptions were not providential interventions by an angry God.

See also Davy, Sir Humphry; Geology; Humboldt, Alexander von; Hutton, James.

References

Hallam, A. *Great Geological Controversies.* 2d ed. Oxford: Oxford University Press, 1989.

Laudan, Rachel. *From Mineralogy to Geology: The Foundations of a Science, 1650–1830.* Chicago: University of Chicago Press, 1987.

Volta, Alessandro Giuseppe Antonio Anastasio (1745–1827)

Alessandro Volta was the late Enlightenment's greatest electrical scientist, best known for his invention of the electric battery. From an Italian clerical family, Volta was originally intended by his family to be a lawyer, but preferred to experiment with electricity. He had little formal training in mathematics or natural philosophy, but was a genius with instrumentation. (He did have a good humanist education, though, and wrote a Latin poem on the discoveries of Joseph Priestley.) In his late teens Volta began corresponding on electrical matters with two of Europe's leading electricians, the abbé Jean-Antoine Nollet and the Turin professor Giambatista Beccaria (1716–1781). The brash youth proposed to Nollet that electricity followed Newtonian laws of attraction, arousing the older man's encouragement, although Nollet warned that the creation of a Newtonian theory of electricity would be difficult yet glorious.

Volta's first great electrical invention was the electrophorus, described in a letter to Priestley in 1775. (The Swede Johann Wilcke [1732–1796] had constructed a similar device in 1762, but Volta's work was independent.) The electrophorus was a device that worked by induction to produce electric sparks "perpetually," once electrified either by friction or by a spark from a Leiden jar. The following year he added to his reputation with the discovery of swamp gas, or methane, winning the chair of experimental physics at the University of Pavia. His discovery that swamp gas could be exploded with an electrical discharge led him to invent a device called an "electric pistol." Volta suggested that the electric pistol could be incorporated into a means of rapid communication over a distance, as a spark could be carried by wires and quickly discharge an electric pistol many miles away. This can be seen as an anticipation of the telegraph.

Volta added to his reputation by a program of correspondence with the leading scientists

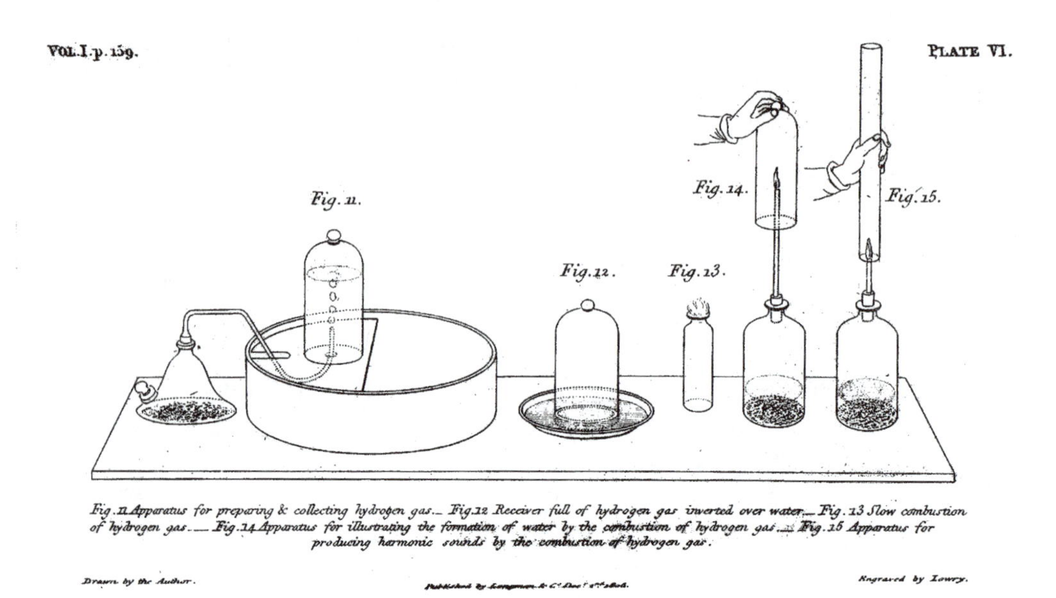

This illustration, from Jane Marcet's popular book Conversations on Chemistry *(1805), shows the use of a voltaic battery to decompose water into hydrogen and oxygen. The use of disembodied hands was a common method of demonstrating experimental procedure. (Edgar Fahs Smith Collection, University of Pennsylvania Library)*

and scientific academies and societies outside Italy. Particularly useful was Tiberio Cavallo (1749–1809), a Neapolitan electrician who had settled in London. Cavallo translated Volta's works into English and served as an agent for him in London. Much of Volta's electrical work was published in the journal of the Royal Society, *Philosophical Transactions.* He was elected a fellow of the society in 1791 and received its Copley Medal in 1794. Volta's quickly growing international reputation helped him procure lavish funding at Pavia, including instruments, a laboratory, paid staff, and a special lecture hall. He also left his position sometimes for a year at a time for travel to foreign centers of science.

In 1782 Volta announced another new device, the *condensatore,* a modification of the electrophorus into an instrument for detecting very weak electrical charges. This device enabled him to measure the minute charges left by a variety of natural processes. During a visit to Paris in 1781–1782, where he was lionized, he worked with Antoine-Laurent Lavoisier and Pierre-Simon de Laplace on a project to establish that the vaporization of water caused an electric charge. The effect Volta identified was actually caused by friction of the evaporating water on its container rather than by vaporization itself.

Volta's controversy with Luigi Galvani over Galvani's discovery of "animal electricity" dominated electrical debate at the close of the century. Volta denied the existence of animal electricity, holding that the twitching frog legs were reacting to an electrical stimulus generated elsewhere, rather than producing an electricity of their own. The frogs' legs functioned as an electrometer, far more sensitive than even those devised by Volta himself. Volta identified the electricity to which the frogs' legs were reacting as caused by the joining under moist conditions of two dissimilar metals. His experiments led him to rank the metals and other substances such as charcoal by their degree of effectiveness, a ranking that later became known as the "electrochemical series."

Volta's experiments with electricity generated by contact led him to the discovery of the electrical battery, or "voltaic pile," announced in a letter to the Royal Society in 1800. The voltaic pile consisted of a stack alternating between silver and zinc disks and moistened cardboard. It differed from all previous electrical generating machines in producing not a quickly discharged spark, but a steady stream, or "current," of electricity. This made it the single most important step in moving electricity out of the physicist's lab into the world of applied technology. In the self-sacrificing tradition of electrical experimenters, Volta used the pile to shock himself and examine the effects of the new electrical current on different parts of the body.

Volta's letter caused huge excitement in the scientific world. Despite the fact that Volta had announced his discovery in a British periodical at a time when Britain and France were at war, the French were as intrigued by the discovery as the rest of Europe. Volta was invited to speak before the Institute of France, the successor to the Royal Academy of Sciences, and his lectures before the Institute in November were graced by the presence of the first consul, Napoléon Bonaparte. Bonaparte and the Institute were much impressed, and Volta was not only admitted as a foreign member but also showered with prizes and rewards, including the title of count bestowed by Napoléon himself. Napoléon, who was inspired by Volta's presentation to announce an enormous prize of 60,000 francs for the next great discovery in electricity, also intervened on occasion to prevent Volta from retiring from Pavia. Volta escaped complete identification with the Napoleonic regime in Italy, however, and had no trouble when the rule of the Hapsburgs was restored—in fact, he was appointed as head of the philosophy faculty at Pavia in 1815, immediately following Napoléon's fall.

The voltaic pile was Volta's last great achievement in science. He finally retired from Pavia in 1819. His closing decades were devoted to his family—he was heartbroken by the death of his son Flaminio, a promising mathematician, at the age of eighteen. His other two sons, Giovanni and Luigi, published some work in industrial science.

See also Electricity; Galvani, Luigi; Napoleonic Science; Nollet, Jean-Antoine.

References

Dibner, Bern. *Alessandro Volta and the Electric Battery.* New York: Franklin Watts, 1964.

Heilbron, J. L. *Electricity in the 17th and 18th Centuries: A Study in Early Modern Physics.* 2d ed. Mineola, NY: Dover, 1999.

Voltaire (1694–1778)

François-Marie Arouet, better known by his adopted pseudonym, Voltaire, was not a scientist, but he was one of the most popular and influential writers on science in the Enlightenment. Voltaire initially became interested in science during a trip to England from 1726 to 1729. There he attended Isaac Newton's funeral and was impressed with how a man of Newton's humble origins was honored by the greatest in the land. Impressed with the relative political freedom and social equality of England, Voltaire, on his return to France, constituted himself as a missionary of English thought, including Newtonianism. Although Newtonianism had already begun to influence French scientific thought, Voltaire described science as the scene of an all-out struggle between French Cartesianism, with its dogmatic rationalism, and English and empirical Newtonianism. Along with Newton, Voltaire praised the emipiricists Francis Bacon (1561–1626) and John Locke (1632–1704) in his *Letters on England* (1732).

The most scientific phase of Voltaire's life was the period he spent with Madame Gabrielle-Émilie du Châtelet at her country estate of Cirey from 1734 to 1750. The two performed experiments, and both submitted entries to a 1736 prize contest sponsored by the Royal Academy of Sciences on the nature of fire. (Neither won; the prize went to a basically Cartesian paper by Leonhard Euler.) Voltaire also wrote a popularization of Newtonianism for the French, *Elements of the Philosophy of Newton* (1738). Although his weak mathematics prevented him from fully grasping Newton's theories,

Voltaire took Newtonian physics seriously enough to be seriously annoyed when Châtelet attempted to blend Newtonianism with Leibnizian mechanics in her *Institutes of Physics* (1740).

During the last years at Cirey and afterward, the focus of Voltaire's scientific interests shifted from physics to the life sciences. He was concerned that the growing vital-materialist tendency that ascribed self-activity to matter would lead to atheism—Voltaire, although no Christian, was a believer in an all-powerful God. For this reason he attacked the English Catholic priest John Turberville Needham (1713–1781), who had claimed that sealed containers of water produced living beings—"spontaneous generation." (It is a mark of Voltaire's unscrupulousness in polemic that he referred to the English secular priest Needham as an "Irish Jesuit.") Voltaire supported and corresponded with Lazzaro Spallanzani, who eventually disproved Needham's claims by demonstrating the flaws in the experiments. Voltaire also believed in a basically static history of the Earth, and attacked those, like Georges-Louis Leclerc de Buffon, who argued that it was marked by great changes. Voltaire's *Dissertation on the Changes of the World* (1746) suggested that the fossil fish found in the Alps were leftovers from the meals of travelers.

Although he never lost his reverence for Newton, the aging Voltaire grew more distant from the science of his day, which he saw as neglecting practical utility in favor of system building and an intellectually barren search for ultimate causes.

See also Châtelet, Gabrielle-Émilie du; The Enlightenment; Materialism; Newtonianism; Popularization; Religion.

References

Perkins, Jean A. "Voltaire and the Natural Sciences." *Studies on Voltaire and the Eighteenth Century* 37 (1965): 161–176.

Sutton, Geoffrey V. *Science for a Polite Society: Gender, Culture, and the Demonstration of Enlightenment*. Boulder: Westview, 1995.

W

War

Although the connection between war and science in the West dated to the scientific revolution, it intensified during the eighteenth century and the Napoleonic era. Eminent scientists who served in militaries or worked closely with militaries or on military projects included Edmond Halley (1656–1742), Luigi Ferdinando Marsigli (1658–1730), Johann Tobias Mayer, Ruggiero Giuseppe Boscovich, Antoine-Laurent Lavoisier, Count Rumford, Jean-Antoine Nollet, Charles-Augustin de Coulomb, Martin Heinrich Klaproth, Claude-Louis Berthollet, Pierre-Simon de Laplace, Gaspard Monge (1746–1818), Lazare-Nicolas-Marguerite Carnot (1753–1823), Jean-Baptiste-Joseph Fourier (1768–1830), and Robert Brown. Scientific exploration was often carried out by militaries, as the voyages of Captain James Cook were by the British navy, or the Egyptian expedition by the French army.

One area where military and scientific concerns overlapped was the artillery. Battlefield artillery was of increased importance during the period, and it was vital that the guns be aimed correctly so as to cause the most damage to enemy forces. This was the subject of some sophisticated mathematical work, notably that of Benjamin Robins (1707–1751), whose *New Principles of Gunnery* (1742) established muzzle velocity rather than range as the key measurement of a gun's effectiveness, and Leonhard Euler. This had only a limited impact on the actual practice of gunnery, which remained heavily influenced by rules of thumb and trial and error, but it did encourage militaries to train their artillery officers in mathematics. Another way in which artillerists required scientific knowledge was the need for gunpowder. In 1775 the leading politician in France, Anne-Robert-Jacques Turgot (1727–1781), requested that the Academy of Sciences consider the problem of the system by which the French army acquired gunpowder. (Turgot was a friend of leading scientists and a great believer in the application of science and reason to administrative problems.) Lavoisier was eventually appointed head of the new commission on powders and spent much time on his new task as state gunpowder administrator. His administration was a great success, enabling France to meet the needs of the war of the American Revolution and later the vastly greater ones of the French Revolution and Napoleonic Wars. This was more due to Lavoisier's administrative than scientific abilities.

The increasing professionalization of the officer corps of the European powers, and later of the United States, led to a need for scientific instructors in military education systems. France, the leader in science, also led in the application of science and mathematics to war. The first European educational institutions in which the curriculum was primarily scientific and technological, outside medicine, were the schools established in the eighteenth century, particularly in France, to train artillery officers. The first and for a long time the most important of these institutions was the French school at La Fère, founded in 1720 and disbanded in 1772. The British followed suit with the foundation of the Royal Military Academy at Woolwich in 1741, devoted to artillery and engineering. Military needs led to the introduction of advanced mathematics in other countries. In early-nineteenth-century America, the most advanced mathematical training was to be found not in the universities, but at the military academy at West Point.

The usefulness of the advanced mathematics taught at artillery schools on the battlefield was not always great. However, common mathematical training helped form artillery officers into a cohesive body. This was particularly important in the French military, as the artillery officers were socially of lesser status than cavalry and infantry officers. The ranking of students by their scores on mathematical exams helped form the meritocratic culture of the artillery, in contrast to other branches of the service where noble birth was more important. Before the French Revolution the artillery corps differed from other branches of the army in admitting nonnoble officers and forbidding the purchase of officers' commissions.

The military also required the services of trained engineers, who formed another military branch to which commoners could be admitted by merit as officers. In 1748 the famous French Royal Engineering School at Mézières was founded as a military institution, replacing the practice of recruiting military engineers on the basis of hereditary succession, informal examination, or apprenticeship. The two-year curriculum at Mézières was more advanced than that of the artillery schools, and instructors and examiners included Nollet and Monge. Its most illustrious scientific graduate was Coulomb, the first product of the military's scientific education to make a major contribution to theoretical science. Monge and Carnot carried much of the Mézières tradition to the founding of the Polytechnic School in 1795. The Polytechnic, originally a generalist rather than a specifically military institution, became Europe's leading institution for mathematical education. It also became increasingly oriented to military needs, with a growing proportion of its graduates entering the army, culminating in Napoléon's militarization of the school in 1804–1805.

Another branch of science in high military demand was medicine, to treat both wounded and diseased soldiers. (Due to poor sanitation and overcrowding military camps were often hotbeds of disease.) This work usually did not attract the most talented practitioners. Most of this work was carried out by surgeons rather than physicians, and often speed at performing amputations and cutting out bullets was more valued than theoretical knowledge. One scientifically minded, military medical man was the Scotsman Sir John Pringle (1707–1782), a graduate of Leiden's medical school and president of the Royal Society. From 1742 to 1758 Pringle was physician general of the British army and the author of several papers communicated to the Royal Society. His most important work was *Observations on the Diseases of the Army* (1752), which emphasized the need for hygiene. The British navy struggled for decades with scurvy, eventually adopting lemon juice in 1795.

The relation of scientists and the military intensified after the French Revolution. The most famous example is Carnot, an engineer and mathematician who became famous as the "organizer of victory" in revolutionary France. Carnot and another graduate of Mézières, Claude-Antoine Prieur (1763–1832), usually referred to as "Prieur of the Côte-d'Or," served as members of the Committee of Public Safety, the body of twelve that effectively ruled France during the Terror. Napoléon carried on the integration of scientists into military organization and culture, notably on the Egyptian expedition.

See also Cook, James; Education; Egyptian Expedition; Exploration, Discovery, and Colonization; French Revolution; Napoleonic Science; Nationalism; Surgeons and Surgery.

References

Alder, Ken. *Engineering the Revolution: Arms and Enlightenment in France, 1763–1815.* Princeton: Princeton University Press, 1997.

Gillispie, Charles Coulston. *Science and Polity in France At the End of the Old Regime.* Princeton: Princeton University Press, 1980.

Porter, Roy. *The Greatest Benefit to Mankind: A Medical History of Humanity.* New York: W. W. Norton, 1998.

Werner, Abraham Gottlob (1750–1817)

Abraham Gottlob Werner, the founder of geology as a science as well as the "neptunist" school of geological thought, like many German scientists had a close connection with the mining industry. He studied at the Freiberg School of Mining in Saxony, and after a period at the University of Leipzig returned to Freiberg to teach in 1775, holding the position for more than forty years until his death. Werner was an inspiring teacher who published relatively little. His first interests were in mineralogy, but he expanded from the identification and classification of rocks and minerals to what he called "geognosy," the science of the mineral body of the whole Earth and the arrangement of different minerals within it.

Werner's theories were mostly set out in his wide-ranging mineralogical lectures at Freiberg, although he did publish a German pamphlet in 1787, *Short Classification and Description of the Various Rocks.* Werner's neptunism (so-called for the ancient Roman sea god, Neptune) drew on previous German writers on stratigraphy to divide the Earth's geological strata into four main groups, later expanded to five. Werner's key innovation was to attach his stratigraphical scheme to a theory of the Earth's history. He identified bodies of rock formed at one time as "formations," thus identifying them by their time of origin rather than their chemical composition. Wernerian neptunism ascribed the arrangement of the Earth's strata to the settling out of different substances from an original universal ocean covering the entire surface of the Earth. (Unlike some of his British disciples, Werner, not particularly religious, did not link this ocean to the biblical story of Noah's Flood.) Initially, chemical precipitates such as granite and various slates formed the deepest strata, the "primitive rocks." Next were limestone and other "transitional" substances, being a transitional stage from chemical precipitates to substances settling by weight. The third layer Werner called "Floetz strata," and was divided into a number of subdivisions. It included salts and basalt, the sedimentary nature of which was to be one of Werner's most controversial assertions. The last two layers had been formed by ongoing processes after the universal ocean had disappeared and were less interesting to Werner. The fourth layer was composed of sands and clays formed by the erosion of the primitive rocks of the first three layers, whereas the fifth was composed of volcanic and similar rocks. Werner originally made little use of fossils in identifying strata, but became increasingly interested in them and encouraged some of his pupils to study them.

Werner's inspiring teaching won his theories a host of exponents throughout Europe and North America, and Werner disciples produced a multitude of geological surveys. However, Werner's neptunism was vulnerable to attack. Some questioned where the primeval ocean had disappeared to, a question Werner's followers dismissed as irrelevant to the validity of the theory. Werner also defended his scheme from a group of mostly French and Italian geologists known as "vulcanists," after the Roman god of fire, Vulcan. They ascribed much greater importance to the action of volcanoes than did Werner and the neptunists, believing basalt to be a product of volcanic eruptions. Some of Werner's prize pupils, such as the Frenchman Jean François d'Aubuisson de Voisons (1769–1819) and Leopold von Buch (1774–1853), became convinced by the evidence for the volcanic origin of basalt, and moved away from Werner's neptunism, although they continued to accept his concept of the rock formation. Alexander von Humboldt was another student of Werner who moved away from some of his assertions. By contrast, Werner's leading student in the English-speaking world, Robert Jameson (1774–1854), held steadily to the Wernerian faith. Jameson, who became Regius Professor of Natural History at the University of Edinburgh on his return from Freiberg in 1804, founded a Wernerian Society in 1808, but Werner's neptunism slowly retreated in the subsequent decades of the nineteenth century.

See also Geology; Hutton, James.

References

Hallam, A. *Great Geological Controversies.* 2d ed. Oxford: Oxford University Press, 1989.

Laudan, Rachel. *From Mineralogy to Geology: The Foundations of a Science, 1650–1830.* Chicago: University of Chicago Press, 1987.

Whiston, William (1667–1752)

The Newtonian William Whiston combined the seventeenth-century world of prophetic interpretation and heretical rebellion against religious authority with the eighteenth-century world of professional science. Whiston's religion was as Newtonian as his science—he agreed with the rejection of the Trinity in the theology of Isaac Newton (1642–1727). But whereas Newton was discreet about his anti-Trinitarianism, Whiston was anything but. Whiston made his scientific reputation with his *New Theory of the Earth* (1696), which associated the Flood of Noah with the near passage of a comet. This brought him to Newton's attention, and when Newton resigned his position as Lucasian Professor of Mathematics at Cambridge, Whiston succeeded him. Although not an original mathematician, his lectures and textbooks played an important role in the dissemination of Newtonian mathematics and astronomy. But his openly expressed heresy meant that he lost the position in 1710.

Although he had been ordained in the Church of England in 1693, Whiston could not deny his opinions, as did some of his associates, and conform in order to receive a position in the church. Bereft of a livelihood, Whiston relocated to London, where he became a scientific entrepreneur, giving lectures at coffeehouses and selling astronomical charts in connection with eclipses and other dramatic celestial events. This was a business that Whiston, along with the London printer John Senex, pioneered. His charts were of high quality. A copperplate engraving of the solar system including cometary orbits, *Scheme of the Solar System* (1712), was reprinted as late as 1760. Although most of his commercial writings on eclipses and other remarkable phenomena do not mention it, he interpreted many of them in providential and apocalyptic terms, as signs of the wrath of God and the approaching end of the world. This was a tendency that increased as Whiston grew older and more isolated from the mainstream of English science. He was one of very few English people to interpret the birth of rabbits to Mary Toft as a sign of the Apocalypse, which let him in for much mockery from satirists. His forthrightness about his religious opinions, which had led him to break with Newton, precluded his admission to the Royal Society. His alienation from the Church of England, particularly intense because he knew some of its leaders privately shared his opinions, grew so bitter that in 1747 he left for the General Baptists, a very marginal group.

Whiston's other scientific activities included giving private lessons and striving to solve the longitude problem. He and Humphrey Ditton (1675–1715) were instrumental in the parliamentary passage of the

Longitude Act in 1714, and the Longitude Board financed his survey of the coasts of England in 1742. He also received patronage from wealthy admirers, including the queen of England, Caroline of Ansbach (1683–1737).

See also Longitude Problem; Newtonianism; Popularization; Religion; Toft Case.

References

Farrell, Maureen. *William Whiston.* New York: Arno, 1981.

Force, James. *William Whiston: Honest Newtonian.* Cambridge: Cambridge University Press, 1985.

Walters, Alice N. "Ephemeral Events: English Broadsides of Early-Eighteenth-Century Solar Eclipses." *History of Science* 37 (1999): 1–43.

Wolff, Christian (1679–1754)

Christian Wolff was the best-known philosopher in Germany between Gottfried Wilhelm Leibniz (1646–1716) and Immanuel Kant, and invented much of the German philosophical vocabulary. His way of doing philosophy was deeply influenced by mathematics. He had started his career at Halle as a professor of mathematics in 1706 and wrote an influential textbook in the field. Wolff hoped to present a strictly deductive philosophy based on mechanistic causality, independent of theology. This stance got him into severe trouble with the Lutheran professors of Halle, who forced him to move to Marburg in 1723. Despite this setback Wolff's philosophy triumphed not only in Germany—Frederick the Great of Prussia (r. 1740–1786) brought him back to Halle in 1740—but also in areas that looked to Germany for intellectual leadership, such as Russia and Sweden. Although he turned down an offer to head the Imperial Academy of Sciences of St. Petersburg, Wolff's influence in selecting the members was very strong there. The fact that he wrote many works in German, whereas previous German philosophers had published mostly in Latin, helped him address a wide audience outside the academy as well.

Wolff's philosophical system was meant to cover every area of philosophy, including natural philosophy. His influence helped spread interest in experimental philosophy throughout the German academic world. A Leibinizian in physics, Wolff shared the master's distrust of Newtonian gravitation as an occult force and hoped to find a strictly mechanical explanation for it. The power of Wolff and his disciples significantly delayed the acceptance of Newtonianism in Germany. Wolff's philosophical system was eventually challenged and overthrown by the "critical philosophy" of Kant beginning in the 1780s.

See also Imperial Academy of Sciences of St. Petersburg; Kant, Immanuel; Physics; Psychology; Universities; University of Halle.

References

Heilbron, J. L. *Electricity in the 17th and 18th Centuries: A Study in Early Modern Physics.* 2d ed. Mineola, NY: Dover, 1999.

Saine, Thomas P. *The Problem of Being Modern; or, The German Pursuit of Enlightenment from Leibniz to the French Revolution.* Detroit: Wayne State University Press, 1997.

Women

Although women were excluded from the universities, learned professions, and scientific societies of Enlightenment Europe, some still managed to participate in science. Women's ability to function as scientists varied by discipline, institution, class, and region. The dominant institutions of eighteenth-century science—the scientific academies and societies—were very nearly completely closed to them. The astronomer Maria Winkelmann (1670–1720), for example, was treated as an embarrassment by the Berlin Academy and, despite her unquestioned ability, was gradually pushed out following the death of her astronomer husband, Gottfried Kirch (1639–1710). A rare exception was the headship of the Russian Imperial Academy of Sciences of St. Petersburg by Princess Yekaterina Dashkova (1744–1810) in 1783. This striking anomaly was connected to the fact that Russia at the time was ruled by a woman, Czarina Catherine the Great (r. 1762–1796). Catherine's despotic power enabled her to make such an appointment, which could not have occurred elsewhere in the scientific world. Universities were also closed to women, whether as faculty or as students. These exclusions were weakest in Italy, whose humanist culture had a tradition of accepting exceptional women. The career of Laure Bassi as a university professor and academician at Bologna could not have occurred in any other European region. Other eighteenth-century Italian women scientists who participated in institutional life include Cristina Roccati, a physics teacher at the Scientific Institute of Rovigo; the mathematician Maria Gaetana Agnesi (1718–1799); and the obstetrician Maria Dalle Donne (1776–1842).

One option for a woman to work as a scientist was as an associate of a male scientist to whom she was related. The outstanding example of such a partnership in this period is that of Caroline Herschel and her brother William Herschel. The Herschels were astronomers, and astronomy had an active tradition of women's participation—many observatories were attached to households and run as family businesses. Caroline worked both as a partner to William and on her own account, discovering several comets and contributing to the mapping of the skies. Caroline Herschel's astronomical participation was increasingly anomalous, and astronomical observatories became more institutionalized and professionalized, and thus closed to women. Another sister of a prominent scientist who participated in his work was Marianna Spallazani, who oversaw the natural history collection Lazzaro Spallanzani gathered at the University of Pavia. A prominent example of a wife who was an intellectual colleague of her scientist husband was Marie-Anne Pierette Paulze (1758–1836), wife of Antoine-Laurent Lavoisier. Marie-Anne assisted Antoine-Laurent in the laboratory and illustrated many of his works, including the famous *Treatise on Chemistry* (1789). She also translated chemical works from English into French for Lavoisier, who did not know English. Marie-Anne's second marriage, to Benjamin Thompson, Count Rumford, may have been inspired by the hope of re-creating such a scientific partnership, but it proved to be a disaster. Like observatories and museums, chemical laboratories were losing their original connections to households and becoming less welcoming to women.

One science that had a number of women practitioners was botany, sometimes identified as a "female science." Women had always been expected to have some botanical knowledge of the healing and other useful properties of available plants. An eighteenth-century innovation was the promotion of botany in its simplest form, the collecting and identification of plants, as a leisure activity suitable for ladies as a way of improving their minds (although some in England had doubts about whether the sexual classification system of Carolus Linnaeus was suited for maidenly ears). Some women went beyond this intellectually passive role to contribute to the science. Noblewomen could combine wealth, large tracts of land, and the leisure to devote to the cultivation of private botanical gardens. Mary Somerset, duchess of Beaufort (1630–1714), owned a large botanical garden and was also a leading breeder of insects. Another English lady who collected plants and made her garden available to botanists was Margaret Harley, duchess of Portland (1715–1785). Kew Gardens owed much to the patronage of women in Britain's royal family. However, European women were cut off from the most glamorous aspect of eighteenth-century botany: the collecting of plants from areas of the world remote from Europe. (One exception was Lady Anne Monson [c. 1714–1776], wife of an officer in the Indian army and associate of the Linnaean circle.)

Another botanical area that saw strong participation by women was illustration. The only woman to be on staff at the French Royal Botanical Garden in the eighteenth century was the illustrator Magdeleine Basseporte (1701–1780). The most technically innovative of eighteenth-century botanical illustrators was an Englishwoman and friend of Margaret Harley, Mary Delany (1700–1788). Delany devised a technique for illustrating plants using colored-paper cutouts, applying hundreds of bits of colored paper to copy living plants, producing illustrations combining beauty with such botanical accuracy that Sir Joseph Banks claimed that they were the only botanical illustrations from which he could unhesitatingly identify the living plant. One woman who broke from the usual pattern of women botanists as observers, illustrators, and popularizers was the Englishwoman Agnes Ibbetson (1757– 1823), whose specialty was plant physiology. Ibbetson published original research papers on the results of her dissections and microscopic observations of plants in general scientific periodicals such as the *Philosophical Magazine.* Although her work was respected by some male botanists, she was never able to publish her book-length manuscript setting forth her theory of how plants worked.

Healing and medical practice was a sphere in which women had been active historically, although opposition from male medical professionals was increasingly constricting women's range of medical operation in the eighteenth century. Although in England obstetrics and gynecology were being taken over by surgeons, against a fierce resistance by midwives, several continental midwives published scientific works on pregnancy, childbirth, and women's health. Lady Mary Wortley Montagu (1689–1762) was partially responsible for the introduction of smallpox inoculation into Europe.

Women were prominent among scientific translators. The most important scientific translation by a woman to appear during the Enlightenment period was Gabrielle-Émelie du Châtelet's translation of *Mathematical Principles of Natural Philosophy,* by Isaac Newton (1642–1727), into French, which appeared in 1759. This was not a mere translation, but expanded Newton's work to take account of developments in mathematical physics since his time. An allied area of cultural transmission was popularization. Jane Marcet (1769–1858), a student of Sir Humphry Davy, was a highly successful popularizer whose *Conversations on Chemistry* (1805) went through many editions in England and America as well as two French translations. She followed it with books on political economy, botany, plant physiology, and natural philosophy. Women were an important audience for popularized science, with many books and journals aimed at them. Although women with an interest in science were often mocked by male satirists as neglectful of their proper feminine duties, other men supported them, or even argued that natural knowledge made them better housewives.

The model of the professional scientist as it emerged in the late eighteenth and early nineteenth centuries offered little scope for female participation. The growing need for formal training, particularly in the mathematical sciences, was a great handicap. Mathematicians Sophie Germain and Mary Somerville (1780–1872) were highly respected by their male professional colleagues, but inability to enter scientific institutions limited their ability to contribute. The slow process by which women won entrance into scientific institutions was the work of the nineteenth and twentieth centuries.

See also Bassi, Laure Maria Catarina; Botany; Châtelet, Gabrielle-Émilie du; Coudray, Angelique Marguerite Le Boursier du; Germain, Sophie; Illustration; Midwives; Popularization; Sexual Difference.

References

Alic, Margaret. *Hypatia's Heirs: A History of Women in Science from Antiquity to the Late Nineteenth Century.* London: Women's Press, 1986.

Linney, Verna. "A Passion for Art, a Passion for Botany: Mary Delany and Her Floral 'Mosaiks.'" *Eighteenth-Century Women* 1 (2001): 203–235.

Shteir, Ann B. *Cultivating Women, Cultivating Science: Flora's Daughters and Botany in England, 1760 to 1860.* Baltimore: Johns Hopkins University Press, 1996.

Z

Zoology

Zoology in the Enlightenment was dominated by the French, including Georges-Louis Leclerc de Buffon in midcentury and the astonishing group of scientific talents at the Museum of Natural History in the early nineteenth century. The first in this line of French zoological savants was René-Antoine Ferchault de Réaumur (1683–1757), who worked mostly on invertebrate creatures. His six-volume *Memoirs on the Natural History of Insects* (1734–1742) was the most elaborate treatment of the subject yet to appear. His close observations established several truths about the social insects. Insects had an advantage over larger animals in that they could be more easily observed and collected, a practice that led to the formation of the first known scientific society devoted to a zoological topic. (Most general natural-history societies were dominated by botany.) This was the shadowy Aurelian Society, a London-based group of butterfly enthusiasts, the first references to which date from the early 1740s. The group was short-lived. A catastrophic fire in March 1748 destroyed its library and collections, and the Aurelians never recovered, although a successor society with the same name was founded around 1762.

Réaumur was relatively uninterested in classification, and that task was left to Carolus Linnaeus. Although primarily a botanist, Linnaeus extended his classification scheme to the animal realm as well. He divided animals into six classes. Four of these classes were vertebrate: mammals (a term Linnaeus invented), birds, a class including modern amphibians and reptiles, and fish. Two were invertebrate: insects including modern arthropods, like spiders, and a huge catchall class called Vermes, or worms. Unlike Linnaeus's famous sexual system of botanical classification, his scheme of animal classification was a "natural" scheme, drawing on a number of different traits, rather than an "artificial" one, arranged around a single characteristic. Linnaeus also originated binomial nomenclature, the identification of animal species by two Latin words, one for the genus and one for the species.

Buffon's *Natural History* was the rival authority to Linnaeus, at least for vertebrate creatures (Buffon ignored the invertebrates), but Buffon was a different kind of zoologist. He was interested in particularity rather than generality, and behavior more than anatomy and conformation. Buffon was suspicious of all classification schemes as imposing a set of human assumptions on the chaos of nature.

Even a species was defined not by conformity to an archetype, but as a population of mutually interfertile individuals. Buffon's and Réaumur's emphasis on behavior would be implicitly rejected by most of their successors, who focused on animal anatomy, working more from the dissection of dead animals than the observations of living ones. Another student of animal behavior whose influence was greater on subsequent literature than on zoology was the English clergyman Gilbert White (1720–1793), whose *Natural History and Antiquities of Selborne,* first published in 1789 and innumerable times thereafter, contains reports of close firsthand observations of many animals.

Although zoology in the eighteenth century was purged of the menagerie of fantastic creatures it had inherited from antiquity, the Middle Ages, and the Renaissance, its range of subjects still vastly increased. Like botany, zoology was transformed by the information brought back from exploration, including accounts of creatures exotic to Europeans, drawings, skins, teeth, skeletons, specimens preserved, and living creatures. Some of these creatures, most notably the platypus, posed thorny problems of classification. Another problematic source of zoological information, but an increasingly prominent one, was the fossil record. The strangeness of some fossilized animals raised the question of extinction: Could a species actually die out to the last member? It was difficult to answer this question with a firm no when much of the Earth's surface remained unknown to European science. Thomas Jefferson (1743–1826) hoped that Lewis and Clark's expedition would find surviving mammoths in the interior of America, and though it failed to do so, it did add several previously unknown species to the stock of Western zoological knowledge. The question of extinction also involved religious issues, such as how a species created by God could simply disappear.

Late-eighteenth-century zoology was dominated by dissection and comparative anatomy rather than the observation of animal behavior. One pioneer in this field was the English surgeon John Hunter, like many early animal dissectors a medical man. The great work of zoological classification was carried out in France, specifically at the Museum of Natural History, by two professors there: Jean-Baptiste-Pierre-Antoine de Monet de Lamarck for the invertebrates and Georges Cuvier for the vertebrates. Both of them established many of the fundamental taxonomic groups used by modern zoologists. Lamarck broke up Linnaeus's enormous and imprecise categories of worms and insects into a larger number of much more precisely defined groups. Cuvier integrated the fossil with the living world and came down firmly on the side of extinction. Both ended by denying the idea of a "great chain of being" in which all species could be arranged in a single hierarchy. Cuvier's skill at dissection and comparative anatomy marks an important stage in the development of zoology as an independent science, as he was not a physician and had no medical training.

The principal theoretical conflicts in early-nineteenth-century zoology were fought at the museum among three professors: Cuvier, Lamarck, and Étienne Geoffroy Saint-Hilaire. Cuvier upheld the fixity of species against Lamarck's belief in evolutionary transformation. Cuvier's approach to animal anatomy was functionalist, emphasizing the perfection with which each animal was adapted to its environment, and he argued that any fundamental change would render the creature unable to survive. Partly through argument and partly through political skill, Cuvier defeated Lamarck, who became increasingly marginalized. Cuvier's quarrel with Geoffroy Saint-Hilaire was about the relationships between different kinds of animals. Cuvier divided "the Animal Kingdom" (the title of his great work of classification) into four entirely distinct classes defined by their nervous systems: the vertebrates, mollusks, *articulata,* and *radiata.* Geoffroy Saint-Hilaire insisted that all animals were variations on one basic form. In 1830 a final confrontation between Geoffroy Saint-Hilaire and Cuvier at the Academy of Sciences ended in a victory on the merits for Cuvier. The victory was a Pyrrhic one for French science, however, as the strength of the Cuvier tradition greatly hindered the spread of evolutionism in nineteenth-century France.

See also Bonnet, Charles; Buffon, Georges-Louis Leclerc de; Cuvier, Georges; Fossils; Geoffroy Saint-Hilaire, Étienne; Lamarck, Jean-Baptiste-Pierre-Antoine de Monet de; Lewis and Clark Expedition; Linnaeus, Carolus; Museum of Natural History.

References

Allen, David Elliston. *The Naturalist in Britain: A Social History.* Princeton: Princeton University Press, 1994.

Corsi, Pietro. *The Age of Lamarck: Evolutionary Theories in France, 1790–1830.* Translated by Jonathan Mandelbaum. Berkeley: University of California Press, 1988.

Chronology

1699 Reorganization of the French Royal Academy of Sciences, with the admission of Cartesians; Thomas Savery's steam engine demonstrated at a meeting of the Royal Society; Gregorian calendar adopted by Denmark and the Protestant states of the Holy Roman Empire.

1700 Founding of the Berlin Academy, with Gottfried Wilhelm Leibniz as its first president; founding of the first Spanish scientific society, the Royal Society of Medicine and Other Sciences of Seville.

1701 Johann Friedrich Böttger held prisoner in Poland pending his making of alchemical gold.

1702 French Royal Academy of Sciences begins to publish annual reports; William Whiston succeeds Isaac Newton as Lucasian Professor of Mathematics at Cambridge University.

1703 Newton becomes president of the Royal Society; Hermann Boerhaave appointed professor of medicine at the University of Leiden.

1704 Publication of Newton's *Opticks;* first appearance of the *Ladies Diary,* English periodical with the purpose of introducing educated women to mathematics and science; publication of the first English dictionary of the arts and sciences, John Harris's *Lexicon Technicum.*

1705 Edmond Halley's *Synopsis of the Astronomy of Comets* predicts the return of the comet of 1682 in 1758.

1706 Founding of the Royal Society of Sciences of Montpellier; publication of a Latin edition of Newton's *Opticks,* with "Queries" that greatly influence eighteenth-century physics.

1707 Publication of Tobias Cohn's Hebrew textbook of medicine and natural philosophy spreads knowledge of modern science among the Jewish community.

1708 Böttger becomes first European to discover the formula of porcelain.

1710 Whiston is expelled from Cambridge and forfeits his chair for heresy.

1712 Halley publishes abbreviated version of John Flamsteed's astronomical data, igniting feud between Flamsteed on one side and Halley and Newton on the other; publication of a report of the Committee of the Royal Society chosen by Newton, backing Newton against Leibniz in the controversy over the origin of the calculus.

1713 Publication of Jakob Bernoulli's *Art of Conjecturing.*

1714 Halley takes over editorship of *Philosophical Transactions* from Sir Hans Sloane, reorienting it from natural history to mathematics, astronomy, and physics; an account of smallpox inoculation appears in *Philosophical Transactions;* approximate date of Daniel Gabriel Fahrenheit's introduction of the temperature scale; foundation of the Bologna Academy of Sciences; British Parliament's Longitude Act establishes the Longitude Board and prizes for successful methods of finding the longitude.

1715 Probable date of the first publication of *Onania,* inaugurating the medical crusade against masturbation.

1715–1716 Leibniz-Clarke controversy.

1717 Amalgamation of four London Masonic lodges to form the London Grand Lodge.

1719 On death of Flamsteed, Halley succeeds him the next year as astronomer royal.

1720 The French government establishes the Department of Maps and Charts in the Ministry of Marine to coordinate geographical and cartographical information; the Japanese shogunate liberalizes the law banning foreign books, exposing Japanese intellectuals to Western science.

1721 Lady Mary Wortley Montagu promotes smallpox inoculation in England, and Cotton Mather promotes it in Boston during an epidemic.

1722 Pierre Poliniere invited to perform his physical demonstrations before Louis XV.

1723 The University of Halle expels Christian Wolff for doubts about his religious orthodoxy.

1724 Foundation of the Imperial Academy of Sciences of St. Petersburg; College of Saint-Come, organization of Paris surgeons, receives from the king the right to give public anatomical courses; John Hutchinson publishes *Moses's Principia,* introducing "Hutchinsonian" natural philosophy.

1725 Publication of Luigi Ferdinando Marsigli's *Physical History of the Sea,* first book devoted to oceanography.

1726 Mary Toft affair in London.

1727 Death of Newton, succeeded as president of the Royal Society by Sloane; publication of Stephen Hales's *Vegetable Staticks;* Leonhard Euler arrives in St. Petersburg to take chair at the Imperial Academy of Sciences; founding of the Hollis Chair of Mathematics and Natural Philosophy at Harvard, with its first incumbent Isaac Greenwood.

1728 Publication of Ephraim Chambers's *Cyclopaedia.*

1729 James Bradley announces the aberration of starlight in a paper before the Royal Society; Royal Society and Imperial Academy of Sciences of St. Petersburg set up first official exchange of publications between scientific societies; founding of the Woodwardian Chair of Geology at Cambridge University; publication of Pierre Boguer's *Optical Treatise on the Gradation of Light,* announcing his invention of the photometer.

1731 Independent invention of the octant by John Hadley in England and Thomas Godfrey in Pennsylvania; founding of the Surgical Academy in Paris; publication of Jethro Tull's *New Horse-Houghing Husbandry,* the most influential agricultural book of the seventeenth century.

1732 Physicist Laure Bassi becomes first woman to receive a degree from the University of Bologna.

1734 Founding of the University of Göttingen.

1735 Charles-Marie de La Condamine's expedition to Ecuador to measure the Earth sets forth; publication of the first edition of Carolus Linnaeus's *System of Nature.*

1736 Pierre-Louis Moreau de Maupertuis's expedition to Lapland procures evidence for the prolate shape of the Earth; the Royal Society admits its first Jewish fellow, Moses da Costa; establishment of the Royal Society's Copley Medal.

1737 Founding of the Edinburgh Philosophical Society.

1738 Publication of Daniel Bernoulli's *Hydrodynamica;* Catholic Church condemns Freemasonry.

1739 Foundation of the Swedish Royal Academy of Sciences; Georges-Louis Leclerc de Buffon appointed director of the French Royal Botanical Garden; first fossil mastodon discovered on the banks of the Ohio River.

1740 Prospero Lambertini, the most important patron of science among eighteenth-century popes, ascends the papal throne under the name Benedict XIV. His pontificate will last until his death in 1758. Halle forced to readmit Wolff.

1741 Euler accepts Frederick the Great's offer to move to Berlin.

1742 Longitude Board finances Whiston's survey of the coasts of Britain; Guillame-François Rouelle introduces Georg Ernst Stahl's phlogiston chemistry to French chemists in his lectures at the Royal Botanical Garden.

1743 Publication of Jean Le Rond d'Alembert's *Treatise on Dynamics.*

1744 Publication of Abraham Trembley's study of polyps.

1746 Invention of the Leiden jar; publication of Jean-Antoine Nollet's *Essay on Electricity,* setting forth his electrical theory.

1747 Publication of Julien Offroy de La Mettrie's *L'Homme-machine.*

1748 Foundation of the French school of military engineering at Mézières, with which many important scientists will be associated; John Needham claims that microscopic organisms emerge spontaneously from decaying broth; publication of Euler's *Introduction to the Analysis of Infinites.*

1749 Imperial Academy of Sciences of St. Petersburg announces its first prize competition, on the relation of lunar motions to Newtonian theory; publication of David Hartley's *Observations on Man, His Frame, His Duty, and His Expectations;* publication of Étienne Bonnot de Condillac's *Treatise on System.*

1750 Publication of Jean-Jacques Rousseau's *Discourse Concerning the Arts and Sciences,* which argues that scientific and technological progress have made humanity more corrupt and unhappy; College of Saint-Come becomes a surgical college.

1751 Alexis-Claude Clairaut wins the St. Petersburg contest demonstrating that Newton's laws account for lunar motions; Charles Le Roy uses a condensation hygrometer to establish the dew point; publication of the first volume of the *Encyclopédie* with d'Alembert's *Preliminary Discourse;* publication of Benjamin Franklin's *Experiments and Observations on Electricity;* publication of Linnaeus's *Botanical Philosophy.*

1752 Franklin's kite experiment; Albrecht von Haller sets forth his theory of the irritability and contractility of fibers in a paper before the Royal Society of Sciences of Göttingen; inauguration of the Leipzig journal *Commentaries on Occurrences in Natural Science and Medicine,* first review journal in the sciences. It will run until 1798.

1753 Transit of Mercury with many astronomical observations; Georg Wilhelm Richmann killed by lightning while experimenting with a lightning rod; publication of Linnaeus's *Species of Plants;* the death of Sloane leads to the establishment of the British Museum to house his enormous collections.

1754 James Lind's experiments on scurvy demonstrate the effectiveness of citrus fruits; Dorothea Erxleben receives from the University of Halle the first M.D. ever given to a woman anywhere.

1755 Destructive earthquake in Lisbon raises interest in earthquakes; Immanuel Kant's *Universal Natural History and Theory of the Heavens* sets forth the nebular hypothesis of the origin of the solar system.

1757–1766 Publication of Haller's *Physiological Elements of the Human Body.*

1758 Robert Symmer's experiments with his socks seem to support "two-fluid" theory of electricity; return of Halley's comet after successful prediction by Clairaut.

1759 Publication of Gabrielle-Émilie du Châtelet's French translation of Newton's *Mathematical Principles of Natural Philosophy;* Angelique Martinique Le Boursier du Coudray receives royal permission to travel France giving midwifery classes; establishment of Kew Gardens by Princess Augusta of Great Britain.

1760 John Michell reads important series of papers on the nature of earthquakes before Royal Society; publication of Samuel-August Tissot's *Onanism.*

1761 First transit of Venus across the face of the Sun with worldwide burst of scientific activity; publication of Giovanni Battista Morgagni's *On the Sites and Causes of Diseases* inaugurates pathological anatomy.

1764 Thomas Bayes's essay in *Philosophical Transactions* introduces notion of inverse probability.

1765 Nevil Maskelyne becomes astronomer royal of Great Britain; publication of Lazzaro Spallanzani's *Accounts of Microscopic Observations Concerning the System of Generation of Needham and Buffon,* setting forth his claims to have experimentally refuted spontaneous generation; founding of the Medical School of the University of Pennsylvania, the first medical school in British America.

1766 Henry Cavendish isolates hydrogen; Euler leaves Berlin for St. Petersburg at the invitation of Catherine the Great; the French captain Louis-Antoine de Bougainville begins a circumnavigation that will last until 1769; publication of Pierre Joseph Macquer's *Dictionary of Chemistry.*

1767 First publication of Maskelyne's *Nautical Almanac and Nautical Ephemeris.*

1768 William Hunter founds school of anatomy in London; foundation of the American Philosophical Society in Philadelphia; James Cook leaves England on the *Endeavour.* His circumnavigation will take until 1771. First appearance of the *Encyclopedia Britannica.*

1769 Second transit of Venus with more observations; Franklin elected president of the American Philosophical Society, a position he will hold until his death.

1771 Paris Observatory made independent of the Royal Academy of Sciences; inauguration of the French journal *Observations on Physics, Natural History, and the Arts,* known as "Rozier's journal"; Mordecai Gumper Schnaber Levinson publishes the first book in Hebrew on Newtonian science and its relevance for Judaism, *A Dissertation upon the Law and Science.*

1772 Daniel Rutherford isolates nitrogen; Louis-Bernard Guyton de Morveau publishes *Academic Digressions,* demonstrating that all metals gain weight when strongly heated in air; Antoine-Laurent Lavoisier's experiments on combustion lead to the formulation of antiphlogiston chemistry; Cook leaves England on his second voyage, which will take until 1775; Peter Simon Pallas investigates a large iron meteorite in Siberia, establishing that it has no connection with local iron deposits; Sir Joseph Banks appointed director of Kew Gardens; committee of the Royal Society created to consider the question of round versus pointed lightning rods; Gustav III's seizure of power in Sweden leads to decline in support for Swedish science; first publication of the French translation of Maskelyne's almanac.

1773 Carl Wilhelm Scheele isolates "oxymuriatic acid," later known as chlorine; dissolution of the Jesuit order precipitates major transformations in European education; an astronomical paper read before the Royal Academy of Sciences precipitates a panic in France, as people fear a comet will strike the Earth; Jean André Deluc and Horace-Bénédict de Saussure introduce new hygrometers; University of Padua is first university to offer a course in veterinary medicine.

1774 Joseph Priestley discovers "dephlogisticated air"; Maskelyne's experiment to measure the "Attraction of Mountains" in Schiehallion; first publication of Charles Messier's catalog of nebulae.

1775 Alessandro Volta announces the invention of the electrophorus; Abraham Gottlob Werner takes teaching position at Freiberg School of Mining; first official meeting of the Lunar Society of Birmingham.

1776 John Hunter secretly supervises the first known case of artificial insemination; Bassi wins chair of experimental physics at the Institute of Bologna; Cook leaves on his last voyage; founding of the Masonic Lodge of the Nine Sisters in Paris.

1778 Banks becomes president of the Royal Society, a position he will hold for more than forty years; foundation of Lorenz Florens Friedrich von Crell's *Chemical Journal for the Friends of Natural Science, Medicine, Domestic Economy, and Manufacturing;* foundation of the French Royal Society of Medicine.

1779 Publication of Jan Ingenhousz's *Experiments on Vegetables.*

1780 Founding of the Boston Academy of Arts and Sciences; founding of the Mannheim Meteorological Society; John Brown's *Elementa Medicinae* sets forth basic principles of "Brunonian" medicine.

1781 William Herschel discovers new planet; founding of the Manchester Literary and Philosophical Society; founding of the Masonic Lodge of True Harmony in Vienna.

1782 Volta announces the invention of the *condensatore,* for detecting very weak electrical charges; founding of the Derby Philosophical Society.

1783 Lavoisier and Pierre-Simon de Laplace announce the invention of the calorimeter and experiments carried out with it in a paper before the Royal Academy of Sciences; Edinburgh Philosophical Society reestablished as Royal Society of Edinburgh; Banks puts down a challenge to his presidency of the Royal Society launched by a group of mathematical scientists; first balloon ascensions; Coudray's last tour of France; Princess Yekaterina Dashkova appointed head of the Imperial Academy of Sciences of St. Petersburg, first woman to head a scientific society; Matthew Baillie inherits William Hunter's anatomy school on Hunter's death.

1784 Crell's journal goes monthly as *Chemical Annals;* reorganization of the General Hospital of Vienna; Erasmus Darwin helps found Derby Philosophical Society; committee of the Royal Academy of Sciences investigates Franz Anton Mesmer's "animal magnetism" and pronounces it worthless; publication of Samuel Thomas von Soemmerring's *On the Physical Differences between the Moor and the European* asserting "scientific racism."

1785 Manchester Literary and Philosophical Society begins to publish series of *Manchester Memoirs;* Lavoisier's paper before the Royal Academy of Sciences openly challenges phlogistonism; James Hutton's geological theories read before the Royal Society of Edinburgh; André Michaux commissioned royal botanist and sent to America to find useful plants for France; an innovative observatory built at the Irish town of Dunsink.

1786 Publication of Kant's *Metaphysical Foundations of Natural Science.*

1787 Charles-Augustin de Coulomb provides evidence for the existence of inverse-square laws governing electrical and magnetic attraction and repulsion in paper read before the Royal Academy of Sciences; Laplace's paper before the academy explains the secular accelerations and decelerations of Saturn and Jupiter; linking of the French and British cartographic systems by Jacques-Dominique de Cassini and William Roy; departure of the Spanish Royal Botanical Expedition; founding of major botanical gardens in Calcutta and St. Helena; founding of the French Linnean Society; publication of the *Method of Chemical Nomenclature* sets out Lavoisier's new scheme for chemical names.

1788 Publication of Joseph-Louis Lagrange's *Analytical Mechanics;* founding of the Linnean Society of London; George III's attack of madness raises interest in the subject. His cure is credited to the mad-doctor Francis Willis. Publication of Gilbert White's literary masterpiece, *Natural History of Selborne.*

1789 Publication of Lavoisier's *Elementary Treatise on Chémie;* foundation of the Lavoisierian journal *Annales de Chémie;* beginning of the French Revolution.

1790 French National Assembly sets forth plan for new, "objective" measurement; David Rittenhouse follows Franklin as president of the American Philosophical Society on Franklin's death.

1791 Luigi Galvani announces that he has made frogs' legs jump by touching them with a bimetallic connector; Nicolas Leblanc patents the "Leblanc process" for making soda; Linnean Society begins to publish *Transactions;* Priestley's house and laboratory in Birmingham attacked by a conservative mob.

1792 Beginning of the measurements to establish the basis of the metric system; Michaux unsuccessfully proposes that the American Philosophical Society sponsor a natural-historical exploration of North America as far as the Pacific; founding of the Royal Mining College of Mexico City, from which Lavoisierian chemistry will be disseminated; Martin Heinrich Klaproth accepts Lavoisierian chemistry, speeding its acceptance in Germany.

1793 Suppression of the Royal Academy of Sciences and the Royal Society of Medicine by the French revolutionaries; establishment of the Museum of Natural History as the successor to the Royal Botanical Garden; establishment of a Register of Accessions at Kew Gardens.

1794 Foundation of the French Polytechnic School; establishment of a menagerie of live animals at the Museum of Natural History; founding of the Conservatory of Arts and Trades in Paris; establishment of first American natural-history museum in the rooms of the American Philosophical Society; French revolutionary government merges surgery and learned medicine; Priestley emigrates to America; publication of Ernst Florenz Friedrich Chladni's *Concerning the Origin of the Mass of Iron Discovered by Pallas and Others Similar* claiming an extraterrestrial origin for meteors and meteorites; publication of Adrien-Marie Legendre's *Elements of Geometry.*

1795 Publication of the Marquis de Condorcet's *Sketch for a History of the Progress of the Human Mind;* founding of the Institute of France as successor to the Royal Academy of Sciences; foundation of the German *Archive for Physiology,* first specialist journal devoted to physiology; position of hydrographer to the British Admiralty established, with its first incumbent Alexander Dalrymple.

1796 Edward Jenner carries out the first smallpox vaccination.

1797 Publication of Friedrich W. J. Schelling's *Ideas for a Philosophy of Nature* inaugurates *Naturphilosophie;* Alois Senefelder invents lithography; Napoléon Bonaparte elected to the First Class of the Institute of France.

1798 End of the metric measurements; foundation of Thomas Beddoes's Pneumatic Institute, at Clifton outside Bristol. It will last until 1802.

1798–1801 French occupation of Egypt, with much scientific activity.

1799 Volta invents the voltaic pile, the first battery, providing a steady source of electricity; publication of Marie-François-Xavier Bichat's *Treatise on Membranes,* first book to analyze the human body as composed of tissues.

1799–1803 Alexander von Humboldt's scientific exploration of the Spanish colonies in America.

1799–1805 Publication of Laplace's *Treatise on Celestial Mechanics.*

1800 Herschel discovers infrared light; Napoléon Bonaparte serves six-month term as head of the Institute; opening of London's Royal Institution under the supervision of Count Rumford.

1801 Giuseppi Piazzi discovers the first known asteroid, Ceres; John Dalton expounds his theory of mixed gases in three papers before the Manchester Literary and Philosophical Society; Johann Wilhelm Ritter demonstrates the existence of ultraviolet light; Sir Humphry Davy moves to London to work at the Royal Institution and uses electricity to decompose water into hydrogen and oxygen.

1801–1805 The *Investigator,* with Robert Brown, voyages to circumnavigate Australia.

1802 Publication of William Paley's *Natural Theology;* establishment of the Bogotá Observatory, the first permanent observatory built in the Americas; late this year or possibly early 1803 Luke Howard sets forth the modern system of cloud classification.

1803 Work begins on the *Description of Egypt,* with financial support from French government; Georges Cuvier becomes permanent secretary of the First Section of the Institute of France.

1804 Closing of the academy of Bologna; Napoléon militarizes the Polytechnic School; publication of Nicolas-Théodore de Saussure's *Chemical Researches on Vegetation.*

1804–1806 Lewis and Clark expedition.

1805 Founding of the first psychiatric journal, Johann Christian Reil's *Journal of Psychological Therapy.*

1807 Foundation of the Society of Arcueil outside Paris, which will last until 1813; founding of the Geological Society of London.

1808 Publication of the first volume of Dalton's *A New System of Chemical Philosophy,* setting forth his theory of chemical atomism; founding of the Wernerian Society of Edinburgh.

1808–1810 Publication of the first edition of Jöns Jakob Berzelius's textbook of chemistry.

1809 Joseph-Louis Gay-Lussac announces the law of the combining volumes of gases; first ovariotomy performed by the American surgeon Ephraim McDowell; founding of the University of Berlin; publication of Lamarck's *Zoological Philosophy;* publication of the first volume of the *Description of Egypt;* Society for Animal Chemistry established as a group within the British Royal Society.

1811 Japanese shogunate opens translation office to promote "Dutch studies."

1812 Publication of Laplace's *Analytical Theory of Probabilities;* publication of Cuvier's *Researches on the Fossil Bones of Quadrupeds.*

1813 Berzelius sets forth modern system of chemical notation; founding of the Analytical Society at Cambridge begins the reintegration of Britain into mathematical mainstream.

1815 Publication of William Smith's geological map of Britain; Sophie Germain wins competitive prize from the Institute of France for her work on elasticity.

1816 René-Théophile-Hyacinthe Laënnec invents the stethoscope; Jean-Baptiste Biot's *Treatise on Experimental and Mathematical Physics* sets forth the program of "Laplacian physics."

1817 Dalton becomes president of the Manchester Literary and Philosophical Society; publication of Cuvier's *Animal Kingdom.*

1818 Berzelius becomes secretary of the Swedish Academy of Science, eventually revitalizing it; publishing of Mary Shelley's *Frankenstein.*

1819 Augustin-Jean Fresnel wins a prize in optics offered by the Academy of Sciences, advancing the wave theory of light.

1820 Hans Christian Ørsted demonstrates the connection of electricity and magnestism; Davy becomes president of the Royal Society on the death of Banks; foundation of the national botanical garden of the United States in Washington, D.C.

1821 Berzelius begins publication of annual yearbook of physical science.

1822 Publication of Joseph Fourier's *Analytical Theory of Heat;* Catholic Church formally permits the teaching of heliocentric astronomy.

1826 First publication of André-Marie Ampère's *Memoir on the Mathematical Theory of Electrodynamic Phenomena,* the basis of electrodynamics; *Description of Egypt* complete in twenty-three volumes; Karl Ernst Adolf von Hoff starts publishing annual list of world earthquakes.

1827 Karl Ernst von Baer discovers the mammalian ovum; Brown discovers "Brownian movement"; Davy resigns as president of the Royal Society.

1828 Friedrich Wöhler announces synthesis of urea, first organic compound to be synthesized; new British Longitude Act disbands the Longitude Board.

1830 Dispute on morphology between Cuvier and Étienne Geoffroy Saint-Hilaire before the Academy of Sciences.

Bibliography

Ackerknecht, Erwin H. *Medicine at the Paris Hospital, 1794–1848.* Baltimore: Johns Hopkins University Press, 1967.

Alder, Ken. *Engineering the Revolution: Arms and Enlightenment in France, 1763–1815.* Princeton: Princeton University Press, 1997.

———. *The Measure of All Things: Seven-Year Odyssey and Hidden Error That Transformed the World.* New York: Free Press, 2002.

Alic, Margaret. *Hypatia's Heirs: A History of Women in Science from Antiquity to the Late Nineteenth Century.* London: Women's Press, 1986.

Allan, D. G. C., and R. E. Schofield. *Stephen Hales: Scientist and Philanthropist.* London: Scolar, 1980.

Allen, David Elliston. *The Naturalist in Britain: A Social History.* Princeton: Princeton University Press, 1994.

Anderson, Lorin. *Charles Bonnet and the Order of the Known.* London: Kluwer, 1982.

Appel, Toby A. *The Cuvier-Geoffroy Debate: French Biology in the Decades before Darwin.* New York: Oxford University Press, 1987.

Atkinson, Dwight. *Scientific Discourse in Historical Perspective: Philosophical Transactions of the Royal Society of London, 1675–1975.* Mahwah, NJ: L. Erlbaum Associates, 1999.

Bacon, John MacKenzie. *The Dominion of the Air: Story of Aerial Navigation.* Philadelphia: David Mackay, 1903.

Baker, Keith M. *Condorcet: From Natural Philosophy to Social Mathematics.* Chicago: University of Chicago Press, 1975.

Beaglehole, J. C. *The Exploration of the Pacific.* London: Adam and Charles Black, 1947.

———. *The Life of Captain James Cook.* Stanford: Stanford University Press, 1974.

Beaucour, Fernand, Yves Laissus, and Chantal Orgogozo. *The Discovery of Egypt.* Paris: Flammarion, 1990.

Bedini, Silvio. *Early American Scientific Instruments and Their Makers.* Rancho Cordova, CA: Landmark Enterprises, 1986.

Beeson, David. *Maupertuis: An Intellectual Biography.* Studies in Voltaire and the Eighteenth Century, no. 299. Oxford: Voltaire Foundation at the Taylor Institution, 1992.

Beringer, Johann Bartholomaus Adam. *The Lying Stones of Dr. Johann Bartholomew Adam Beringer, Being His "Lithographia Wirceburgensis."* Translated and annotated by Melvin E. Jahn and Daniel J. Woolf. Berkeley: University of California Press, 1963.

Blunt, Wilfrid. *The Compleat Naturalist: A Life of Linnaeus.* New York: Viking, 1971.

Bowen, James. *A History of Western Education.* Vol. 3, *The Modern West: Europe and the New World.* London: Methuen, 1981.

Boyer, Carl B. *A History of Mathematics.* 2d ed. Revised by Uta C. Merzbach. New York: John Wiley and Sons, 1991.

Bradbury, S. *The Evolution of the Microscope.* Oxford: Pergamon Press, 1967.

Brinitzer, Carl. *A Reasonable Rebel: Georg Christoph Lichtenberg.* Translated by Bernard Smith. New York: Macmillan, 1960.

Brock, William H. *The Norton History of Chemistry.* New York: W. W. Norton, 1993.

Broman, Thomas. "J. C. Reil and the 'Journalization' of Physiology." In *The Literary Structure of Scientific Argument,* edited by Peter Dear, 13–42. Philadelphia: University of Pennsylvania Press, 1991.

Brooke, John Hedley. *Science and Religion: Some Historical Perspectives.* Cambridge: Cambridge University Press, 1991.

Brown, Sanford C. *Benjamin Thompson, Count Rumford.* Cambridge: Massachuseets Insitute of Technology Press, 1979.

Buhler, Walter Kaufmann. *Gauss: A Biographical Study.* Heidelberg: Springer-Verlag, 1981.

Burke, John G. *Cosmic Debris: Meteorites in History.* Berkeley: University of California Press, 1986.

Burkhardt, Richard W., Jr. *The Spirit of System: Lamarck and Evolutionary Biology.* Cambridge: Harvard University Press, 1977.

Bynum, W. F., and Roy Porter, eds. *Brunonianism in Britain and Europe.* Medical History, Supp. 8. London: Wellcome Institute for the History of Medicine, 1988.

Cantor, G. N. *Optics after Newton: Theories of Light in Britain and Ireland, 1704–1840.* Manchester: Manchester University Press, 1983.

Cardwell, Donald. *The Norton History of Technology.* New York: W. W. Norton, 1994.

Carter, G. N., and M. J. S. Hodge, eds. *Conceptions of Ether: Studies in the History of Ether Theories, 1740–1900.* Cambridge: Cambridge University Press, 1981.

Clark, Ronald W. *Benjamin Franklin: A Biography.* London: Weidenfeld and Nicolson, 1983.

Coleby, L. J. M. *The Chemical Studies of P. J. Macquer.* London: Allen & Unwin, 1938.

Coleman, William. *Georges Cuvier, Zoologist: A Study in the History of Evolution.* Cambridge: Harvard University Press, 1964.

Collison, Robert. *Encyclopedias: Their History throughout the Ages.* New York and London: Harper, 1966.

Corsi, Pietro. *The Age of Lamarck: Evolutionary Theories in France, 1790–1830.* Translated by Jonathan Mandelbaum. Berkeley: University of California Press, 1988.

Crabtree, Adam. *From Mesmer to Freud: Magnetic Sleep and the Roots of Psychological Healing.* New Haven: Yale University Press, 1993.

Crocker, Lester G. *Diderot: Embattled Philosopher.* New York: Free Press, 1966.

Crosland, Maurice. *Gay-Lussac: Scientist and Bourgeois.* Cambridge: Cambridge University Press, 1978.

———. *Historical Studies in the Language of Chemistry.* Rev. ed. Mineola, NY: Dover, 1978.

———. *The Society of Arcueil: A View of French Science at the Time of Napoleon I.* Cambridge: Harvard University Press, 1967.

Cunningham, Andrew, and Nicholas Jardine, eds. *Romanticism and the Sciences.* Cambridge: Cambridge University Press, 1990.

Darnton, Robert. *Mesmerism and the End of the Enlightenment in France.* Cambridge: Harvard University Press, 1968.

Daston, Lorraine. *Classical Probability in the Enlightenment.* Princeton: Princeton University Press, 1988.

———. "Nationalism and Scientific Neutrality under Napoleon." In *Solomon's House Revisited: Organization and Institutionalization of Science,* edited by Tore Frangsmyr, 95–119. Canton, MA: Science History Publications, 1990.

Daumas, Maurice. *Scientific Instruments of the Seventeenth and Eighteenth Centuries.* New York: Praeger, 1972.

Davison, Charles. *The Founders of Seismology.* 1927. Reprint, New York: Arno, 1978.

Deacon, Margaret. *Scientists and the Sea, 1650–1900: A Study of Marine Science.* London: Academic Press, 1971.

Delaporte, François. *Nature's Second Kingdom: Explorations of Vegetality in the Eighteenth Century.* Translated by Arthur Goldhammer. Cambridge: Massachuseets Insitute of Technology Press, 1982.

Dematteis, Philip B., and Peter S. Fosl, eds. *British Philosophers, 1500–1799.* Dictionary of Literary Biography, no. 252. Detroit: Gale, 2002.

De Ridder-Symoens, Hilde, ed. *Universities in Early Modern Europe.* Cambridge: Cambridge University Press, 1996.

De Terra, Helmut. *Humboldt: Life and Times of Alexander von Humboldt, 1769–1859.* New York: Alfred A. Knopf, 1955.

Dibner, Bern. *Alessandro Volta and the Electric Battery.* New York: Franklin Watts, 1964.

Donnelly, Marian Card. *A Short History of Observatories.* Eugene: University of Oregon Books, 1973.

Earnest, Ernest. *John and William Bartram, Botanists and Explorers.* Philadelphia: University of Pennsylvania Press, 1940.

Engstrad, Iris H. W. *Spanish Scientists in the New World: Eighteenth-Century Expeditions.* Seattle: University of Washington Press, 1981.

Farrell, Maureen. *William Whiston.* New York: Arno, 1981.

Findlen, Paula. "Science As a Career in Enlightenment Italy: The Strategies of Laura Bassi." *Isis* 84 (1993): 441–469.

Fletcher, Harold R., and William H. Brown. *The Royal Botanic Garden Edinburgh, 1670–1970.* Edinburgh: Her Majesty's Stationery Office, 1970.

Forbes, Eric G. *Tobias Mayer (1723–62): Pioneer of Enlightened Science in Germany.* Göttingen: Vandenhoeck and Rupprecht, 1980.

Force, James. *William Whiston: Honest Newtonian.* Cambridge: Cambridge University Press, 1985.

Ford, Brian J. *Images of Science: A History of Scientific Illustration.* New York: Oxford University Press, 1993.

Frangsmayr, Tore, ed. *Linnaeus: Man and His Work.* Berkeley: University of California Press, 1983.

Friedman, Michael. *Kant and the Exact Sciences.* Cambridge: Harvard University Press, 1992.

Frisinger, H. Howard. *The History of Meteorology: To 1800.* New York: Science History Publications, 1997.

Fry, Howard Tyrrell. *Alexander Dalrymple (1737–1808) and the Expansion of British Trade.* London: Cass for the Royal Commonwealth Society, 1970.

Gager, Charles Stuart. "Botanic Gardens of the World: Materials for a History." 2d ed. *Brooklyn Botanic Garden Record* 27 (1938): 151–406.

Gay, Peter. *The Enlightenment: An Interpretation, Volume II: The Science of Freedom.* New York: Alfred A. Knopf, 1969.

Gelbart, Nina Rattner. *The King's Midwife: A History and Mystery of Madame du Coudray.* Berkeley: University of California Press, 1998.

Gerbi, Antonello. *The Dispute of the New World: History of a Polemic, 1750–1900.* Translated by Jeremy Moyle. Rev. ed. Pittsburgh: University of Pittsburgh Press, 1973.

Gibbs, F. W. *Joseph Priestley: Revolutions of the Eighteenth Century.* Garden City, NY: Doubleday, 1967.

Gillispie, Charles Coulston. *The Edge of Objectivity: An Essay in the History of Scientific Ideas.* Princeton: Princeton University Press, 1960.

———. *Science and Polity in France at the End of the Old Regime.* Princeton: Princeton University Press, 1980.

Gillispie, Charles Coulston, with the assistance of Robert Fox and Ivor Grattan-Guiness. *Pierre-Simon Laplace, 1749–1827: A Life in Exact Science.* Princeton: Princeton University Press, 1997.

Gilmor, C. Stewart. *Coulomb and the Evolution of Physics and Engineering in Eighteenth-Century France.* Princeton: Princeton University Press, 1971.

Gleeson, Janet. *The Arcanum: The Extraordinary True Story.* New York: Warner, 1999.

Golinski, Jan. *Science As Public Culture: Chemistry and Enlightenment in Britain, 1760–1820.* Cambridge: Cambridge University Press, 1992.

Graupe, Heinz Moshe. *The Rise of Modern Judaism: An Intellectual History of German Jewry, 1650–1942.* Translated by John Robinson. Huntington, NY: Krieger, 1978.

Green, J. Reynolds. *A History of Botany in the United Kingdom from the Earliest Times to the End of the 19th Century.* London: J. Dent and Sons, 1904.

Greenberg, John L. *The Problem of the Earth's Shape from Newton to Clairaut.* Cambridge: Cambridge University Press, 1995.

Greene, John C. *American Science in the Age of Jefferson.* Ames: Iowa State University Press, 1984.

Grove, Richard H. *Green Imperialism: Colonial Expansion, Tropical Island Edens, and the Origins of Environmentalism, 1600–1860.* Cambridge: Cambridge University Press, 1995.

Guerlac, Henry. *Essays and Papers in the History of Modern Science.* Baltimore: Johns Hopkins University Press, 1977.

Guerrini, Anita. *Obesity and Depression in the Enlightenment: The Life and Times of George Cheyne.* Norman: University of Oklahoma Press, 2000.

Hahn, Roger. *The Anatomy of a Scientific Institution: The Paris Academy of Sciences, 1666–1803.* Berkeley: University of California Press, 1971.

Hall, Thomas S. *History of General Physiology, 600 B.C. to A.D. 1900, Volume Two: From the Enlightenment to the End of the Nineteenth Century.* Chicago: University of Chicago Press, 1969.

Hallam, A. *Great Geological Controversies.* 2d ed. Oxford: Oxford University Press, 1989.

Hankins, Thomas L. *Jean d'Alembert: Science and the Enlightenment.* Oxford: Oxford University Press, 1970. Reprint, New York: Gordon and Breach, 1990.

———. *Science and the Enlightenment.* Cambridge: Cambridge University Press, 1985.

Hannaford, Ivan. *Race: The History of an Idea in the West.* Washington, DC: Woodrow Wilson Center Press, 1996.

Harman, P. M. *Energy, Force, and Matter: The Conceptual Development of Nineteenth-Century Physics.* Cambridge: Cambridge University Press, 1982.

Hayes, Roslynn D. *From Faust to Strangelove: Representations of the Scientist in Western Literature.* Baltimore: Johns Hopkins University Press, 1994.

Hearnshaw, L. S. *The Shaping of Modern Psychology.* London: Routledge and Kegan Paul, 1987.

Heilbron, J. L. *Electricity in the 17th and 18th Centuries: A Study in Early Modern Physics.* 2d ed. Mineola, NY: Dover, 1999.

Hine, Ellen McNiven. *A Critical Study of Condillac's "Traite des Systems."* The Hague: Martinus Nijhoff, 1979.

Hofmann, James R. *André-Marie Ampère.* Cambridge: Cambridge University Press, 1996.

Holmes, Frederic Lawrence. *Antoine Lavoisier—The Next Crucial Year; or, The Sources of His Quantitative Method in Chemistry.* Princeton: Princeton University Press, 1998.

———. *Lavoisier and the Chemistry of Life: An Exploration of Scientific Creativity.* Madison: University of Wisconsin Press, 1985.

Horne, David Bayne. *A Short History of the University of Edinburgh, 1556–1889.* Edinburgh: University Press, 1967.

Howell, Benjamin F. *An Introduction to Seismological Research: History and Development.* Cambridge: Cambridge University Press, 1990.

Howse, Derek. *Nevil Maskelyne: The Seaman's Astronomer.* Cambridge: Cambridge University Press, 1989.

Hufbauer, Karl. *The Formation of the German Chemical Community, 1720–1795.* Berkeley: University of California Press, 1982.

Hughes, David W. "The History of Meteors and Meteor Showers." *Vistas in Astronomy* 26 (1982): 325–45.

Jacob, Margaret. *Scientific Culture and the Making of the Industrial West.* New York: Oxford University Press, 1997.

Jardine, N., J. A. Secord, and E. C. Spary, eds. *Cultures of Natural History.* Cambridge: Cambridge University Press, 1996.

Jessop, L. "The Club at the Temple Coffee House: Facts and Supposition." *Archives of Natural History* 16 (1989): 263–274.

Johns, Alessa, ed. *Dreadful Visitations: Confronting Natural Catastrophe in the Age of Enlightenment.* New York: Routledge, 1999.

Jungnickel, Christina, and Russell McCormmach. *Cavendish: Experimental Life.* Rev. ed. Lewisburg, PA: Bucknell, 1999.

Kafker, Frank A., ed. *Notable Encyclopedias of the Late Eighteenth Century: Eleven Successours of the "Encyclopédie."* Studies in Voltaire and the Eighteenth Century, no. 315. Oxford: Voltaire Foundation, 1994.

———. *Notable Encyclopedias of the Seventeenth and Eighteenth Centuries: Nine Predecessors of the "Encyclopédie."* Studies in Voltaire and the Eighteenth Century, no. 194. Oxford: Voltaire Foundation, 1981.

Kafker, Frank A., in collaboration with Serena L. Kafker. *The Encyclopedists As Individuals: A Biographical Dictionary of the Authors of the "Encyclopédie."* Studies in Voltaire and the Eighteenth Century, no. 257. Oxford: Voltaire Foundation, 1988.

Katz, David. "The Occult Bible: Hebraic Millenarianism in Eighteenth-Century England." In *The Millenarian Turn: Millenarian Contexts of Science, Politics, and Everyday Anglo-American Life in the Seventeenth and Eighteenth Centuries,* edited by James E. Force and Richard H. Popkin, 119–132. Dordrecht, Netherlands: Kluwer, 2001.

Khrgian, Aleksandr Khrsioforovich. *Meteorology: A Historical Survey.* 2d ed., revised. Edited by Kh. P. Pogosyan. Jerusalem: Israel Program for Scientific Translations, 1970.

King, Henry C. *The History of the Telescope.* London: Charles Griffin, 1955.

King-Hele, Desmond. *Doctor of Revolution: Life and Genius of Erasmus Darwin.* London: Faber and Faber, 1977.

Knight, David. *Humphry Davy: Science and Power.* Oxford: Blackwell, 1992.

———. *Ideas in Chemistry: A History of the Science.* New Brunswick: Rutgers University Press, 1992.

Kronick, David A. *A History of Scientific and Technical Periodicals: Origins and Development of the Scientific and Technical Press, 1665–1790.* 2d ed. Metuchen, NJ: Scarecrow Press, 1976.

Lafuente, Antonio. "Enlightenment in an Imperial Context: Local Science in the Late-Eighteenth-Century Hispanic World." *Osiris,* 2d ser., 15 (2000): 155–173.

Laqueur, Thomas. *Making Sex: Body and Gender from the Greeks to Freud.* Cambridge: Harvard University Press, 1990.

Laudan, Rachel. *From Mineralogy to Geology: Foundations of a Science, 1650–1830.* Chicago: University of Chicago Press, 1987.

Leith, James A., ed. *Facets of Education in the Eighteenth Century.* Studies in Voltaire and the Eighteenth Century, no. 167. Oxford: Voltaire Foundation, 1977.

Lenoir, Timothy. *The Strategy of Life: Teleology and Mechanics in Nineteenth-Century German Biology.* Dordrecht, Netherlands: D. Reidel, 1982.

Lesch, John E. *Science and Medicine in France: The Emergence of Experimental Physiology, 1790–1855.* Cambridge: Harvard University Press, 1984.

Limoges, Camille. "The Development of the Muséum d'Histoire Naturelle of Paris, c. 1800–1914." In *The Organization of Science and Technology in France, 1808–1914,* edited by Robert Fox and George Weisz, 211–240. Cambridge: Cambridge University Press, 1980.

Lindberg, David C., and Ronald L. Numbers, eds. *God and Nature: Historical Essays on the Encounter between Christianity and Science.* Berkeley: University of California Press, 1986.

Lindeboom, G. A. *Hermann Boerhaave: Man and His Work.* London: Methuen, 1968.

Linney, Verna. "A Passion for Art, a Passion for Botany: Mary Delany and Her Floral 'Mosaiks.'" *Eighteenth-Century Women* 1 (2001): 203–235.

Lodge, Sir Oliver. *Pioneers of Science and the Development of Their Scientific Theories.* 1926. Reprint, Mineola, NY: Dover, 1960.

Lyons, Henry. *The Royal Society, 1660–1940: A History of Its Administration under Its Charters.* Cambridge: Cambridge University Press, 1944.

Mabberley, D. J. *Jupiter Botanicus: Robert Brown of the British Museum.* Brunswick: Verlag von J. Cramer, 1985.

MacGregor, Arthur, ed. *Sir Hans Sloane: Collector, Scientist, Antiquary, Founding Father of the British Museum.* London: British Museum Press, 1994.

MacLeod, Roy, ed. *Nature and Empire: Science and the Colonial Enterprise. Osiris,* 2d ser., 15 (2000).

Maistrov, L. E. *Probability Theory: A Historical Sketch.* Translated and edited by Samuel Kotz. London: Academic Press, 1974.

Mason, Stephen F. *A History of Sciences.* Rev. ed. New York: Collier, 1962.

Mazzolini, Renato G., ed. *Non-verbal Communication in Science prior to 1900.* Florence: Leo S. Olschki, 1993.

McClellan, James E., III. *Science Reorganized: Scientific Societies in the Eighteenth Century.* New York: Columbia University Press, 1985.

McKie, Douglas. *Antoine Lavoisier: Scientist, Economist, Social Reformer.* New York: Harper and Row, 1952.

Meade, Teresa, and Mark Walker, eds. *Science, Medicine, and Cultural Imperialism.* New York: St. Martin's, 1991.

Melhado, Evan M. *Jacob Berzelius: The Emergence of His Chemical System.* Madison: University of Wisconsin Press, 1981.

Melhado, Evan M., and Tore Frangsmyr, eds. *Enlightenment Science in the Romantic Era: The Chemistry of Berzelius and Its Cultural Setting.* Cambridge: Cambridge University Press, 1992.

Middlekauf, Robert. *The Mathers: Three Generations of Puritan Intellectuals.* Oxford: Oxford University Press, 1971.

Middleton, W. E. Knowles. *A History of the Theories of Rain and Other Forms of Precipitation.* London: Oldbourne, 1966.

———. *Invention of the Meteorological Instruments.* Baltimore: Johns Hopkins University Press, 1969.

Miller, David Phillip, and Peter Hans Reill, eds. *Visions of Empire: Voyages, Botany, and Representations of Nature.* Cambridge: Cambridge University Press, 1996.

Miller, Edward. *That Noble Cabinet: A History of the British Museum.* London: Andre Deutsch, 1973.

Morselli, Mario. *Amedeo Avogadro: A Scientific Biography.* Dordrecht, Netherlands: D. Reidel, 1984.

Morton, A. G. *History of Botanical Science: An Account of the Development of Botany from Ancient Times to the Present Day.* London: Academic Press, 1981.

Murray, David. *Museums: Their History and Use with a Bibliography and a List of Museums in the United Kingdom.* 1904. Reprint, Staten Island: Pober, 2000.

Musson, A. E., and Eric Robinson. *Science and Technology in the Industrial Revolution.* Toronto: University of Toronto Press, 1969.

Nicolson, Marjorie Hope. *Newton Demands the Muse: Newton's "Opticks" and the Eighteenth-Century Poets.* Princeton: Princeton University Press, 1946.

North, John. *The Norton History of Astronomy and Cosmology.* New York: W. W. Norton, 1995.

O'Brian, Patrick. *Joseph Banks: A Life.* Boston: David R. Godine, 1993.

Oldroyd, David. "An Examination of G. E. Stahl's *Philosophical Principles of Universal Chemistry.*" *Ambix* 20 (1973): 36–52.

Ørsted, Hans Christian. *Selected Scientific Works of Hans Christian Ørsted.* Edited and translated by Karen Jelved, Andrew D. Jackson, and Ole Knudsen. Princeton: Princeton University Press, 1998.

Outram, Dorinda. *Georges Cuvier: Vocation, Science, and Authority in Post-Revolutionary France.* Manchester: Manchester University Press, 1984.

Paul, Charles B. *Science and Immortality: "Eloges" of the Paris Academy of Sciences (1699–1791).* Berkeley: University of California Press, 1980.

Perkins, Jean A. "Voltaire and the Natural Sciences." *Studies on Voltaire and the Eighteenth Century* 37 (1965): 161–176.

Persaud, T. N. *A History of Anatomy: Post-Vesalian Era.* Springfield, IL: Charles C. Thomas, 1997.

Petitjean, Patrick, Catherine Jami, and Anne Marie Moulin, eds. *Science and Empires: Historical Studies about Scientific Development and European Expansion.* Boston Studies in the Philosophy of Science, no. 136. Dordrecht, Netherlands: Kluwer, 1992.

Petrovich, Vesna Crnjanski. "Women and the Paris Academy of Sciences." *Eighteenth-Century Studies* 32 (1999): 383–390.

Pinto-Correia, Clara. *The Ovary of Eve: Egg and Sperm and Preformation.* Chicago: University of Chicago Press, 1997.

Porter, Roy. *The Creation of the Modern World: Untold Story of the British Enlightenment.* New York: W. W. Norton, 2000.

———. *The Greatest Benefit to Mankind: A Medical History of Humanity.* New York: W. W. Norton, 1998.

———. *The Making of Geology: Earth Science in Britain, 1660–1815.* Cambridge: Cambridge University Press, 1977.

———. *Mind-Forg'd Manacles: A History of Madness in England from the Restoration to the Regency.* Cambridge: Harvard University Press, 1987.

Porter, Roy, and Mikulas Teich, eds. *The Scientific Revolution in National Context.* Cambridge: Cambridge University Press, 1992.

Porter, Roy, and W. F. Bynum, eds. *William Hunter and the Eighteenth-Century Medical World.* Cambridge: Cambridge University Press, 1985.

Pullman, Bernard. *The Atom in the History of Human Thought.* Translated by Axel Reisinger. New York: Oxford University Press, 1998.

Pyenson, Lewis, and Susan Sheets-Pyenson. *Servants of Nature: A History of Scientific Institution, Enterprises, and Sensibilities.* New York: W. W. Norton, 1999.

Roberts, Lissa. "A Word and the World: Significance of Naming the Calorimeter." *Isis* 82 (1991): 199–222.

Roe, Shirley. *Matter, Life, and Generation: Eighteenth-Century Embryology and the Haller-Wolff Debate.* Cambridge: Cambridge University Press, 1981.

Roger, Jacques. *Buffon: A Life in Natural History.* Translated by Sarah Lucille Bonnefoi. Edited by L. Pearce Williams. Ithaca: Cornell University Press, 1997.

Rothschuh, Karl E. *History of Physiology.* Translated and edited by Guenter B. Risse. Huntington, NY: Robert E. Krieger, 1973.

Rousseau, G. S., and Roy Porter, eds. *The Ferment of Knowledge: Studies in the Historiography of Eighteenth-Century Science.* Cambridge: Cambridge University Press, 1980.

Ruderman, David B. *Jewish Enlightenment in a New Key: Anglo-Jewry's Construction of Modern Jewish Thought.* Princeton: Princeton University Press, 2000.

———. *Jewish Thought and Scientific Discovery in Early Modern Europe.* New Haven: Yale University Press, 1995.

Rudwick, Martin J. S. *The Meaning of Fossils: Episodes in the History of Palaeontology.* 2d ed. New York: Science History Publications, 1976.

Ruestow, Edward G. *Physics at Seventeenth- and Eighteenth-Century Leiden: Philosophy and the New Science in the University*. The Hague: Nijhoff, 1973.

Russell, Edward John. *A History of Agricultural Science in Great Britain, 1620–1954*. London: Allen & Unwin, 1966.

Saine, Thomas P. *The Problem of Being Modern; or, The German Pursuit of Enlightenment from Leibniz to the French Revolution*. Detroit: Wayne State University Press, 1997.

Savage, Henry, Jr., and Elizabeth Savage. *André and François André Michaux*. Charlottesville: University Press of Virginia, 1986.

Schiebinger, Londa. *Nature's Body: Gender in the Making of Modern Science*. Boston: Beacon, 1993.

Schofield, Robert E. *The Lunar Society of Birmingham: A Social History of Provincial Science and Industry in Eighteenth-Century England*. Oxford: Clarendon, 1963.

———. *Mechanism and Materialism: British Natural Philosophy in an Age of Reason*. Princeton: Princeton University Press, 1970.

Shaffer, Elinor S., ed. *The Third Culture: Literature and Science*. Berlin: Walter de Gruyter, 1998.

Shea, William R., ed. *Science and the Visual Image in the Enlightenment*. Canton, MA: Science History Publications, 2000.

Shteir, Ann B. *Cultivating Women, Cultivating Science: Flora's Daughters and Botany in England, 1760 to 1860*. Baltimore: Johns Hopkins University Press, 1996.

Slaughter, Thomas P. *The Natures of John and William Bartram*. New York: Alfred A. Knopf, 1996.

Sobel, Dava. *Longitude: The True Story of a Lone Genius Who Solved the Greatest Scientific Problem of His Time*. New York: Walker, 1995.

Sorrenson, Richard. "Towards a History of the Royal Society in the Eighteenth Century." *Notes and Records of the Royal Society* 50 (1996): 29–46.

Stansfield, Dorothy A. *Thomas Beddoes, M.D., 1760–1808: Chemist, Physician, Democrat*. Dordrecht, Netherlands: D. Reidel, 1984.

Stearns, Raymond Phineas. *Science in the British Colonies of America*. Urbana: University of Illinois Press, 1970.

Stengers, Jean, and Anne van Neck. *Masturbation: The History of a Great Terror*. Translated by Kathryn A. Hoffman. New York: Palgrave, 2001.

Stephenson, R. H. *Goethe's Conception of Knowledge and Science*. Edinburgh: Edinburgh University Press, 1995.

Stevens, Peter F. *The Development of Biological Systematics: Antoine-Laurent de Jussieu, Nature, and the Natural System*. New York: Columbia University Press, 1994.

Stoye, John. *Marsigli's Europe, 1680–1730: Life and Times of Luigi Ferdinando Marsigli, Soldier and Virtuoso*. New Haven: Yale University Press, 1994.

Sugimoto, Masayoshi, and David L. Swain. *Science and Culture in Traditional Japan, A.D. 600–1854*. Cambridge: Massachusetts Insitute of Technology Press, 1978.

Sutton, Geoffrey V. *Science for a Polite Society: Gender, Culture, and the Demonstration of Enlightenment*. Boulder, CO: Westview, 1995.

Tapper, Alan. "The Beginnings of Priestley's Materialism." *Enlightenment and Dissent* 1 (1982): 73–82.

Taton, René, and Curtis Wilson, eds. *Planetary Astronomy from the Renaissance to the Rise of Astrophysics, Part B: Eighteenth and Nineteenth Centuries*. Cambridge: Cambridge University Press, 1995.

Thackray, Arnold. *John Dalton: Critical Assessments of His Life and Science*. Cambridge: Harvard University Press, 1972.

———. "Natural Knowledge in Cultural Context: Manchester Model." *American Historical Review* 79 (1974): 672–709.

Todd, Dennis. *Imagining Monsters: Miscreations of the Self in Eighteenth-Century England*. Chicago: University of Chicago Press, 1995.

Toulmin, Stephen, and June Goodfield. *The Architecture of Matter.* Chicago: University of Chicago Press, 1962.

Truesdell, Clifford. *Essays in the History of Mechanics.* New York: Springer-Verlag, 1968.

Turner, Gerard L'Estrange. *Scientific Instruments and Experimental Philosophy, 1550–1850.* Aldershot, UK: Variorum, 1990.

Turner, R. Steven. "University Reformers and Professorial Scholarship in Germany, 1760–1806." In *The University in Society II: Europe, Scotland, and the United States,* edited by Lawrence Stone, 495–531. Princeton: Princeton University Press, 1974.

Van Berkel, Klaas, Albert Van Helden, and Lodewijk Palm, eds. *A History of Science in the Netherlands: Survey, Themes, and Reference.* Leiden: Brill, 1999.

Vartanian, Aram. *Diderot and Descartes: A Study of Scientific Naturalism in the Enlightenment.* Princeton: Princeton University Press, 1953.

Vizthum, Richard C. *Materialism: An Affirmative History and Definition.* Amherst, NY: Prometheus, 1995.

Walters, Alice N. "Ephemeral Events: English Broadsides of Early Eighteenth-Century Solar Eclipses." *History of Science* 37 (1999): 1–43.

Watkins, Eric, ed. *Kant and the Sciences.* Oxford: Oxford University Press, 2001.

Wellman, Kathleen. *La Mettrie: Medicine, Philosophy, and Enlightenment.* Durham: Duke University Press, 1992.

Whyte, Lancelot Law, ed. *Roger Joseph Boscovich, S.J., F.R.S., 1711–1787: Studies of His Life and Work on the 250th Anniversary of His Birth.* London: Allen & Unwin, 1961.

Wilde, C. B. "Hutchinsonianism, Natural Philosophy, and Religious Controversy in Eighteenth-Century Britain." *History of Science* 18 (1980): 1–24.

Wolf, Abraham. *A History of Science, Technology, and Philosophy in the Eighteenth Century.* 2d ed. Revised by D. McKie. Gloucester, MA: Peter Smith, 1968.

Woolf, Harry. *The Transits of Venus: A Study in Eighteenth-Century Science.* Princeton: Princeton University Press, 1959.

Yeo, Richard. *Encyclopaedic Visions: Scientific Dictionaries and Enlightenment Culture.* Cambridge: Cambridge University Press, 2001.

Yeomans, Donald K. *Comets: A Chronological History of Observations, Science, Myth, and Folklore.* New York: John Wiley and Sons, 1991.

Yolton, John. *Thinking Matter: Materialism in Eighteenth-Century Britain.* Minneapolis: University of Minnesota Press, 1983.

Zinsser, Judith P. "Translating Newton's Principia: Marquise du Châtelet's Revisions and Additions for a French Audience." *Notes and Records of the Royal Society of London* 55 (2001): 227–245.

Useful Web Sites

Curtis's Botanical Magazine:
http://www.nal.usda.gov/curtis/index.shtml

The United States National Agricultural Library hosts this database of texts and images from a popular British botanical magazine, published from 1787 to 1807.

Edgar Fahs Smith Collection:
http://dewey.library.upenn.edu/sceti/smith/

A collection of images from the history of chemistry, including many from the eighteenth and early nineteenth centuries, housed at the Schoenberg Center for Electronic Texts and Images at the University of Pennsylvania.

Images from the History of Medicine:
http://wwwihm.nlm.nih.gov/

From the National Library of Medicine, contains thousands of images relevant to Enlightenment medicine and other sciences including portraits, frontispieces, and diagrams.

The Incompleat Chymist:
http://dbhs.wvusd.k12.ca.us/Chem-History/Obsolete-Chem-TermsTOC.html

Jon Eklund's dictionary of eighteenth-century chemical terms.

Internet Library of Early Journals:
http://www.bodley.ox.ac.uk/ilej/

Housed at Oxford's Bodelian Library, this database includes a run of twenty years of the Royal Society's journal, *Philosophical Transactions,* from 1757 to 1777.

Jesuits and the Sciences, 1540–1995:
http://www.luc.edu/libraries/science/jesuits/index.html

Contains a number of reproductions of early modern Jesuit scientific books, as well as discussion of Jesuit scientists.

Lamarck.net:
http://www.Lamarck.net

Professor Pietro Corsi's site devoted to Jean-Baptiste Lamarck and his students. In English and French.

The MacTutor History of Mathematics Archive:
http://www-groups.dcs.st-andrews.ac.uk/~history/

From the School of Mathematics at the University of St. Andrews, a large and growing database of mathematical biographies and treatments of specific problems in the history of mathematics.

The Messier Catalog:
http://www.seds.org/messier/

Students for the Exploration and Development of Space maintains this Web site with information about Charles Messier and his catalog of celestial objects. The site also includes information on William Herschel and his catalog, as well as others.

Museum of the History of Science, Oxford:
http://www.mhs.ox.ac.uk/

The museum site includes on-line exhibitions and thousands of images of items from its collections.

Panopticon Lavoisier:
http://moro.imss.fi.it/lavoisier/

The Institute and Museum of the History of Science in Florence, Italy, maintains this site devoted to Antoine-Laurent Lavoisier.

Index

Note: Page numbers in italic type refer to illustrations; page numbers in bold type refer to main entries.

About the Author

William E. Burns is a historian and history teacher in Washington, D.C. He received his Ph.D in 1994 from the University of California at Davis, with a specialization in early modern Europe. His dissertation was published as *An Age of Wonders: Prodigies, Politics and Providence in England, 1657–1727* (2002). Dr. Burns has also written *The Scientific Revolution: An Encyclopedia* (2001). In addition to the history of science, his interests include Restoration England, the history of astrology, and witches and witchhunting.